Fossil Frogs and Toads of North America

Fossil Frogs and Toads of North America

J. Alan Holman

INDIANA UNIVERSITY PRESS
BLOOMINGTON AND INDIANAPOLIS

*Publication of this book was assisted by
the Friends of Indiana University Press.*

This book is a publication of

Indiana University Press
601 North Morton Street
Bloomington, IN 47404-3797 USA

http://iupress.indiana.edu

Telephone orders   800-842-6796
Fax orders   812-855-7931
Orders by e-mail   iuporder@indiana.edu

The paper used in this publication meets the minimum requirements of American
National Standard for Information Sciences—Permanence of Paper for Printed Library
Materials, ANSI Z39.48-1984.

Manufactured in the United States of America

Library of Congress Cataloging-in-Publication Data

Holman, J. Alan, date
  Fossil frogs and toads of North America / J. Alan Holman.
      p. cm. — (Life of the past)
Includes bibliographical references and indexes.
  ISBN 0-253-34280-5 (cloth : alk. paper)
  1. Anura, Fossil—North America. I. Title. II. Series.
  QE868.A5H65 2003
  567'.8—dc21

                        2003008459

1 2 3 4 5 08 07 06 05 04 03

Dedicated to my principal mentors:
Pierce Brodkorb
Claude Hibbard
Sherman Minton
Hobart Smith
Mike Voorhies

All the frogs of which you are about to read have croaked.

CONTENTS

Because frog and toad populations appear to be dwindling on Earth at an appalling rate, they are finally getting the attention that they deserve after many years of neglect. Going back to the 18th century, we find the great naturalist Carolus Linnaeus (1758) flatly stating in his summary in *Systema Naturae*, "Most amphibia are abhorrent because of their cold body, pale colour, cartilaginous skeleton, filthy skin, fierce aspect, calculating eye, offensive smell, harsh voice, squalid habitation, and terrible venom; and so their Creator has not exerted his powers (to make) them many of them."

Frogs have been savored as a gourmet dish for generations and to that effect have been persecuted to extirpation in many regions. Moreover, frogs by the millions have been dissected in elementary courses in biology and zoology as a "representative vertebrate," even though they are not typical vertebrates at all. As Romer (1959) stated, "The choice of the frog as a favorite laboratory animal is due to several factors. One item (not a minor one) is the fact that it is common, readily available, and hence inexpensive." In Spicer County, Minnesota, in 1930 Wright and Wright (1949) recorded the exploitation of the then abundant Leopard Frog (*Rana pipiens*). Referring to a specific frog gatherer, they wrote, "He has seen them moving in such numbers at this season (first frost) as to form a band two rods wide and one-half mile long where no one could step without crushing frogs," and "One night he and two other men picked up 750 pounds by hand." In those days, small Leopard Frogs were used as fish bait, and large ones were shipped to biological supply houses.

The present intense human interest in frog and toad depletion, I believe, falls into two categories. On the one hand, we have been informed by the media that the demise of these animals indicates great trouble in the environment. Needless to say, this is a situation of grave concern, which, of course, if true, will affect our lives quite adversely. On the other hand, many of us just like having frogs and toads around. Contrary to the opinion of the narrow-minded Linnaeus, they are mainly attractive creatures, whose voices enrich the outdoors, not to mention the tremendous biological interest in them from the morphological, physiological, behavioral, ecological, and evolutionary standpoints. This book deals with frogs and toads from still another dimension, that of North American vertebrate paleontology. Oddly, no such comprehensive study exists.

The introductory chapter begins with an overview of the anurans, including a discussion of the definition of the order Anura and of the specializations found in the group. A short history of fossil anuran studies follows. Next, the chapter presents a general account of the anuran skeleton, as well as detailed accounts of individual anuran bones used in paleontological studies. A discussion of the early evolution of the Anura is presented. This is followed by a definition of the chronological terms used in the book.

The heart of this book (Chapter 2) consists of detailed systematic accounts of the known fossil frogs and toads (anurans) of North America, as well as the localities where each taxon occurs. Extinct fossil frogs and toads are fully discussed and illustrated, and in some cases they are rediagnosed and redescribed on the basis of additional information and fossil material that has recently come to light. Moreover, the taxonomic status of some forms in the literature has either been questioned or changed.

Fossil anuran taxa that are living today are presented in a somewhat different way. The modern characteristics, ecological attributes, and modern ranges of genera and species are given. Criteria for the identification of fossils of these animals are presented, and illustrations of diagnostic skeletal elements are included for most modern species.

Chapter 3 presents a detailed, epoch-by-epoch discussion of Mesozoic, Tertiary, and Pleistocene anurans, stressing the changes in taxonomic content of each epoch, and with a special overview of the Pleistocene.

Finally, an epilogue deals with the state of the art of the classification and phylogeny of the anurans as related to the fossil record. A comprehensive list of references follows.

I have tried to write this book as clearly as possible and have tried to avoid convoluted arguments, scientific jargon, and hazy semantics whenever possible. From time to time I have resorted to a slight bit of levity. My hope is that this book will be useful to neoherpetologists, paleoherpetologists, general paleontologists, biologists, zoologists, and of course anyone that has an interest in frogs and toads.

JAH
East Lansing, Michigan

ACKNOWLEDGMENTS

I gratefully acknowledge those special people that have shared their interest in North American fossil frogs and toads with me. This book could not have been written without them. These people are Walter Auffenberg, Bayard Brattstrom, Charles Chantell, the late Richard Estes, Leslie Fay, Kenneth Ford, James Fowler, James Gardner, Frederick Grady, James Harding, the late Max Hecht, Ami Henrici, the late Claude Hibbard, Jim I. Mead, Dennis Parmley, J.-C. Rage, Ronald Richards, Zbynek Roček, Karel Rogers, Elise Schroeder, Anthony Stuart, Robert Sullivan, Zbigniew Szyndlar, the late Joseph Tihen, George Van Dam, Thomas Van Devender, Michael Voorhies, Rachael Walker, Richard Wilson, the late Vincent Wilson, Alisa Winkler, and J.-P Zonneveld.

I thank Robert Sloan, Tony Brewer, and the other staff members of Indiana University Press for their effort in the production of this book. James Farlow was the special editor, and I thank him for his comments and suggestions. I am grateful for the thorough copyediting of the manuscript by Sharon Stewart, which greatly improved the book. I especially thank Merle Graffam for his series of illustrations of modern frog bones and James H. Harding and Richard D. Bartlett for photographic images of modern frogs. Other providers of illustrations are acknowledged in the Figure Credits section or on the legends of the photographic images.

The National Geographic Society has provided grants through the years that have funded much of my work on fossil frogs. The Michigan State University Museum has provided office and research space for this work.

ABBREVIATIONS

AMNH	American Museum of Natural History
CM	Carnegie Museum of Natural History
DINO	Dinosaur National Monument
FAM	Frick Laboratory, American Museum of Natural History
FGS	Florida Geological Survey
FMNH	Field Museum of Natural History
KU	Museum of Natural History, University of Kansas
LACM	Los Angeles County Museum of Natural History
MCZ	Museum of Comparative Zoology, Harvard University
MNA	Museum of Northern Arizona
MSUHS	Michigan State University, Museum Herpetological Skeletal Collection
MSUVP	Michigan State University, Museum Vertebrate Paleontology Collection
MU	Midwestern University
ND	Notre Dame University
PU	Princeton University Museum (its vertebrate fossils are now at YPM)
ROM	Royal Ontario Museum
SDSM	South Dakota School of Mines
SMNH	Saskatchewan Museum of Natural History
SMPSMU	Shuler Museum of Paleontology, Southern Methodist University
UALP	University of Arizona Laboratory of Paleontology
UALVP	University of Alberta Laboratory for Vertebrate Paleontology
UC	University of Chicago
UCMP	University of California Museum of Paleontology
UF	Florida Museum of Natural History
UMMPV	University of Michigan Museum of Paleontology, Vertebrate Collection
UNSM	University of Nebraska State Museum
USNM	National Museum of Natural History, Smithsonian Institution
YPM	Yale Peabody Museum, Yale University

FOSSIL FROGS AND TOADS
OF
NORTH AMERICA

1

Introduction

AN OVERVIEW OF FROGS AND TOADS

This section provides a definition of the Amphibia, the Lissamphibia, and the Anura (frogs and toads), as well as a discussion of the specializations of the Anura. Hereafter, when the word *frog* is used in the general sense, it indicates the Anura. The term *Amphibia* is used in the traditional Linnaean sense, rather than in a cladistic sense.

Definition of Amphibia, Lissamphibia, Anura. The members of the class Amphibia may be defined as tetrapods or derivatives of tetrapods that are characterized by a non-amniote egg (in general an "aquatic egg") and commonly by an aquatic larval stage that metamorphoses into at least a partially terrestrial adult. Many ancient, extinct amphibians, however, were large, highly ossified, flat-bodied forms, obviously morphologically quite different from the living ones. These ancient groups are included in the Amphibia mainly because the fossil evidence indicates they laid their eggs in water, had aquatic larval stages, and were at least partially terrestrial as adults or had evolved from other groups that were terrestrial as adults.

Modern amphibians belong to three distinct orders: Caudata (salamanders), Anura (frogs and toads), and Gymnophiona (caecilians). As of 1992, extant species of salamanders numbered 357; of frogs and toads, 3967; and of caecilians, 163. Judging by the increase in the naming of anuran species from 1985 to 1992 (Duellman, 1993, p. 13), I would estimate that at least 9% more species of modern amphibian orders are presently recognized. The modern orders are usually grouped together as the Lissamphibia, a taxonomic term that has been widely accepted by recent authors (e.g., Trueb and Cloutier, 1991; Cannatella and Hillis, 1993; Trueb, 1993; Sanchiz, 1998; Carroll, 2000b). The term was outlined first by Parsons and Williams (1963), who studied the anatomy and functional anatomy of the living groups and suggested that they must have had a common ancestry from some specific lineage of ancient Paleozoic amphibians. Osteological characters of Parsons and Williams that separate the lissamphibians from the ancient groups include (1) the unique teeth, which consist of a pedicel and a crown (pedicellate teeth); (2) the operculum–plectrum complex in the ear; (3) the odd fenestration in the posterolateral part of the skull roof; (4) the loss of posterior bones of the skull roof; (5) the open palate; (6) the presence of two occipital condyles; (7) the characteristically shaped cervical vertebra (atlas). But documenting the phylogenetic relationships between the ancient and modern amphibian groups has been a daunting task, as such studies are severely hampered by the scarcity of fossils that bridge the gap.

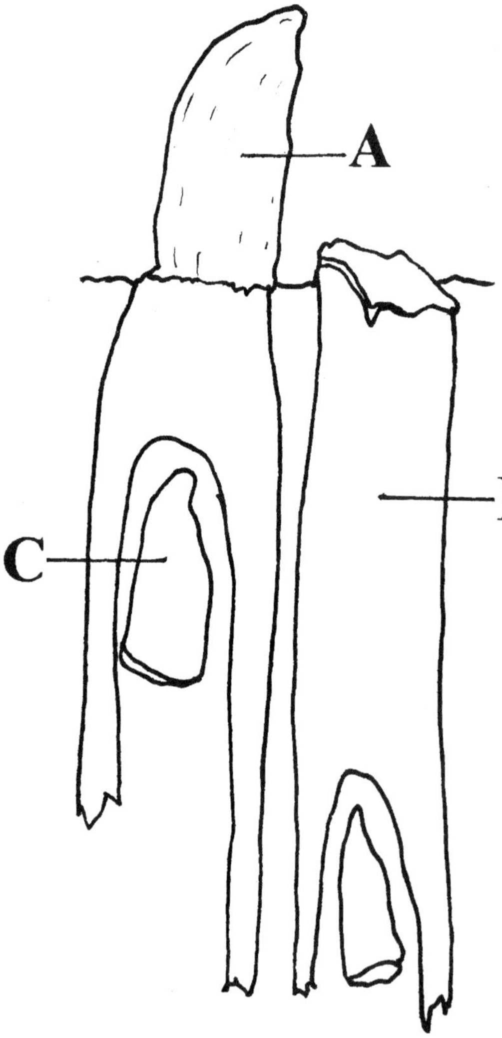

FIGURE 1. Two pedicellate teeth (semidiagrammatic) drawn from *Thaumastosaurus* sp., an anuran from the Late Eocene of England. (A) Erupted functional tooth crown. (B) Pedicel with the crown broken off. (C) Replacement crown growing within pedicel.

The presence of pedicellate teeth is probably the strongest character that unifies the three orders of modern amphibians. These teeth (Fig. 1) are composed of two portions, a basal pedicel and a distal pointed crown. The crown is usually shorter than the pedicel. In frogs, the pedicel is divided from the crown either by uncalcified dentine or by fibrous connective tissue.

In salamanders and caecilians the two tooth parts are divided only by fibrous connective tissue. In most modern amphibians the crown is capped by enamel or other hard substances. The division between the pedicel and the crown lies near the gum-free margin of the jaw.

During the developmental stage, both tooth parts form a continuous mass of tissue that early becomes separate from the jaw. During the life of anurans, crowns break off or become separated from their pedicels in various ways, so varying numbers of crownless teeth are usually present in the adults. When skeletons of anurans are prepared by the maceration method (bacterial action in water), the crowns usually become detached and are often lost.

New crown buds form in the base of pedicels and grow upward within this structure to take the place of previously detached crowns. In fossil and skeletal anurans, a medial opening on the basal part of the lingual surface of the pedicel, indicating the position of a former crown bud, is often present in many teeth. In a few anurans the boundary between pedicels and crowns may be obscure, and in the leptodactylid genus *Ceratophrys* undivided teeth develop from divided ones.

Most modern amphibian groups have two small bones that together transmit sounds to the inner ear. The operculum is the proximal element, and the plectrum (columella) is the distal one. Sometimes the two bones are fused. The operculum develops in association with the fenestra ovalis (oval window) in all amphibians. In a few frogs the plectrum is lost, but in the others this bone is usually joined to the tympanum.

In ancient amphibians the posterolateral portion of the skull is bounded by the jugal and postorbital bones that separate the eye socket (orbit) from the temporal region. But in the modern amphibian orders both the jugal and the postorbital bones are lacking, thus forming an odd fenestration in this region. Of note is the fact that some frogs have formed dermal bone in this region, thus secondarily reducing this fenestration. These anurans include some Burrowing Treefrogs and toads in which this area is elaborated for support and protection.

All of the modern orders lack certain bones in the posterior part of the skull roof that were present in the ancient amphibians. These elements include the supratemporal, intertemporal, tabular, and postparietal bones.

An open palate is characteristic of modern amphibians, in contrast to the ancient groups, which have much more closed palates. The modern taxa have pterygoid bones that are reduced and widely separated and that normally articulate with the brain case. Also, unlike the situation in the ancient primitive groups, the parasphenoid tends to be broad and to form a significant part of the palate.

In modern amphibians, two occipital condyles on the back of the skull articulate with the vertebral column, unlike the ancient amphibians, which had only one occipital condyle. Moreover, in modern amphibians the first presacral vertebra (cervical vertebra) bears two cotyles for the reception of the two occipital condyles on the skull and lacks transverse processes or ribs.

The specialized modes of swimming and jumping are the primary characters that differentiate the frogs from the other two orders of the Lissamphibia. Feeding adaptations, especially the ability to swallow relatively large prey species whole, are also important. Characters that can be used in the study of the relationships of ancient and modern amphibians are mainly associated with the skull, the vertebral column, and the pelvic girdle and hind limbs.

The skull (Figs. 2, 3) is very large in comparison with the body, is short and wide, and usually has huge orbits. The vertebral column (see Fig. 2) is short and contains eight or nine presacral vertebrae; and the tail is missing, as the caudal vertebrae are fused to form a single slender element, the urostyle (coccyx), which is functionally associated with the pelvic girdle. The sacrum is composed of a single vertebra that has developed a single pair of stout transverse processes that articulate with the pelvic girdle. The ilia are elongate and have become the origin of primary muscles associated with the locomotion of the animal. The femur is very long, and long, fused tibiofibulae articulate distally with a calcaneum and an astragalus that are fused proximally and distally in

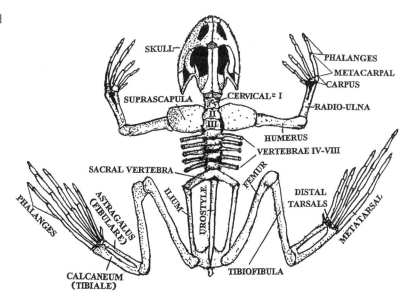

FIGURE 2. Generalized advanced anuran skeleton (semidiagrammatic) in dorsal view, with major bones and groups of bones on the postcranial skeleton labeled.

advanced anurans. The bones of these three segments of the hind limb are conspicuously hollow and thin. The enlarged hind feet are supported by very long digits.

Anuran Specializations. Frogs are not "typical" vertebrates in any sense of the word, and the dissection of frogs by the millions in general biology laboratories was a wanton misuse based on the former abundance and low cost of the animals. I shall discuss several of the special nonskeletal adaptations of anurans in the following paragraphs.

THE SKIN

As with other lissamphibians, the skin of anurans is thin and "naked," lacking the covering of scales, feathers, or hair found in other vertebrate groups. Frog skin is unique in being loosely attached, basically because it is fastened to the body wall only in five discrete places (Duellman and Trueb, 1994). In a world populated with humans, this has been a disadvantage to frogs, as the skin is easy to tear and frogs are without a doubt the easiest vertebrates for humans to skin, whether in the biology lab or in preparation for the gourmet table. Frog skin is permeable to water and thus functions in osmotic regulation, respiration, and to a somewhat limited extent, temperature regulation.

The external appearance of frogs is greatly affected by the action of special structures in the skin, as well as its color and pattern, which are regulated by the activity of chromatophores (pigment bodies). Many frogs, including the highly poisonous ones, are brightly colored, and others have the ability to change their colors and patterns by changing the distribution of chromatophores.

The outermost layer of the epidermis, the stratum corneum, is periodically shed in both frogs and salamanders. In both, the stratum corneum first splits along the midline at the top of the head, and then the splitting proceeds posteriorly. Most frogs use their limbs to help remove the molt, either in one piece or in sections. The molted skin is normally

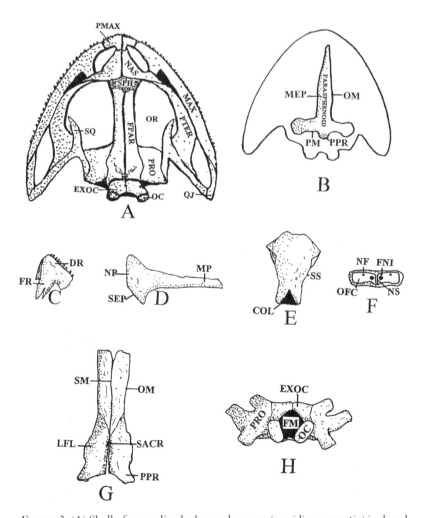

FIGURE 3. (A) Skull of generalized advanced anuran (semidiagrammatic) in dorsal view: EXOC, exoccipital; FPAR, frontoparietal; MAX, maxilla; NAS, nasal; OC, occipital condyle; OR, orbit; PMAX, premaxilla; PRO, prootic; PTER, pterygoid; QJ, quadratojugal; SPH, sphenethmoid; SQ, squamosal. (B) Position of parasphenoid in skull of generalized advanced anuran in ventral view: MEP, medial process (also called cultriform process); OM, orbital margin; PM, posterior margin; PPR, posterior process. (C) Right premaxilla in lateral view: DR, dental region; FR, facial region. (D) Right nasal in lateral view: MP, maxillary process; NP, nasal process; SEP, sphenethmoid process. (E) Sphenethmoid in dorsal view: COL, cavity for olfactory lobe; SS, supraorbital shelf. (F) Sphenethmoid in anterior view: FNI, foramen for nerve I; NF, nutrient foramen; NS, nasal septum; OFC, olfactory cavity. (G) Frontoparietal in dorsal view: LFL, lateral flange; OM, orbital margin; PPR, posterior process; SACR, sagittal crest; SM, sagittal margin. (H) Occipital complex in posterior view: EXOC, exoccipital bone; FM, foramen magnum; PRO, prootic bone.

eaten by the shedding individual. Some estivating frogs form cocoons by accumulating multiple layers of shed skin.

Mucous glands and larger structures called granular glands are important in frogs. They form in the epidermis but become imbedded in the upper part of the dermis (stratum spongiosum). Both types of glands are saclike but differ in their distribution in the stratum spongiosum, in their secretory regulation, and in their function. Mucous glands are more numerous and more widely distributed in the skin than granular glands, which tend to be confined to specific areas. But in many frogs, clusters of mucous or granular glands form discrete structures called macroglands (Duellman and Trueb, 1994). Macroglands that are formed from mucous glands spontaneously secrete mucopolysaccharides, which keep the skin slimy and moist. Macroglands that are formed from granular glands secrete a large variety of poisons in response to nervous and hormonal stimuli. These poisons, some of which are extremely toxic, protect frogs from a variety of predators.

Two other types of skin glands occur in frogs. Lipid glands, found in the Treefrog *Phyllomedusa*, secrete an impervious coating on the skin, which keeps the animal from drying out. Lipid glands are often somewhat larger than granular glands. Breeding glands can be found in the frog family Microhylidae. These glands are about the same size as granular glands, occur in the chest region, and produce a sticky secretion that fastens the breeding males to the dorsal part of the females during mating.

Since frog fossils consist almost entirely of hard parts, such as bones and teeth, the presence of bones in the dermis of some frogs and toads is especially pertinent to the present book. The dermis of the dorsal part of the body in some Treefrogs (family Hylidae) contains small dermal plates that are capped by bony spines that penetrate the epidermis. Moreover, in the widespread, mainly Neotropical leptodactylid frogs (family Leptodactylidae), the genera *Caudiverbera*, *Ceratophrys*, and *Lepidobatrachus* have large dermal plates in the dorsal body skin. In several groups of frogs, especially some toads of the genus *Bufo* and several Treefrogs, the dermis becomes fused (coossified) with the underlying bone, forming a readily identifiable character in fossils. Non-bony warts (verrucae) and pointed projections (coni) in the skin of some frogs are hardened by apical keratin.

FOOD PROCUREMENT

Adult frogs are carnivores, but the majority of larval forms are herbivorous. Most adult frogs sit and wait for their prey rather than actually foraging for it. In frogs, visual location is the most commonly documented method of finding prey. But olfactory location may be used in fossorial (burrowing) forms (e.g., *Rhinophrynus*); and auditory detection of insect sounds or of calls of other frogs sought as prey has been documented in several anuran species.

With the exception of the aquatic pipids, frogs use the tongue to capture their prey. The usual method in advanced species consists of a flip of the tongue forward in which the posterodorsal surface of the retracted tongue becomes the anteroventral surface of the extended tongue. This anteroventral surface adheres to the prey, which is drawn into the open mouth when the tongue is retracted. The hyoid apparatus, or hyo-branchium, is a well-integrated musculoskeletal system in the floor of the mouth that supports the tongue and takes part in tongue activity. These skeletal elements are for the most part slender and fragile and are not often found in the fossil record.

In the more primitive species, the tongue does not have much free surface and cannot be moved far, or it may even be fused to the floor of the mouth. Some frog groups have conditions of the tongue or tongue skeleton that have not yet been studied from a functional standpoint. In *Ceratophrys*, the mandibles are strongly fused and armed with dorsally directed spikes (odontoids) for holding large prey.

Anuran feeding mechanisms that differ markedly from tongue flipping include those of the permanently aquatic pipids and those of fossorial taxa. In the pipids, food is drawn into the mouth by hyobranchial pumping that is essentially the same as that used by frog larvae. Many fossorial frogs feed underground on ants, termites, and worms. The tongue of the most specialized of these, *Rhinophrynus dorsalis*, is a rodlike tube that is protruded through a small buccal groove at the end of the rostrum. Termite prey is captured by a cuplike tip on the end of the tongue, which is drawn slowly into the mouth.

VOCALIZATION

Vocalization in animals is mainly used to indicate their presence (advertising) to others of the same species. It is often found in flying animals (e.g., insects and birds) or hopping animals (anurans) that do not leave continuous trails that may be followed by chemosensory means. Most frogs have well-developed vocal systems that produce sounds that attract mates, advertise territories, or indicate distress.

The interactions involved in frog sound production and reception are quite complex (see Duellman and Trueb, 1994, pp. 89–97). In essence sound production is based on the interaction of nares, laryngeal vocal cords, body wall, and specialized vocal sacs that expand and act mainly as resonating chambers for sound produced by vibrations of the vocal cords.

Structurally, there are three types of vocal sacs in anurans. The most generalized type consists of a single median sac on the bottom of the throat (median subgular sac). Derived structures include paired subgular sacs that are either completely or incompletely separated from each other and paired lateral sacs that are completely separated from each other and occur on either side of the head.

Frogs have a unique sound-reception system. The primary sound receptors are the external ears, but the forelimbs also take part in this activity. High-frequency sounds travel from the tympanum, which forms the external ear, through the plectrum (columella) of the middle ear to the inner ear, where the acoustic information is processed by a ventral inner ear structure called the papilla basilaris. Low-frequency sounds are transmitted through the opercular muscle (operculum) to a structure called the papilla amphibiorum, which processes this acoustic information. The papilla amphibiorum occupies a more dorsal position in the inner ear. This so-called operculum–plectrum complex in the ear is one of the unique characters in the Lissamphibia. Moreover, the papilla amphibiorum is also unique in the group. The middle-ear structure of a primitive Jurassic frog has been interpreted as indicating that vocalization may have been present in these early anurans (Estes and Reig, 1973).

Several types of vocalizations occur in anurans. Advertisement calls are designed to attract conspecific females or to announce occupied territories to males of the same or other species. Reciprocation calls are given by receptive females in reaction to the advertisement calls of conspecific males. Reciprocation calls occur in only a few species. Release calls are vocalizations accompanied by body vibrations made by males or unwilling females in response to reproductive seizure (amplexus) by males. Finally, distress calls are loud sounds produced by either sex, usually with the mouth open, in response to seizure or disturbance by predators.

COURTSHIP AND MATING
Courtship and mating activities in frogs are varied and highly interesting. Vocalizations play an important role in guiding anurans to breeding sites, as do olfactory (odor) and visual stimuli. Detection of land slope and associated water bodies can also be important. Some species use several of these clues at once. Odor is probably the major factor in orientation and movement to the breeding site in the Leopard Frog (*Rana pipiens*), as blinded individuals orient to and move toward such sites over distances up to 800 m (Dole, 1968). Attachment to particular breeding sites has been demonstrated in anurans, including individuals of the Common European Toad (*Bufo bufo*) that returned to areas where former breeding ponds had been destroyed (Heusser, 1960).

Secondary sexual characters occur in most modern amphibians, including anurans. Female frogs are usually larger than males, with the exception of a few species whose males are of equal size or even slightly larger than females. Spines are present in males of many species. These include spines on the ends of the digits of the front feet (prepollical spines); projecting spines from the proximal ends of the humeri (humeral spines); and spines on outer surface of upper lip (labial spines). Spines are sometimes used in combat. In the Gladiator Frogs of the Treefrog genus *Hyla*, prepollical spines may produce fatal wounds in males grappling with other males to defend their nests.

There are indications that the humeral spines in several species of centrolenid frogs may be used by males to hook each other in grappling matches. A Papuan hylid genus (*Nyctimystes*) also has humeral spines and may engage in combat with this weapon. The Spadefoot genus *Leptobrachium* of the Oriental faunal region has a cornified row of labial spines on the outer surface of the upper lip. Sharp tusks (odontoids) on the lower jaws occur in both sexes in several genera of frogs (e.g., *Ceratobatrachus*, *Hemiphractus*, and *Pyxicephalus*). In frogs of other genera (*Phyllodytes* and *Adelotus*), the males may have much larger tusks than the females.

The most prominent external secondary sexual characters in frogs, other than vocal sacs, are structures called nuptial excrescences that are composed of modified dermal and epidermal portions of the skin. These structures are most highly developed in species that breed in water. They mainly consist of roughened patches and spines or spiny areas that are usually on the first finger or sometimes on the forearm and are used in amplexus, associated with mating.

Osteological secondary sex characters that occur on the humerus of males of many frogs take the form of flanges and roughened areas on the distal shaft of the humerus. These structures serve as attachments for muscles associated with the gripping of females that occurs during amplexus. These humeral characters are important in determining the sex of fossil anurans; and in North America, the humeral flanges of males of the larger species of fossil *Rana* are quite distinctive.

In many taxa, various glands occur on the ventral surface of breeding males. Sometimes these glands secrete a sticky substance that helps the males maintain amplexus. Such glands occur in the North American genus *Gastrophryne* (Microhylidae). The function of other ventral glands in male frogs is unknown, but it is thought that because these glands are in contact with the female during amplexus, they produce secretions that stimulate ovulation or oviposition. Glands on the throat (mental or gular glands) occur in several taxa of male tropical frogs and are usually present throughout the year. Other male tropical frogs have rounded glands on the chest (pectoral glands) or on the ventral surface of the thighs (femoral glands).

Sexual dimorphism in skin texture occurs in some toads of the genus *Bufo* (Bufonidae): in those species the males have more numerous cornified tubercles on the dorsal surface of the back than females. On the other hand, in the Common European Toad (*Bufo bufo*), males have rather smooth skin with low tubercles and females have rougher skins with more robust tubercles. A most striking modification of the skin occurs in the African frog *Trichobatrachus robustus* (Ranidae): during the breeding season, males have long hairlike filaments consisting of vascularized epidermis on the thighs and flanks. These males sit on clutches of eggs on stream beds, and it is reasoned that the filaments increase the cutaneous respiration of the animal during these episodes.

The most striking modification of the cloaca in frogs is associated with sexual dimorphism in the genus *Ascaphus* (Leiopelmatidae), of the streams of the northwestern United States. In this taxon, a cloacal extension, or "tail," of the male is inserted into the cloaca of the female during the mating process. Internal fertilization has also been suggested for the small African bufonid, *Mertensophryne micranotis*, as the male has protruding spiny vents that putatively fit into grooves in the female vent.

In male anurans of many species, the vocal sac changes color during the breeding season. In many small species of the Treefrog genus *Hyla*, the throat becomes yellow, whereas in toads of the genus *Bufo* the throat becomes gray or black. In some species of the genus *Rana*, the tympanum of males is significantly larger than that of females. This is true of the common North American taxa, the American Bullfrog, *Rana catesbeiana*, and the Green Frog, *Rana clamitans*. In most anurans, however, the tympanum is about the same size in both sexes or may be slightly larger in females.

Actual courtship behavior in anurans begins with the advertisement calls made by males. Normally, the males reach the breeding site first, to begin the ritual. When the sexes are ready to mate the male usually grabs the female dorsally to achieve amplexus (clasping by one or both pairs of limbs). Inguinal amplexus, where the male clasps the female around the waist, bringing his groin near hers, occurs in primitive frogs. This method is not considered to be as efficient for the fertilization of eggs as the more anterior position, called axillary amplexus, which brings the vents closer together. In typical axillary amplexus the forelimbs of the female are clasped by the male, who inserts his thumb solidly into the axilla (armpit) of the female. Most advanced anurans use axillary amplexus, but there are some notable exceptions.

In some short-limbed anurans with fat, round bodies, axillary amplexus is not possible. Thus, in the African frog *Breviceps* (Microhylidae) the males become glued to the posterior part of the female's back. Some frogs of the family Dendrobatidae use cephalic amplexus: the male grabs the underside of the female's throat. In a few ranids from Madagascar the male straddles the female across the head and shoulders. True amplexus is absent in some frogs, such as the Granular Poison-arrow Frog *Dendrobates granuliferus* (Dendrobatidae); instead, the male and the female face opposite directions (butt to butt) and press their cloacal areas together.

In frogs that use amplexus, the female usually selects the oviposition (egg-placing) site; but in the terrestrial dendrobatids that lack amplexus, the male usually selects the site. In these same dendrobatid species there is usually a vigorous defense of territories by either the male or the female. In those species where males aggressively defend territories, the females approach the males, but when females defend territories the opposite occurs. Courtship of females by males includes such activities as jumping on the back, prodding and poking, and even performing a short cephalic amplexus. Visual activities include hopping in front of the female, a performance called a "toe dance," or a darkening of coloration.

Courtship behavior occurs during amplexus in a few anurans. In the permanently aquatic anurans of the family Pipidae, the male uses arm-pumping behavior to inhibit rejection activity of the female. In the terrestrial Common European Spadefoot Toad *Pelobates fuscus* (Pelobatidae), the male not only uses arm pumping but makes up and down movements of the head and strokes the female's cloaca with his hind feet.

Both fertilization and oviposition occur during amplexus in most frogs and toads. When the female selects an oviposition site, abdominal contractions accompany movement of the eggs down the oviduct. It has been suggested that this may signal the male that oviposition is about to occur. Oviposition and fertilization are accompanied by synchronized movements in the mating pair. These movements range from rather simple ones to highly acrobatic movements in the permanently aquatic pipid frogs.

The Tailed Frog, *Ascaphus truei*, is the only anuran that has true internal fertilization in that it has a phallodeum formed by a posterior extension of the cloaca. When the phallodeum becomes turgid and is inserted in the cloaca of the female, copulation, which can last for 30 hours, takes place. Internal fertilization occurs in a few tropical frogs of the family Leptodactylidae by simple cloacal apposition, and it has been suggested that other modifications of the cloacal region in a few poorly known species of frogs indicate internal fertilization.

Eggs and Larvae

Anuran eggs are small and fairly simple and generally similar to those of other lissamphibians. The small amount of yolk in these eggs does not allow for a long incubation period, so the hatchling larvae are relatively small, in contrast to the much larger hatchlings that develop in shelled eggs with large amounts of yolk, such as are found in reptiles and birds. Larvae that develop rapidly in non-permanent aquatic situations also tend to be especially small.

In frogs and other modern amphibian groups a strong, thin, vitelline membrane encloses the egg proper. This membrane is surrounded by a series of capsules, the number of which varies between species. Frogs, however, generally have fewer capsules than salamanders. The American Bullfrog (*Rana catesbeiana*), for instance, has only one capsule. In general, anurans that bypass the tadpole stage and have direct development have larger eggs than species with a larval stage. Frog and toad eggs in sites exposed to sunlight usually have melanin (dark pigment) deposits over the portion of the egg (animal hemisphere) containing the developing embryo. It has been suggested that this is to protect the embryos from ultraviolet radiation.

In the Holarctic it is common for frogs to breed and lay their eggs in relatively shallow, still, permanent bodies of water, where the larvae hatch, feed, and metamorphose into tiny individuals. The relatively large newly metamorphosed individuals of some large species of *Rana* are a notable exception. On the global scene, however, there are many of kinds of egg-deposition sites as well as modes of tadpole development, which have led to some of the most interesting studies of anuran biology (see Duellman and Trueb, 1994). In fact, the very specific modes of reproduction in some anuran species make them extremely vulnerable to extinction.

Egg deposition. Egg-deposition sites for anuran genera with aquatic eggs include those in still water (*Bufo* [Bufonidae], *Pseudacris* [Hylidae], and *Rana* [Ranidae]); in running water (*Ascaphus* [Leiopelmatidae] and *Atelopus* [Bufonidae]); in tree holes and bromeliads (*Anotheca* [Hylidae] and *Mertensophryne* [Bufonidae]); in foam nests in ponds (*Physalaemus* [Leptodactylidae]) or stream pools (*Megistolotis* [Myobatrachidae]); or on the back of the female in small excavations (*Pipa* [Pipidae]).

Deposition sites for anurans with non-aquatic eggs include a variety of terrestrial nests (*Centrolene* [Centrolenidae], *Dendrobates* [Dendrobatidae], *Eleutherodactylus* [Leptodactylidae], *Hemisus* [Ranidae], and *Pseudophryne* [Myobatrachidae]); in leaves hanging over water (*Cochranella* [Centrolenidae] and *Hyla* [Hylidae]); in walls of tree holes (*Acanthixalus* [Hyperoliidae] and *Chirixalus* [Rhacophoridae]); in foam nests in burrows (*Adenomera* [Leptodactylidae] and *Heleioporus* [Myobatrachidae]); in trees (*Chiromantis* [Rhacophoridae]); on the legs of males (*Alytes* [Discoglossidae]); in a dorsal pouch on the female (*Flectonotus, Gastrotheca,*

Hemiphractus [Hylidae]); or as eggs retained in oviducts (*Eleutherodactylus* [Leptodactylidae] and *Nectophrynoides* [Bufonidae]).

Tadpole development. In anurans with aquatic eggs, tadpole development may occur in temporary or permanent ponds where feeding takes place (*Bufo* [Bufonidae]; *Physalaemus* [Leptodactylidae], a foam-nest builder; *Pipa carvalhoi* [Pipidae], whose eggs are initially deposited on the female's back; and *Rana* [Ranidae]). Anuran tadpole development may also occur in streams where feeding takes place. These taxa include *Ascaphus* [Leiopelmatidae]; *Atelopus* [Bufonidae]; and *Megistolotis* [Myobatrachidae], a stream-pool foam-nest builder. Tadpole development may occur in tree holes or bromeliads where the larvae feed in the available water (*Anotheca* [Hylidae] and *Mertensophryne* [Bufonidae]). Larval development may also occur in tree holes and leaf axils where the animals do not feed (*Anodonthyla* and *Syncope* [Microhylidae]); or on the female's back, where direct development takes place (*Pipa pipa* [Pipidae]).

In anurans with non-aquatic eggs, tadpoles may develop in the water after being carried there from terrestrial nests (*Dendrobates* [Dendrobatidae]); in a terrestrial nest where the animals do not feed (*Leiopelma* [Leiopelmatidae]); in vocal sacs (*Rhinoderma* [Rhinodermatidae]); on the skin of the back (*Sooglossus* [Sooglossidae]); in the walls of tree holes where the animals feed (*Acanthixalus* [Hylidae] and *Chirixalus* [Rhacophoridae]); or directly in either the skin of the back (*Hemiphractus* [Hylidae]) or back pouches (*Stefania* [Hylidae]). Tadpoles may also develop by absorbing nutrients provided by yolk in the oviduct of females (*Eleutherodactylus* [Leptodactylidae]); or by absorbing nutrients provided by specialized secretions of the oviduct (*Nectophrynoides* [Bufonidae]).

Tadpole structure and habits. Each frog species with a larval form is essentially two discrete animals, each with a totally different structure and set of habits. This is especially striking in taxa with specialized tadpoles. Many anuran families (e.g., Bufonidae, Hylidae, Microhylidae, Ranidae) have larvae of the so-called generalized pond type. The bodies of these animals are ovoid, and the tail is about twice as long as the body. But among genera with this body form, a variety of feeding patterns occur. Tadpoles of the genus *Rana* in North America are generalized pond grazers. On the other hand, tadpoles of the genus *Microhyla* (Microhylidae) are surface feeders, having a funnel-mouth adapted for such activities. Mid-water feeders, such as North American *Gastrophryne* (Microhylidae), tend to be less rounded and more compressed. Other pond types are bottom feeders.

The generalized stream-type tadpole tends to have a depressed body and a long tail. Those forms that inhabit riffle areas of streams tend to have down-turned mouths like certain species of fishes in the sucker and minnow families. Other stream-type tadpoles cling to stones by means of specialized suctorial discs on the ventral surface. Lee-side tadpoles that cling to stones tend to have upturned mouths and suckers. Tadpoles that live in litter or gravel beds in streams tend to have an elongate, almost serpentine body, with few fins.

Odd types of tadpoles include semiterrestrial wrigglers with reduced fins and snakelike bodies. Others are specialized feeders with mouths adapted for such activities.

Tadpoles that live in bromeliads usually have depressed bodies and elongate tails. Some of the most interesting tadpoles are carnivorous forms that are specialized for eating other tadpoles.

In the leptodactylid genus *Ceratophrys*, the adult is characterized by its daggerlike teeth and preys on other adult frogs. *Ceratophrys* tadpoles are also carnivorous, having strong jaws and beaks for feeding on other tadpoles. On the other hand, the adult rhinophrynid, *Rhinophrynus*, preys on ants by extruding its long, sticky tongue, but its tadpoles prey on other frog larvae.

Needless to say, much more can be learned about the morphology and ecology of tadpoles, especially those from remote areas on the planet.

HISTORY OF STUDIES OF NORTH AMERICAN FOSSIL ANURANS

Only a few people have written about North American fossil anurans. There are two reasons for this: the fragile, hollow bones and cartilaginous portions of the anuran skeleton often do not fossilize well (and when they do fossilize, they are usually fragmentary); and large vertebrate fossils (e.g., dinosaurs and mammoths) have been considered more desirable for exhibition and study.

Studies of fossil anurans began much earlier in Europe than in North America. European writings on anuran fossils are known from as early as 1580 and 1600 (see Roček and Rage, 2000b). Andreae (1776) discussed a fossil anuran from the Middle Miocene of Ohningen in southern Germany, and the fossil was later described as *Palaeophrynos gessneri* Tschudi, 1839. It has been suggested that this specimen is probably the earliest fossil frog reported that is still available for study (Roček and Rage, 2000b). Other notable early European zoologists with paleontological interests who dealt with fossil anurans include Goldfuss (1831), Agassiz (1835), who moved to America after being a professor at Neuchâtel from 1832 to 1846, Pomel (1844, 1853), Lartet (1851), von Meyer (1852, 1860), and Gervais (1859).

The legendary Edward Drinker Cope (1840–1897) founded the study of fossil frogs in North America. His writings on frogs figure among his numerous other works in modern ichthyology and herpetology, as well as his prodigious number of publications in vertebrate paleontology.

Cope's greatest contributions to vertebrate zoology and vertebrate paleontology resulted for the most part from his mastery of the study of internal morphology, especially the skeleton, which he used in the classification of both fossil and modern amphibians and reptiles. With respect to the frogs, his work on anuran classification (Cope, 1865) was especially important. Other publications by Cope that deal with skeletal morphology of modern and fossil anurans include his works in 1864, 1866, and 1884. Cope (1865) named the only extinct anuran family that is presently recognized, the Palaeobatrachidae. After Cope's work, it was not until the middle 1900s that fossil frog studies revived in a meaningful sense in North America.

At this point it seems prudent to discuss historical differences between European and North American fossil frog studies. In essence, European anuran fossils are often more complete that those in North America, which tend to be represented by unassociated, individual elements. Roček and Rage (2000b) recognized more than 60 fossil anuran species in the Tertiary of North America. Fewer are known from the European Tertiary, although the study of fossil anurans began much earlier in the region. This disparity is said to be based on the fact that many North American fossil frog identifications (especially of bufonids, hylids, and ranids) have been based on isolated ilia and sacra, and some of these should be considered suspect (Roček and Rage, 2000b).

On the other hand, many European "complete" anuran fossils tend to be impressions in rocks, rather than being composed of three-dimensional bones. These impressions not only lack depth, but usually show little morphological detail. Compare, for instance, the sketchy line drawings of "complete" European fossil frogs (Sanchiz, 1998; Roček, 2000; Roček and Rage, 2000b) with the morphologically detailed figures of individual elements representing North American anuran taxa in these publications.

I attribute the true beginning of the study of fossil anurans in North America to a single event, the perfection of a simple technique for washing and screening (sieving) very small vertebrate fossils, perfected shortly after World War II by Claude W. Hibbard (1949) (see also Holman, 2000a for details on Hibbard's work). This technique and the ensuing publications on smaller species of vertebrates sparked interest in what is now called microvertebrate paleontology, the study of small fossils representing such groups as small fishes, birds, mammals (such as shrews, bats, and mice), as well as amphibians and reptiles.

Edward H. Taylor, an outstanding herpetologist who described many modern herpetological species during his long tenure at the University of Kansas, came in contact with Hibbard before Hibbard moved to the University of Michigan. In the 1930s and 1940s Taylor published studies on new fossil anurans from the Miocene of Nevada and the Miocene and Pliocene of Kansas. His most detailed works dealt with the genus *Rana* (e.g., Taylor, 1938, 1942). His Kansas material was collected mainly by Hibbard.

Joseph A. Tihen, who spent most of his professorial career at the University of Notre Dame, began his career in herpetology with studies on modern lizards. Tihen was also influenced by Hibbard and not only published seminal works on fossil anurans, mainly in the 1950s and 1960s, but became an active vertebrate paleontologist. His publications dealt with fossil anurans ranging from the Miocene to the Pleistocene and involved work at sites in Florida, Kansas, Nebraska, Oklahoma, and Texas. His classic work on New World Bufonidae included an osteological study on modern New World *Bufo* species (Tihen, 1962a) and another on New World fossil *Bufo* (Tihen, 1962b).

Walter Auffenberg was influenced by Tihen and also by Pierce Brodkorb, of the University of Florida, who was using Hibbard's techniques in the 1950s to collect fossil bird bones. Auffenberg, who, like Brodkorb, spent almost all of his professional career at the University of Florida, began his studies in herpetological paleontology in the 1950s. Later he switched to the study of modern varanid lizards and became the world's expert on the subject. Auffenberg is most well known paleontologically for his studies on fossil snakes and turtles, but he also produced excellent short papers on anurans from the Miocene and Pliocene of Florida and the Pleistocene of Barbuda, Leeward Islands (Auffenberg 1956, 1957, 1958).

Bayard H. Brattstrom of California State University, Fullerton, was a pioneer in the study of small herpetological remains in California and Nevada in the 1950s and 1960s. In the course of analyzing Pleistocene herpetofaunas, Brattstrom identified several anuran species. His publication on the amphibians and reptiles from the famous Rancho La Brea Pleistocene site in Los Angeles (Brattstrom, 1953a), which included anurans, is recognized as a classic. Brattstrom later published mainly on physiological and ecological adaptations in modern amphibians and reptiles.

J. A. Holman first published on fossil amphibians and reptiles in the 1950s while studying fossil birds with P. Brodkorb for his Ph.D. degree. Holman was strongly influenced by W. Auffenberg, J. A. Tihen, and B. H. Brattstrom. After graduation from the University of Florida in 1961, Holman spent most of his academic career at Michigan State University, where he learned about fossil collecting in Kansas from C. W. Hibbard and in Nebraska from M. Voorhies. Holman has published mainly on Tertiary and Pleistocene frogs, turtles, and snakes.

Max K. Hecht, of Queens College, New York, and a long-time associate of the American Museum of Natural History, had a long career in paleoherpetology and evolutionary biology, but he published most of his work on fossil anurans in the 1960s. His studies on fossil frogs encompass mainly the early history of frogs from the Jurassic to the early Tertiary, but he has published on anurans from the Miocene as well. His studies of fossil frogs have often been of an evolutionary nature (e.g., Hecht, 1962, 1963, 1969), rather than a purely descriptive nature, although his descriptive works (e.g., Hecht, 1959, 1960, 1970) have been very detailed.

Charles J. Chantell, a Ph.D. student of J. A. Tihen at the University of Notre Dame, published several detailed works on fossil anurans in the 1960s and early 1970s. Chantell, who spent his academic years at the University of Dayton (Ohio), was a specialist in fossil frogs and published several detailed papers on these animals from the Tertiary of Colorado, Idaho, and Nebraska. Chantell is known not only for his studies of fossil frogs (Chantell, 1964, 1965, 1966, 1970), but also for his osteological studies of the frog genera *Acris* and *Pseudacris* (Chantell, 1968a, b), which have been useful in fossil interpretations.

John D. Lynch (Ph.D., University of Kansas; career, University of Nebraska) is best known for his comprehensive studies of modern leptodactylid frogs (especially the modern species of the huge genus *Eleutherodactylus*). He published several papers on fossil frogs in the 1960s, mostly dealing with the distribution of Pleistocene frogs (Lynch, 1964, 1965, 1966). The anuran osteological knowledge gained by Lynch in these early studies was well used in his later comprehensive works on anurans (e.g., Lynch, 1971, 1978).

Richard Estes received his Ph.D. from the University of California and spent his academic career at San Diego State University. Here he produced numerous publications and was a mentor to some outstanding students in paleoherpetology. Although he is better known for his comprehensive works on salamanders and lizards, Estes actively published on fossil frogs from the 1960s through the 1980s. His works on fossil frogs mainly dealt with those from the Cretaceous and early Cenozoic, and he published not only on North American taxa but also on those from South America, Europe, Africa, and Australia. A summary of Late Cretaceous and Cenozoic herpetofaunas of North and South America with regard to faunal interchange (Estes and Báez, 1985) and a major paper on the early fossil record of frogs (Estes and Reig, 1973) are paleoherpetological classics.

Let us turn now to modern North American anuran fossil studies. Given the hundreds of contemporary herpetologists in the United States alone, the small number of people who have recently published in fossil frogs is surprising. Ami C. Henrici of the Carnegie Museum of Natural History is one of the leaders, with publications on rhinophrynids and pelobatids from the Eocene of Wyoming and Montana (Henrici, 1991; Henrici and Fiorillo, 1993) and on a new pelobatid from the Arikareean of Montana (Henrici, 1994). Karel L. Rogers, of Grand Valley State University (Michigan), has provided records of anurans from the Pliocene of Nebraska and Texas and ecologically significant records of anurans from the Pleistocene of Kansas and Colorado (Rogers, 1976, 1982, 1984, 1987).

Leslie P. Fay of Rock Island College, Illinois, has provided interesting information on the ecology and distribution of Appalachian Pleistocene anurans (Fay, 1984, 1986, 1988). Jim I. Mead of Northern Arizona University, in Flagstaff, with several junior authors has contributed ecologically important records of Pleistocene and Holocene anurans from both the Colorado Plateau and the Great Basin (1982–1985). Thomas R. Van Devender, Mead's counterpart in the Sonoran and Chihuahuan desert regions, along with his junior authors, has also provided some ecologically important Quaternary records of anurans from those regions. Neil Shubin, of the University of Pennsylvania, and Farish A. Jenkins, Jr., have reported an important jumping frog from the Jurassic of Arizona (Shubin and Jenkins, 1995; Jenkins and Shubin, 1998). Finally, James B. Gardner, presently at the Royal Tyrrell Museum of Palaeontology in Alberta, Canada, and a student of Cretaceous amphibians, is a bright light on the horizon of fossil frog studies. Hopefully, more herpetologists will turn to the study of this fossil group.

THE ANURAN SKELETON

The study of fossil frogs is based almost entirely on the study of individual fossilized bones or small assemblages of bones. Unfortunately, anuran teeth, although specialized, are almost useless in making specific identifications. This is a far cry from the situation in some other vertebrate groups, where pharyngeal teeth (e.g., in suckers and minnows) and molar teeth (in many mammalian taxa) may be identified to the specific level. The paragraphs below provide an introduction to the anuran skeleton and some of the terminology associated with individual bones, as well as comments on the usefulness of various elements in paleontological studies. For the most part, the terminology used here is an anglicized version of that of Sanchiz (1998).

The frog body may be divided into three general regions: the head, the neck, and the trunk. Fishes essentially have no neck, but in adult amphibians, with the disappearance of internal gills, a neck, albeit still a very short one, appears. Actually, little movement of the head upon the trunk occurs in frogs. Moreover, in frogs, the trunk itself is much shorter than in most other vertebrates, and a true tail is absent in adults. On the other hand, the hind legs are developed to a great degree in anurans. These adaptive specializations in the anuran body are mainly associated with the hopping and leaping gait of the animals and are obvious in the skeleton.

Figure 2 is a semidiagrammatic depiction (in dorsal view) of an advanced frog. The skull is depressed. Enlarged nasal openings (solid black) occur in the anterior part of the skull, and very large orbits (solid black) occur in its posterior part. The large eyeballs, situated within the orbits, are associated with sight feeding, which is characteristic in frogs. Moreover, in some frogs (e.g., *Rana*), the large eyes, equipped with strong muscles, are used to help push the prey down into the throat. The curved upper jaw is armed with numerous small pedicellate teeth that are used to hold onto struggling prey while it is being swallowed.

The very short vertebral column is composed of nine vertebrae in which the cervical vertebra or atlas (I) connects the remainder of the vertebral column to the skull, and the robust sacral vertebra (technically the ninth vertebra) connects the vertebral column to the pelvic girdle and hind limbs. Intermediate vertebrae II–VIII have rounded transverse processes (not labeled in Figure 2), although in vertebrae II and III these processes are covered by the suprascapula, a thin element derived from cartilage. The front limbs connect to the sternum ventrally by means of the proximal end of the humerus to the scapula and coracoid (this obscured in the drawing by the suprascapula).

The humerus connects distally to the radio-ulna, which consists of the fused radius and ulna. The elongate metacarpals and phalanges and the small semirounded bones of the carpus make up the bones of the forefeet. In locomotion, the forelimbs are used to orient and position the head and body, and they also contribute to the launch phase of the jump. In the mating process the forefeet are used to grasp the female. Moreover, during the feeding process, many anurans use the forefeet along with the eyeballs to push food down the throat.

The pelvic girdle and hind limbs of anurans are elongate and specialized and are among the most highly modified of any vertebrate group. All the elements involved are functionally associated with the animal's ability to hop, leap, crawl and swim; but they also interact, like a set of springs, to cushion the body from shock when it makes contact with the substrate at the end of a hop or leap. The long, compressed urostyle (coccyx) evolved by means of the fusion of posterior vertebrae. It not only strengthens the pelvic girdle but diminishes the landing impact in leaping frogs. The ilia are also elongate, and as will be discussed later, are important single elements used in the identification of fossil frogs.

The extent of the enlargement and elongation in anuran hind limbs and their associated muscles is illustrated by the fact that, from a practical standpoint, "frog legs" are the only frog parts many modern humans choose to eat. The hind limbs of anurans, rather than having three functional joints as in most tetrapod vertebrates, have four. This is made possible by the elongation and enlargement of the calcaneum and astragalus, which are often fused proximally and distally. In effect, this has created a huge, double-jointed hind foot. The femur is rounded in cross section, long, hollow, and slightly twisted. The equally long tibiofibula is more compressed in cross section and represents the fusion of the tibia and the fibula of most other tetrapods. Proximal and distal grooves in this element indicate this fusion. Appended to the distal (usually fused) ends of the calcaneum (tibiale) and astragalus (fibulare) are very small distal tarsals, oddly, sometimes absent from commercially prepared articulated frog skeletons. Both the metatarsals and phalanges are elongate, except for the terminal phalanges.

Turning to bones of the anuran skull that are used in the identification of fossil anuran remains, we examine a dorsal view of an anuran skull with important bones identified below (Figs. 3 and 4). Again, we note the very large orbits. Tooth-bearing bones around the outside of the skull include, at its anterior tip, the small premaxillae. The long, exter-

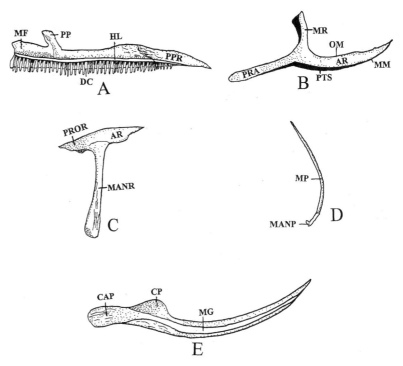

FIGURE 4. Additional individual elements of generalized advanced anuran skull. (A) Right maxilla in medial view: DC, dental crest; HL, horizontal lamina; MF, maxillary fossa; PP, palatine process; PPR, posterior process. (B) Right pterygoid in dorsal view: AR, anterior ramus; MM, maxillary margin; MR, medial ramus; OM, orbital margin; PRA, posterior ramus; PTS, pterygoidal sulcus. (C) Right squamosal in lateral view: AR, anterior ramus; MANR, mandibular ramus; PROR, prootic ramus. (D) Right quadratojugal in dorsal view: MANP, mandibular process; MP, maxillary process. (E) Right angulosplenial in lateral view: CAP, cranial articular process; CP, coronoid process; MG, mandibular groove.

nally curved maxillary bones make up most of the rest of the outside rim of the skull, articulating with the small quadratojugals posteriorly. The roughly T-shaped squamosal bone articulates with the quadratojugal laterally, the curved pterygoid ventrally, and the squarish prootic medially. The squamosal bone does not actually contribute to the wall of the orbit because it rides over the pterygoid and prootic bones, which form part of the anterior and lateral (pterygoid) and posterior (prootic) wall of the orbit.

The roughly triangular nasal bones lie posterior to the premaxillae and form the medial walls of the nasal openings (external nares). Posterior to the nasals one finds a dorsal exposure of the sphenethmoid bone. The major structural component of the dorsal roof of the skull consists of a pair of long, flat elements, the frontoparietals, each composed of two fused dorsal skull bones, the frontals (anterior) and the parietals (posterior). The frontoparietals themselves are often fused at the midline. The two short bones posterior to the frontoparietals are the exoccipitals. These bones, joined together, surround a cavity called the foramen magnum that contains and protects the spinal cord. Each exoccipital has a posterior, rounded occipital condyle that articulates with the first vertebra (called cervical, or atlas), a double-socketed element. Some of these individual skull bones will be discussed individually later, depending on their abundance and use in the identification of fossil anurans.

Moving now to the ventral side of the skull, we feature only one large palatal bone, the T-shaped parasphenoid that forms much of the floor of the palate (see Fig. 3B) and is the only palatal bone that has been used much in the identification and description of fossil anurans. Significant areas on the bone include the medial process, orbital margin, posterior margin, and posterior process. Any of these areas may differ in separate anuran groups, but enough intraspecific variation exists that one needs an adequate sample of recent and fossil material to determine the usefulness of characters on this bone.

Returning to individual elements labeled in the dorsal view of the skull (see Fig. 3A), we can see that the right premaxilla in lateral view (Fig. 3C) is toothed and has two regions, the facial and the dental, that may be examined for osteological characters. The premaxilla has had a somewhat minimal use in the identification of fossil species. A right nasal bone in lateral view (Fig. 3D) can be divided into three processes, nasal, sphenethmoid, and maxillary. Anuran fossil nasal bones are usually rather fragmentary and have been of limited use in paleontological studies.

A sphenethmoid in dorsal view (Fig. 3E) shows the cavity for the olfactory lobe of the brain; a supraorbital shelf also exists. A sphenethmoid in anterior view (Fig. 3F) shows several characters. A nasal septum divides the bone into two olfactory cavities. Somewhat large foramina for cranial nerve I may be observed just lateral to the nasal septum. Smaller nutrient foramina occur lateral and somewhat dorsal to the nerve I foramina. The sphenethmoid is a relatively sturdy bone of the anuran skull and thus has been preserved more often in the fossil record than several other cranial elements. It can been used with moderate success in anuran paleontological identifications.

The frontoparietal, a paired bone (Fig. 3G, dorsal view), may also be used with some degree of success in the study of fossil frogs. Fused fron-toparietals form a sturdy median roof in the anuran skull; thus, these elements occur fairly often in paleontological sites. The degree of fusion varies somewhat, but a general rule is that the two bones are more firmly fused in the adults than in immature specimens. Important structures include the sagittal and orbital margins, the lateral flanges, a median sagittal crest, and a posterior process.

The fused occipital complex, composed of the prootic and exoccipital bones (Fig. 3H, posterior view), is a sturdy structure in the posterior region of the anuran skull. It is another element that is found relatively often in frog paleontological sites and that can be used with moderate success in the identification of taxa. Characters include the configuration and shape of the laterally projecting processes of the prootic, as well as the size and shape of the foramen magnum and the size and position of the occipital condyles.

Other bones of the skull that have supplemented fossil studies are indicated in Figure 4. A right maxilla in medial view is depicted in Figure 4A. Several important structures occur on this element. Although individual teeth are not particularly diagnostic, the number, size, and position of the teeth may be helpful in the identification of anuran taxa if the fossil bone is complete and adequate series of modern and fossil skeletons are available, which unfortunately is usually not the case. Other features of the maxilla include the dental crest, the horizontal lamina, the posterior process, the palatine process, and the maxillary fossa. A complete maxilla can be one of the most useful elements in fossil anuran studies.

Figure 4B shows a right pterygoid in dorsal view. This is one of the more fragile bones in the frog skull, and it usually turns up in a fragmentary condition at fossil sites. Important structures include the medial ramus, the posterior ramus, the pterygoidal sulcus, the anterior ramus, the maxillary margin, and the orbital margin. The T-shaped squamosal bone (Fig. 4C, right side, lateral view) is also prone to breakage during the fossilization process and thus is also not often described in fossil studies. The important parts of the bone are the prootic ramus, anterior ramus, and mandibular ramus. A right quadratojugal is shown in dorsal view (Fig. 4D). This bone is thin and fragile and seldom turns up as a fossil in identifiable form unless it is a part of an articulated skull, and as I have previously indicated, fossil frog skulls are exceedingly rare. Two noteworthy portions of the quadratojugal are the mandibular process and the maxillary process.

Turning to the mandible (lower jaw), which is usually called the angulosplenial in paleontological studies, we find a very sturdy bone (Fig. 4E, right side, lateral view). In fact, this element is probably the most dense bone of the anuran skull, as it often exists in fossil deposits in a relatively complete state. It is also a moderately good bone to work with, as it has rather consistent characters within species. The most prominent characters include the cranial articular process, the coronoid process, whose size and shape may be interspecifically variable, and the mandibular groove.

We now move on to axial postcranial elements (Fig. 5). The cervical vertebra, or atlas (Fig. 5A, in dorsal view), technically the first (I) of the

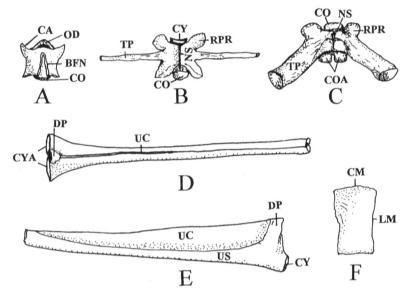

FIGURE 5. Axial postcranial elements of generalized advanced anuran skeleton. (A) Cervical vertebra (atlas) in dorsal view: BFN, bifid neural spine; CA, articular facet for left occipital condyle; CO, condyle; OD, odontoid process. (B) Mid-region presacral vertebra in dorsal view: CO, condyle; CY, cotyle; NS, neural spine (undivided); RPR, right prezygapophysis; TP, transverse process. (C) Sacral vertebra in dorsal view: COA, condyles; (double) CY, cotyle; NS, neural spine; RPR, right prezygapophysis; TP, transverse process. (D) Urostyle (coccyx) in dorsal view: CYA, cotyles (double); DP, dorsal prominence; UC, urostylar crest. (E) Urostyle in lateral view: CY, right condyle; DP, dorsal prominence; UC, urostylar crest; US, urostylar shaft. (F) Posterior sternal plate in dorsal view: CM, coracoidal margin; LM, lateral margin.

nine vertebrae in the vertebral column, functions to join the column to the skull (see Fig. 2). This cervical is rather strongly built and has been used with some success in the identification of fossil anurans. Noteworthy structures include paired, concave condylar articular facets on the anterior end of the bone in a dorsal position and a single ventral odontoid process. These elements articulate with the fused occipital bones at the rear of the skull (see Fig. 2). This anuran articular complex, where two horizontal condyles from the skull articulate with two horizontal, concave condylar articular surfaces of the cervical vertebra, allows up and down motion of the head upon the body but restricts side to side motion. Thus, to orient the head sideways, anurans must move the head and body as a unit. On the posterior surface of the cervical, however, is a single condyle that articulates with a single, concave cotyle on presacral vertebra II. This allows the head and the cervical together to move as a rotary unit. A bifid neural spine occurs on the dorsal surface of the cervical vertebra.

A presacral vertebra from the middle part of the frog body is shown in Figure 5B. Anuran presacral vertebrae are sturdy elements and tend to occur frequently at fossil sites. Nevertheless, isolated presacral vertebrae have not been used much in fossil anuran studies. Important structures include the concave cotyle on the anterior end of the vertebra and the convex condyle on its posterior surface. These articular structures join adjacent vertebrae and allow some motion within the column. Other articular surfaces that join adjacent vertebrae include the prezygapophyses, with their articular surfaces facing upward, and the postzygapophyses (not labeled), with their articular surfaces facing downward. Elongate transverse processes function as immovable ribs in the anurans. The neural spine on the dorsal surface of presacral vertebrae in anurans is normally not bifid like it is on the cervical vertebra.

The terminal vertebra (Fig. 5C) in the column forms the entire sacrum in frogs. It is a dense, strongly built bone that frequently occurs in fossil deposits and one that has been rather widely used in paleontological studies. Unfortunately, it is somewhat intraspecifically variable, and an adequate series of both fossil and recent sacra is needed for accurate studies.

Noteworthy structures include the backswept, thickened transverse processes, whose distal ends articulate with the proximal ends of the ilia (see Fig. 2) and the prezygapophyses. A condyle in this individual frog is anterior, and additional paired condyles occur posteriorly for articulation with paired cotyles in the urostyle. Later we shall see that condylar and cotylar positions in the vertebral column proper and sacrum are quite variable in anurans and can be useful in separating taxa (e.g., *Rana* from *Bufo*). Moreover, the transverse processes may be structurally quite different in different anuran groups.

The urostyle (Figs. 5D, 5E) is an elongate, rather fragile bone that plays an important role in the suspensorium formed by the pelvic girdle. In dorsal view, noteworthy structures include the paired cotyles on the proximal end. These concave structures articulate with the condyles of the sacral vertebra. A dorsal prominence lies in the midline above and just posterior to the cotyles, and a urostylar crest is visible. In lateral view, this crest is visible as an extensive but thin structure that lies above the shaft of the urostyle. The structure of the dorsal prominence and its position relative to the cotylar portion of the bone may be seen in this view. Bits of the urostyle, especially pieces of the proximal end, are not uncommon in paleontological sites, but they have not been used extensively in fossil anuran studies. The posterior sternal plate (Fig. 5F) turns up occasionally in fossil localities, but it is not a particularly diagnostic frog bone. Its coracoidal and lateral margins are indicated in the figure.

Elements of the frog pectoral girdle and forelimb are shown in Figure 6. The bony portion of the mainly cartilaginous suprascapula is shown in Figure 6A. This bone usually appears as thin scraps in fossil sites and is seldom used in paleontological studies. The anterolateral and posteromedial margins of this bone are indicated in the figure. The scapula (Fig. 6B) plays an important role in the structure of the pectoral girdle complex, as it articulates with four other bones in the complex. Moreover, the scapula can be a diagnostic bone in studies of fossil anurans at both higher and lower taxonomic levels. Fortunately, it is a fairly dense, sturdy bone that is often found at fossil sites, sometimes rather completely preserved. Features of the scapula include the suprascapular articular surface; the mesial crest, which varies in its development and extent among anuran taxa; the humeral articular facet; the coracoidal articular surface; and the clavicular articular surface.

FIGURE 6. Pectoral girdle and forelimb of generalized advanced anuran skeleton. (A) Bony portion of right suprascapula in dorsal view: ALM, anterolateral margin; PMM, posteromedial margin. (B) Right scapula in medial view: CAS, coracoidal articular surface; CLAS, clavicular articular surface; HAF, humeral articular facet; MC, mesial crest; SSA, suprascapular articular surface. (C) Right coracoid in dorsal view: AM, anterior margin; HAF, humeral articular facet; IAS, inter-coracoidal articular surface; SAF, scapular articular facet; SAM, sternal articular margin. (D) Right clavicle in dorsal view: CG, coracoidal groove; LCAF, lateral coracoidal articular facet; MCAF, medial coracoidal articular facet; SAF, scapular articular facet; SAM, sternal articular margin. (E) Right humerus in ventral view: CCH, calcified cartilaginous head of humerus; CF, cubital fossa; DC, distal condyle; HS, humeral shaft; MC, mesial crest; RE, radial (lateral) epicondyle; UE, ulnar (medial) epicondyle; VC, ventral crest. (F) Right radio-ulna in lateral view: FHC, facet for humeral condyle; CCA, calcified cartilaginous area for articulation with carpus; LRG, lateral radio-ulnar groove.

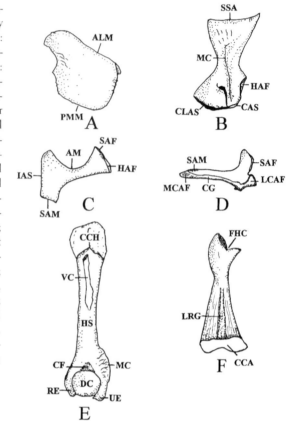

The coracoid itself (Fig. 6C) is not an uncommon frog fossil, but it is a simpler and thus less diagnostic bone at the lower taxonomic levels than the scapula. It is an important part of the pectoral complex. Characters on the coracoid include the inter-coracoidal articular surface, the anterior margin, the scapular articular facet, the humeral articular facet, and the sternal articular margin. The clavicle (Fig. 6D), another important element of the pectoral complex, is relatively thin; thus, it is usually found only as scattered fragments at fossil sites and has seldom been used in paleontological studies. Noteworthy characters of the clavicle include the medial coracoidal articular facet, the sternal articular margin, the scapular articular facet, the lateral coracoidal facet, and the extensive coracoidal groove.

One of the most common anuran bones used in paleontological identifications is the humerus (Fig. 6E), a common bone at anuran-bearing fossil sites but often difficult to use at the specific level because of intraspecific and sexual variations. The mesial crest flares in many male anurans (as in Fig. 6E), a character associated with amplexus, and tends to be diminished in females. The rounded distal condyle (sometimes incorrectly called the ball) may be rounded or oval in anuran taxa, but there is some intraspecific variation in this character. The humeral shaft tends to be slender in some anurans and stout in others. Other important humeral characters include the calcified cartilaginous head of the humerus, which, unfortunately, is often missing in fossil specimens; the ventral crest; the cubital fossa; the radial epicondyle, and the ulnar epicondyle, which lies below the mesial crest.

The radio-ulna (which reminds James O. Farlow of "a tiny ice cream scoop") is a common element, especially its proximal end, which is strongly built for its articulation with the distal condyle of the humerus (Fig. 6F). Unfortunately, there appear to be few structures of diagnostic value on the radio-ulna. Important parts of the radio-ulna include the facet for the humeral condyle, the lateral radio-ulnar groove, and the calcified cartilaginous area for articulation with the carpus.

Finally, we turn to elements of the pelvic girdle and the hind limb. The puboischium (Fig. 7A), a bone that represents the fusion of the pubis and ischium of the pelvic girdle, contains the posterior portion of the acetabular fossa, which forms the posterior part of the articular surface for the head of the femur. This puboischium is sometimes recovered as an anuran fossil, but it has seldom been used in the identification of fossil frogs. Structures of note on the puboischium include the calcified cartilaginous posterior border, the posterior portion of the acetabular fossa, and the ilial articular surface. In some adult anurans the puboischium may fuse rather firmly with the two ilia.

The ilium (Fig. 7B) is the bone most often relied on for identifications in North American fossil frog studies. The element is common in anuran fossil deposits, although in such situations, the bone usually has one-fourth to one-half of its anterior portion broken off. Nevertheless, frog ilia have many characters that are useful to the vertebrate paleontologist. This is not to say that intraspecific variations do not occur, though, so fossil ilia should be studied with an ample series of comparative modern skeletons available. Parts of the anuran ilium that are often mentioned in fossil studies include the dorsal acetabular expansion (sometimes called the supra-acetabular expansion); the dorsal prominence, which sometimes has a distinct rounded or knoblike structure on it that has been called the dorsal protuberance or tubercle; the dorsal ilial crest, a part of the ilial shaft; the preacetabular fossa; the ventral acetabular expansion (sometimes called the subacetabular expansion); and the anterior portion of the acetabular fossa. Ilia will be referred to time and again in the systematic accounts of this book.

The femur is a long, sigmoidally curved, hollow bone (Fig. 7C) that in most frogs is capped by calcified cartilaginous articular surfaces at both ends. Although pieces of anuran femora are relatively abundant in fossil deposits, they have not been particularly useful in fossil studies because the femur bears practically no prominent structures or muscle scars. For this reason it is even difficult to distinguish left anuran femora from right ones. The three prominent features of the anuran femur are the proximal calcified cartilaginous surface that articulates with the acetabulum of the pelvis; the sigmoidally curved shaft; and the distal calcified cartilaginous surface that articulates with the tibiofibula. The degree of calcification of these femoral articular surfaces increases with age, so they may be preserved in the fossil femora of older individuals. Usually, however, anuran fossil femora are hollow at both ends.

The anuran tibiofibula (Fig. 7D) is a slightly curved (not sigmoidally) bone that is about the same length as the femur, but it has a prominent

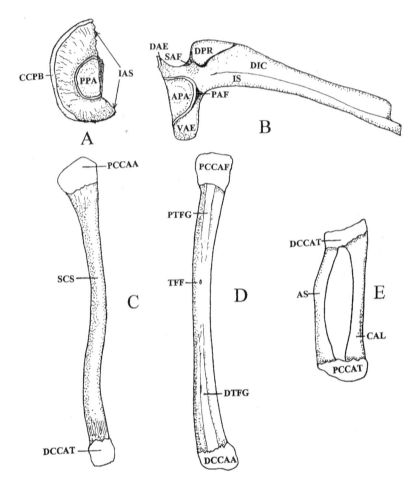

FIGURE 7. Pelvic girdle and hind limb of generalized advanced anuran skeleton. (A) Right puboischium in lateral view: CCPB, calcified cartilaginous posterior border; IAS, ilial articular surface; PPA, posterior portion of acetabular fossa. (B) Right ilium in lateral view: APA, anterior portion of acetabular fossa; DAE, dorsal acetabular expansion (supra-acetabular expansion); DIC, dorsal ilial crest (ilial crest, ilial blade); DPR, dorsal prominence; IS, ilial shaft; PAF, preacetabular fossa; SAF, supra-acetabular fossa; VAE, ventral acetabular expansion (subacetabular expansion). (C) Right femur in lateral view: DCCAT, distal calcified cartilaginous articular surface for tibiofibula; PCCAA, proximal calcified cartilaginous articular surface for pelvic acetabulum; SCS, sigmoidally curved shaft. (D) Right tibiofibula in dorsal view: DCCAA, distal calcified cartilaginous articular area for tarsus; DTFG, distal tibiofibular groove; PCCAF, proximal calcified cartilaginous articular surface for femur; PTFG, proximal tibiofibular groove; TFF; tibiofibular foramen. (E) Right astragalus (AS) and calcaneum (CAL) in dorsal view: DCCAT, distal calcified cartilaginous area for articulation with distal tarsals; PCCAT, proximal calcified cartilaginous area for articulation with tibiofibula.

tibiofibular foramen on its dorsal surface and is grooved at both the proximal and the distal ends. These grooves indicate the dual origin of the bone—the fusion of the tibia and the fibula. Tibiofibulae are moderately common elements in fossil frog sites and are somewhat more taxonomically useful than the femur in anuran paleontological studies. The four most prominent structures on the tibiofibula are the proximal calcified cartilaginous articular surface that articulates with the femur, the proximal tibiofibular groove, the tibiofibular foramen, the distal tibiofibular groove, and the distal calcified cartilaginous articular surface that articulates with the astragalus and the calcaneum. As in the femur, the degree of calcification of the proximal and distal articular surfaces increases with age.

The astragalus and calcaneum (Fig. 7E) (often called the tibiale and fibulare) are elongated tarsal elements that add a third joint to the anuran hind limb. Both bones are slightly curved and are joined both proximally and distally by calcified cartilaginous articular surfaces. These bones usually appear as separate elements in paleontological digs and have infrequently been used in fossil anuran studies; they are fused throughout their entire lengths to form a compound bone in only two anuran families, the Centrolenidae and the Pelodytidae. The proximal calcified cartilaginous surface articulates with the tibiofibula. The distal calcified cartilaginous surface articulates with the distal tarsals. These articular surfaces both become increasingly calcified with age.

Early Evolution of the Anura

Unlike the situation with the salamander and caecilian orders, the evolutionary events leading to the frogs may be discussed rather thoroughly, primarily because the fossil record is better for the anurans than for the other modern amphibian orders. Roček and Rage (2000a) have broken down the "why" of the origin of the anurans into addressable categories. Paedomorphosis (the retention of juvenile characters of ancestral forms by adults) is considered the principal evolutionary factor in the origin of the Anura, with "progressive evolution," changes associated with locomotion, premetamorphic transformations, and ecological impetus being considered important as well.

Several paedomorphic changes could have occurred: A shift to small size, a process sometimes called miniaturization, could have led to a retention of the juvenile skull shape. Fossil evidence has shown that some ancient amphibian skulls changed during postmetamorphic growth from being short and wide with large orbits in the middle portion of the skull to being long and narrow with small posterior orbits (see Roček and Rage, 2000a, fig. 3, p. 1277). Moreover, a shift of the pineal organ to an anterior position in primitive living anurans, in which this structure is still preserved, and the presence of a pineal foramen in the transitional froglike amphibian *Triadobatrachus* (Figs. 8A, 8B) correspond to the larval rather than the adult condition in the ancient amphibians.

Reduction in size and the loss of dermal bones, as well as a reduction in the size of the palate, may be related to paedomorphic processes. It has been suggested that a shortening of the ossification sequence itself would produce an anuranlike animal in ancient amphibians. In support of this is the fact that bones such as the parasphenoid, pterygoid, palatine, maxilla, nasal, frontal, parietal, and dentary, which appear in the premetamorphic stage in some ancient amphibians (branchiosaurs), also occur in anurans. On the other hand, some bones, such as the ectopterygoid, prefrontal, postfrontal, postorbital, and jugal, appear in late larval or postmetamorphic stages in branchiosaurs but are absent or cartilaginous in the anurans.

Bolt (1977, 1979) pointed out that in the dissorophoids (an ancient amphibian group), pedicellate bicuspid teeth were present in juvenile individuals, whereas the adults developed labyrinthine teeth, a common tooth type in many ancient amphibians. I have already discussed the fact that pedicellate teeth are characteristic of adult frogs. It has been argued that the rare appearance of non-pedicellate teeth in a few adult anurans is the result of incomplete tooth development resulting from paedomorphosis.

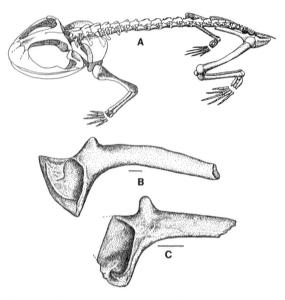

FIGURE 8. The Triassic proanurans *Triadobatrachus massinoti* and *Czatkobatrachus polonicus*. (A) Restoration of the whole skeleton of *T. massinoti*. (B) Right ilium of *T. massinoti*. (C) Right ilium of *C. polonicus*. Each scale bar = 1 mm.

Other anuran characteristics that have been attributed to paedomorphic events include the loss of the interclavicle; reduction and final loss of the adductor mandibulae externus muscle; the shift of the position of the tympanum, basipterygoid articulation, and otic process; the presence of a shallow otic notch; and the fact that the oval foramen faces ventrolaterally, as in the early developmental stages of the temnospondyls, an ancient amphibian group.

Changes thought to be due to "progressive evolution" by Roček and Rage (2000a), that is, changes that result in more derived structures than are found in the ancient amphibians, include such characters as the fusion of the frontal and parietal bones of the skull; the miniaturization of the stapes that ultimately led to the conversion of large-amplitude, low-intensity vibrations to small-amplitude, high-intensity vibrations; the development of a nearly monospondylous (single unit) vertebra derived from the pleurocentrum (one of the two main developmental units of the vertebral centrum of primitive amphibians); and the reduction of ribs and their incorporation into the transverse processes.

Changes associated with the specialized mode of locomotion in anurans include the reduction in the number of body vertebrae and incorporation of vertebrae into the urostyle (an event that began in the proanuran stage and reached its maximum development at a later time); the posterior elongation of the pelvis, brought about by the anterior elongation of the ilia; and the shortening of the tail.

Changes that occurred in the larval stages (premetamorphic stages) during the evolution of the anurans were profound. Distinct external gills are present in the developmentally earliest known larvae of the ancient temnospondyl amphibians. In the frogs, external gills of the same type as found in the temnospondyls develop first. But with the ontogenesis of opercular folds, external gills are reduced and internal gills are developed under the folds. These new gills are not homologous with either those of fishes or those of salamanders. Another evolutionary innovation in the larvae of anurans is the change from that of predation, as in larval temnospondyls, to that of feeding on plants (although some modern anuran tadpoles have secondarily become predators).

Turning now to an ecological impetus for the origin of anurans, it has been suggested that the appearance of transitional forms between temnospondyls and anurans in the Early Triassic may have reflected some profound climatic change at the close of the Permian. Such an event is considered to be responsible for the massive extinction in the world biota that included 75% of the amphibian and 80% of the reptilian families (Nevo, 1995). Studies of modern neotenic populations of salamanders suggest that either an increase in aridity or a drop in temperature could have led to such populations of transitional forms. Therefore, either or both of these factors could have been a major ecological impetus for the paedomorphosis leading to the origin of the anurans.

TRIADOBATRACHUS AND CZATKOBATRACHUS

Two key fossils from the Early Triassic, *Triadobatrachus* from Madagascar and *Czatkobatrachus* from Poland (Evans and Borsuk-Białynicka, 1998), morphologically bridge the gap between ancient Paleozoic amphibians and anurans. *Triadobatrachus* (see Figs. 8A, 8B) is remarkably complete, as it was found in the form of a natural mold in a nodule, whereas *Czatkobatrachus* (Fig. 8C) is represented by disarticulated vertebrae, humeri, and ilia. Such intermediate forms are not yet available in the salamanders and caecilians, although a caecilian with legs is known from the Early Jurassic of Arizona (Jenkins and Walsh, 1993; Carroll, 2000b).

The skull of *Triadobatrachus* is remarkably similar to that of anurans (see Figs. 2, 3). In the skull, the orbits are very large, and the frontoparietal forms the main part of the skull roof. The palate is reduced, the parasphenoid is T-shaped, and the palatine bones are transversely placed. The pterygoids are triradiate. In the postcranial skeleton, the ilia are elongate and the pubis is cartilaginous.

On the other hand, primitive characters not found in the anurans occur in *Triadobatrachus*. The skull may have a prefrontal bone, an element that is absent in the frogs. In the pectoral girdle, a cleithrum, an element present in primitive tetrapods but absent in anurans, is present. In the forelimb, the radius and the ulna are separate bones, not fused as in the frogs. The vertebral column is long and all the ribs are separate from the transverse processes, whereas in the frogs the vertebral column is short and only a few short anterior ribs occur in primitive forms. In *Triadobatrachus*, the pelvic girdle (even though having a long, anteriorly directed ilium), is of fairly normal size, rather than being greatly expanded as in anurans. Moreover, the hind limbs are of normal proportions, rather than being greatly lengthened as in the frogs. Expanded rather than reduced tarsal bones in *Triadobatrachus* indicate that the extra joint in the hind limb of anurans was not present in the Triassic fossil. Thus, considering the situation of the pelvic girdle and hind limb, it is safe to say that *Triadobatrachus* did not have a hopping and leaping (saltatorial) mode of locomotion as in the frogs and toads, but probably crawled around rather slowly.

Triadobatrachus (see Figs. 8A, 8B) was discovered in Madagascar in 1936; thus, the discovery of another "missing link" genus in Poland is eagerly embraced by the vertebrate paleontological community. *Czatko-batrachus* (see Fig. 8C) was found in karstic (irregular limestone) deposits of the Early Triassic (Scythian) at the Czatkowice locality (Evans and Borsuk-Białynicka, 1998). Disarticulated postcranial bones—vertebrae, humeri, and ilia—occurred at the Polish locality.

The new genus is about 5 million years (Ma) younger than *Triadob-atrachus* and is smaller than the Madagascar fossil, having an estimated snout–vent length of only 50 mm or so. It is also more derived than *Triadobatrachus* in characters of the vertebrae and elbow joint. Of interest is that the sacrum of *Czatkobatrachus* appears to be intermediate between that of *Triadobatrachus* and that of the frogs.

The First Anurans

Frogs and caecilians are known from the Early Jurassic, about 190 Ma before present (BP), and the salamanders appear in the Middle Jurassic about 165 Ma BP (Carroll, 2000a). The first anurans occur about 40 Ma after the intermediate fossil *Triadobatrachus* and about 35 Ma after *Czat-kobatrachus*. Based on the fossil record, it is obvious that the evolution of the Anura was a slow process but that once the frog body plan was achieved, little basic structural change took place in the 190 Ma that followed. I cannot resist saying here that it would be a unthinkable to allow this highly specialized and fascinating group of vertebrates to disappear from the Earth in a mere instant of geologic time because of human neglect and exploitation of the environment.

The earliest known fossil anuran is *Prosalirus bitis* Shubin and Jenkins, 1995 from the Early Jurassic Kayenta Formation, in Arizona. This important fossil will be briefly discussed here, and because it is of North American origin, will be detailed in the systematic accounts section of the book. This frog occurs in the same deposits as the earliest caecilians, which are limb-bearing rather than limbless, unlike the modern representatives of the group. Although *Prosalirus* bears no clear relationship to any fossil or modern anuran groups known, it is unquestionably a frog and was capable of jumping.

The second most ancient frog presently known is *Vieraella herbsti* Reig, 1961 from a somewhat younger deposit in the Early Jurassic than *Prosalirus*, namely, the plant-bearing Rosa Blanca Formation in Santa Cruz Province in southern Patagonia, Argentina. Part of the specimen was found in 1961, and its counterpart was discovered in 1965. *Vieraella* was a quite small frog with a snout–vent length of about 30 mm. *Vieraella* was essentially a modern frog (see Roček, 2000, fig. 3, p. 1301), and there is no question that it could jump. This taxon retained some primitive features: free ribs in two of the anterior vertebrae; and 11 rather than 9 total vertebrae in the vertebral column.

Because of such features as the lack of ossified epiphyses, the unossified sphenethmoidal part of the postnasal wall, and possibly the free ribs it has been suggested that *Vieraella* was not a mature frog (Roček, 2000). Relative to the phylogenetic position of this taxon, Estes and Reig (1973) surmised that it can be considered a structural ancestor to two primitive modern frog families, the Leiopelmatidae and the Discoglossidae. Since it has many characters shared with the *Leiopelma*, a living representative of the Leiopelmatidae, they assigned it to this family, although Sanchiz (1998) placed it with the Anura incertae sedis ("of uncertain taxonomic position").

Appearing about 20 Ma later, *Eodiscoglossus oxoniensis* Evans, Milner, and Musset, 1990 is the first fossil frog that has been unquestionably assigned to a modern family, the Discoglossidae, the Disc-tongued Toads that presently occur in Eurasia, North Africa, Southeast Asia, and the Philippines. This taxon was first reported from the Kirlington Mammal Bed, of Middle Jurassic age, in Oxfordshire, England. *Eodiscoglossus* closely resembles the modern genus *Discoglossus*: it has 15–18 premaxillary teeth and about 50 maxillary teeth; the ilium has a dorsal crest and dorsal ilial prominences; the coronoid process is smooth and convex, with no notches; the vertebral centra have the condyle directed anteriorly and the cotyle directed posteriorly (the opisthocoelous condition); and the urostyle has small, anterior transverse processes. Thus, we can say that truly modern frogs were hopping around among Middle Jurassic dinosaurs and have changed little in basic skeletal characters since then.

Another early frog, *Callobatrachus sanyanensis* Wang and Gao, 1999, from the Yixian Formation of China, has been assigned to the Discoglossidae. Unfortunately, two dates from the site conflict: one indicates a Late Jurassic age; the other, an Early Cretaceous age (Gao and Wang, 2001).

EARLY CRETACEOUS TADPOLES

Any fossil tadpole discoveries are remarkable, but those from Early Cretaceous deposits of the Taysir Volcanics in the Shomron region of Israel are of special interest. Not only do they reveal the skeletal structure of these early anuran larvae, but many of the specimens preserve nerves and blood vessels, as well as other parts of the soft anatomy. These ancient tadpoles, described by Estes et al. (1978), were given the name *Shomronella jordanica*. *Shomronella* was referred to the modern family Pipidae (Clawed and Surinam Frogs) based on soft and skeletal anatomy, including a large mouth, the presence of an angular bone in the lower jaw, and a narrow, lance-shaped parasphenoid in the palate. Large frogs usually have relatively big tadpoles, and as tadpoles go, those of *Shomronella* are indeed large, having heads up to about 30 mm wide. *Shomronella* differs from other known pipids in that the cleithrum has a prominent but thin anterior process and a small posterior process and the parasphenoid has a large, diamond-shaped posterior end.

Comments on the Identification of Fossil Anurans. An entire book should probably be written on the subject of the identification of fossil anurans, as it a difficult and often frustrating task. Here, I shall make a few simple suggestions for the beginner. First, obtain a mounted, articulated frog skeleton from one of the biological supply houses. These skeletons usually represent one of the large frogs of the genus *Rana*. You need to become familiar with all the bones in this skeleton. You can find laboratory manuals on the subject, and supply-house specimens often come with a drawing of the specimen that identifies the individual bones. Next, procure the dead body of a large frog or toad (roadkill is abundant in many areas during the frog and toad breeding season). Put the frog in a large jar filled with water, and screw the lid on tight. Let the jar sit in some warm, protected place, at room temperature, for a few weeks. Bacterial decay of the soft tissue of the specimen will then occur, unless some inhibiting mold or fungus is accidentally introduced into the jar.

Wearing rubber gloves, unscrew the top of the jar and pour the thickened liquid through a large tea strainer or other type of sturdy sieve. You should do this under a laboratory hood or outdoors, far away from people. A disarticulated frog or toad skeleton will soon appear in the strainer or sieve. Wash these bones thoroughly, and put them back in the jar for about two weeks at room temperature. At the end of the two weeks, repeat the pouring-off procedure. Be extremely careful not to spill any of the liquid on yourself or on your clothes, or you will not be welcome at home or anyplace else.

After the second pouring off, the disarticulated bones are usually ready to be put in a tray to dry. Some people prefer to soak the bones in a weak solution of ammonia for a few days before they are dried, to cut down on any lingering odor on the skeletal remains. Unlike many of the disarticulated skeletons produced by enzymes or chemicals, these bacterially macerated specimens should last indefinitely if stored in a reasonably cool, dry place.

Place the clean, dry, disarticulated specimen in a sturdy box, and attach labels inside and outside, identifying the species. The inside label should also give, at least, the snout–vent length (SVL) of the original specimen, the sex, the locality, the date, and the collector. At this point, compare the individual bones of your disarticulated skeleton with those of the articulated one. The individual bones of the body are relatively easy to identify, but learning to identify the individual bones of the skull is a challenging task.

When you are able to identify the individual body elements and most of the individual elements of the skull of your first disarticulated anuran, it is time to gather more roadkill of the same species and prepare a few more skeletons. When the additional skeletons are ready, lay out a series of the same bone from the different individuals, for comparison. First, determine which bones have a simple structure and which have a complicated one. It will probably be obvious that bones such as the femur are relatively simple, whereas the scapulae, humeri, and ilia are more complex. In most cases, you will find the complex bones more useful for identifying species than the simpler ones.

Next, study the series of bones to determine the amount of intraspecific (within a species) variation that exists. You will notice that a cavity or prominence in the humerus of one individual of the same species may be round, while in another individual the same structure is oval. Or, in a particular skull bone, there may be three foramina (small openings) in a specific area in one individual and one or two in the same area in another. Generally, you will be surprised at how much individual variation there is within any given anuran species.

After the above exercise, obtain a series of disarticulated skeletons of another anuran species to compare with the first one. In beginning studies such as this, it is useful to compare a series of a species of *Rana* with that of a toad (*Bufo*). A rather quick study of the ilia in these two anuran genera will show that these elements in *Rana* are rather easily distinguished from those in *Bufo*. The same is true of several other bones. On the other hand, if you compare the ilia (or other bones) of species within the genus *Rana* (or within the genus *Bufo*), you will quickly see that the individual bones appear to be similar or even identical. Thus, intensive studies are usually needed to differentiate individual elements in this category. This can be frustrating. In fact, you might come to the conclusion that certain bones in species within the same genus cannot be differentiated.

When you are identifying fossil anurans, it is important to keep a modern anuran skeletal collection at hand. You should certainly not try to identify fossil frogs and toads on the basis of the written descriptions and figures in this book, or for that matter, any other book. Intraspecific, intracolumnar, ontogenetic, and pathological conditions occur in anuran bones, both fossil and modern—without comparative series of skeletons, these might be misinterpreted as characters useful in the description of new species. Such mistakes have often been made in the past and are still being made.

The first step when you are identifying a fossil frog bone is to determine what element in the skeleton the fossil represents. The next logical step is to try to determine the anuran family or genus represented by the element, based on comparative skeletons available. If your comparative collection is small, you'll need to check the literature.

Identifying an individual anuran bone (or even a fairly complete fossil anuran skeleton) to the specific level is a daunting task and will take hours, days, and sometimes even weeks to accomplish. A good rule of thumb is to find good characters first and then substantiate them by measurements and ratios later. It is a mistake to "give up" early in the identification process and resort to the interpretation of "blind" measurements alone, as you are sure to miss much more reliable characters. A leading anuran osteologist once told me, "You can tell a beginning student right off because they start measuring things before they even know all of the bones of the skeleton."

Once you do decide to use measurements and ratios to distinguish anuran bones from one another, you'll soon find that such elements have curves and irregularities that make them difficult to measure accurately. For this reason, you should establish landmarks that can be reference points for comparative measurements. Esteban et al. (1995) illustrate landmarks on frog bones, and Sanchiz et al. (1993) indicate how they used some simple landmarks and statistics to establish the taxonomic relationships of an Oligocene frog. Actually, few morphometric (related to the characterization of form for quantitative analysis) or even simple statistical studies have been reported in papers dealing with the identification of fossil anurans, and the need for this type of work is apparent. I assure the reader these kinds of studies are important, but they will be challenging to say the least.

CHRONOLOGICAL TERMS USED IN THE BOOK

Here I shall use the globally standardized era, period, epoch, and age (for the Mesozoic era) units of the *1999 Geologic Time Scale* provided by the Geological Society of America (Palmer and Geissman, 1999). The three post-Precambrian eras (oldest to youngest) are Paleozoic, Mesozoic, and Cenozoic (Figs. 9, 10). I shall discuss only the Mesozoic and Cenozoic here because both proanurans and anurans are restricted to these units. Periods are divided into epochs, which may be divided either into two subdivisions, Early and Late, or three subdivisions, Early, Middle, and Late. Unlike the daily time units that we use, which can be broken down into subdivisions of equal duration, classical geologic subdivisions are of unequal duration because they are defined on the basis of the relative sequence of rocks and fossils that make up geologic time, rather than on the basis of actual time.

Age systems, especially those of the Cenozoic, may differ locally. Here I shall use the internationally recognized age units for the Mesozoic (see Table 1), but I shall use the North American land-mammal age (NALMA) for the Cenozoic (see Fig. 10), as the internationally recognized age units are mainly based on marine sequences. The NALMAs were originally based on the concept that large land mammals were not only wide ranging, but often restricted to narrow units of time (Wood et al., 1941). Also, somewhat understated at the time, was the fact that most of the Cenozoic vertebrate fossils that had been collected were remains of large mammals.

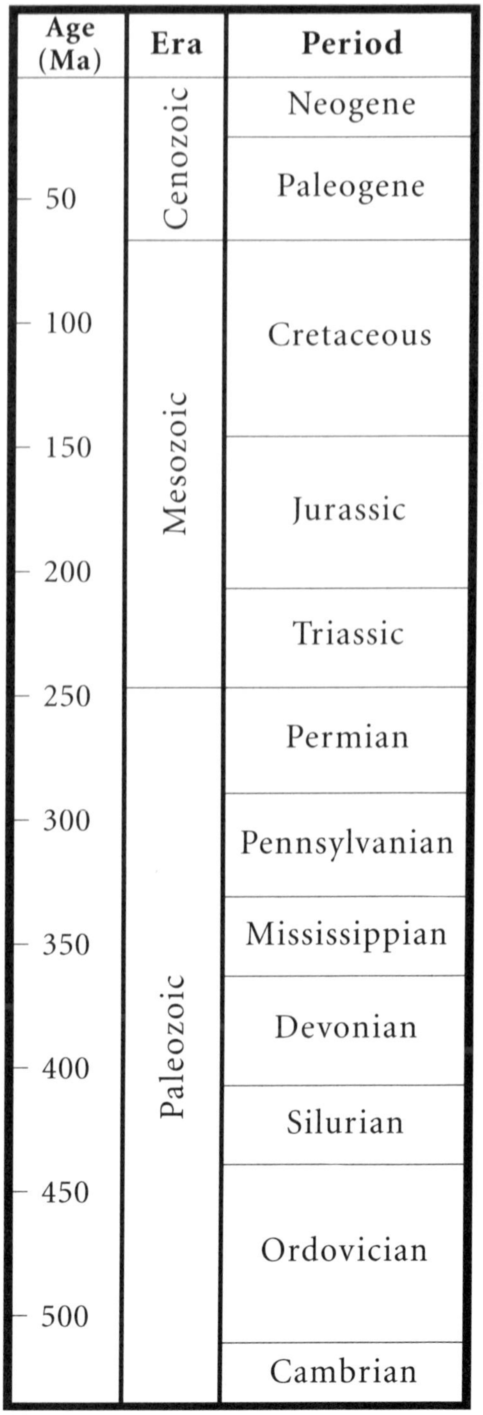

Age (Ma)	Era	Period
	Cenozoic	Neogene
50		Paleogene
100	Mesozoic	Cretaceous
150		Jurassic
200		Triassic
250	Paleozoic	Permian
300		Pennsylvanian
350		Mississippian
		Devonian
400		Silurian
450		Ordovician
500		Cambrian

FIGURE 9. Eras and periods (other than the Quaternary) of the geological time scale. The Quaternary period, which represents about the last two million years (2 Ma) of the Cenozoic era, appears in Figure 10.

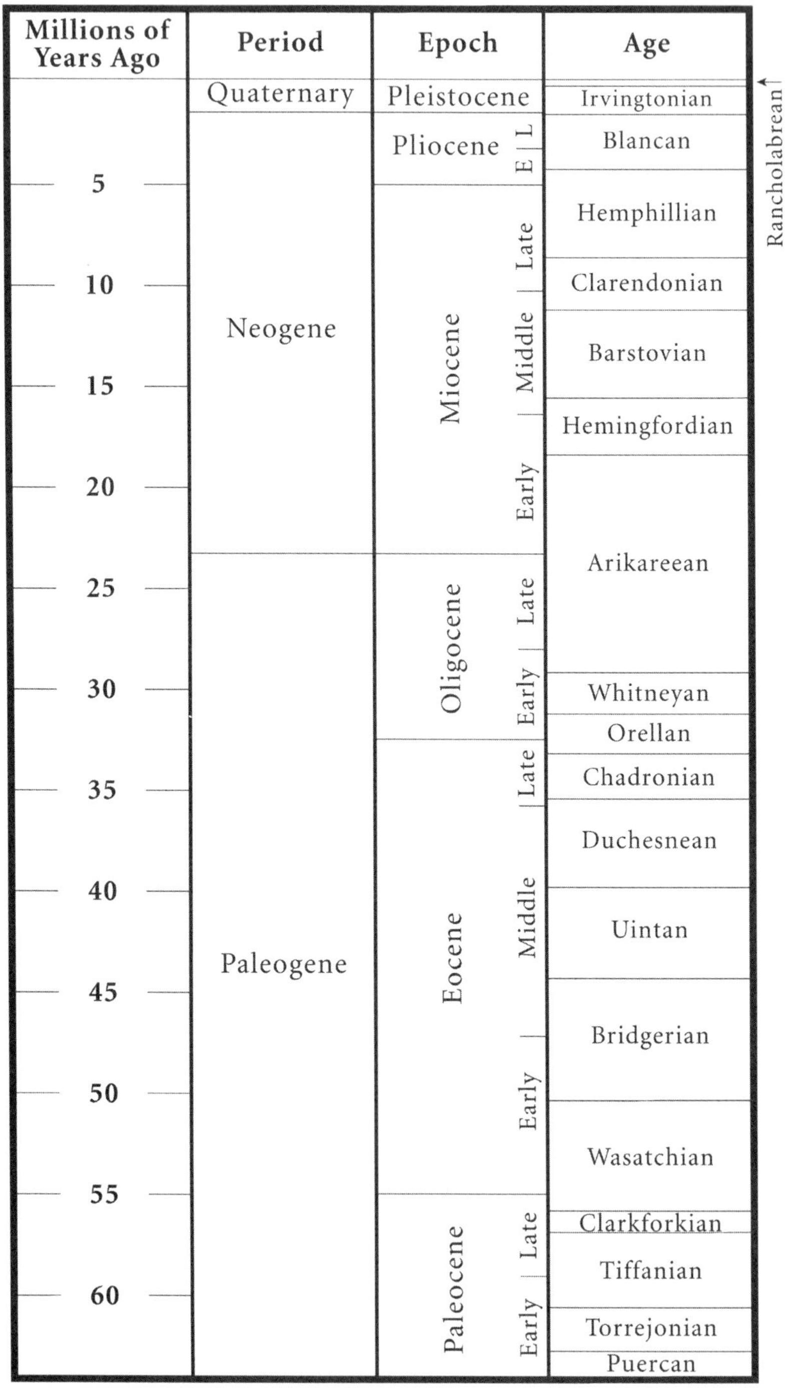

FOSSIL FROGS AND TOADS

FIGURE 10. Periods, epochs, and North American land-mammal ages (after Hulbert, 2001).

Millions of Years Ago	Period	Epoch		Age	
	Quaternary	Pleistocene		Irvingtonian	Rancholabrean
		Pliocene	L	Blancan	
5			E		
	Neogene	Miocene	Late	Hemphillian	
10				Clarendonian	
			Middle	Barstovian	
15				Hemingfordian	
20			Early	Arikareean	
25	Paleogene	Oligocene	Late		
30			Early	Whitneyan	
				Orellan	
35		Eocene	Late	Chadronian	
				Duchesnean	
40			Middle	Uintan	
45				Bridgerian	
50			Early	Wasatchian	
55		Paleocene	Late	Clarkforkian	
				Tiffanian	
60			Early	Torrejonian	
				Puercan	

TABLE 1. Internationally recognized ages of the Mesozoic.

Epoch/Age	Dates (Ma BP)	Epoch/Age	Dates (Ma BP)
Late Cretaceous		Middle Jurassic	
Maastrichtian	71.3–65.0	Callovian	164–159
Campanian	83.5–71.3	Bathonian	169–164
Santonian	85.8–83.5	Bajocian	176–169
Coniacian	89.0–85.8	Aalenian	180–176
Turonian	93.5–89.0	Early Jurassic	
Cenomanian	99.0–93.5	Toarcian	190–180
Early Cretaceous		Pliensbachian	195–190
Albian	112–99.0	Sinemurian	202–195
Aptian	121–112	Hettangian	206–202
Barremian	127–121	Late Triassic	
Hauterivian	132–127	Rhaetian	210–206
Valanginian	137–132	Norian	221–210
Berriasian	144–137	Carnian	227–221
Late Jurassic		Middle Triassic	
Tithonian	151–144	Ladinian	234–227
Kimmeridgian	154–151	Anisian	242–234
Oxfordian	159–154	Early Triassic	
		Olenekian	245–242
		Induan	248–245

Note: Oldest is at the bottom. Ma BP, million years before present.
Source: After 1999 *Geologic Time Scale* (Palmer and Geissman, 1999).

THE NORTH AMERICAN PLEISTOCENE

Although there are more fossil records of anurans in the North American Pleistocene than in all the other North American geologic units put together, this record has been understated and poorly elaborated in recent comprehensive fossil anuran studies (e.g., Sanchiz, 1998; Heatwole and Carroll, 2000). To rectify this, I am giving the anurans of this epoch special attention in this book. The terminology used to depict Pleistocene events may be somewhat confusing to the non-geologist.

Two systems of dividing the North American Pleistocene into temporal units presently exist. The older system, the Pleistocene glacial- and interglacial-age system, relies principally on the glacial and interglacial sedimentary record. The second system, the NALMA system, briefly discussed above, relies on the biochronology (dating of biological events using biostratigraphic or other objective paleontological date) of land mammals.

Turning first to the older system, we find that before the 1840s, North American scientists were prone to attributing what was actually glacial deposition to the effects of flooding. But later, Louis Agassiz's work showed that ancient glacial landforms and deposits existed far south of existing glaciers and that this indicated former cold climates. Stratigraphic studies then found that layers distinguished by weathered zones of organic soils and containing plant remains existed between layers of glacial sediments, such as sands and gravels. These early scientists postulated that the organic layers formed during unglaciated intervals and that the glaciers must have advanced and retreated several times.

Soon after the turn of the 19th century, four major glacial drift sheets were identified, each of which was separated from the other layers on the basis of organic layers or fossils that indicated an interglacial environment. The classical North American glacial and interglacial sequence that evolved from these studies is presented below (oldest at the bottom).

Wisconsinan glacial age
Sangamonian interglacial age
Illinoian glacial age
Yarmouthian interglacial age
Kansan glacial age
Aftonian interglacial age
Nebraskan glacial age

At present, the classical glacial and interglacial ages that occurred before the late part of the Illinoian are considered poorly defined and highly questionable. At best, strata of the Nebraskan and Kansan glacial ages are difficult to identify because they have been exposed to long periods of weathering and erosion, as well as the scouring effects of later glacial activity, especially that of the Wisconsinan. Since the Wisconsinan is the youngest glacial age, its strata contain the most detailed record of Pleistocene events.

The Wisconsinan glacial age is the best documented Pleistocene stage in North America. It has been shown that several warmer intervals, or interstadials, within the generally cold Wisconsinan led to the temporary withdrawal of the ice. These intermittent warm spells have been given names, especially in areas where Pleistocene glaciers were active and are well documented. In the Toronto, Ontario, area, for instance, two interstadials are recognized in the Wisconsinan: the Port Talbot interstadial, which has yielded a biological assemblage dating somewhat earlier than 54–45 thousand years (ka) BP, and the Plum Point interstadial, which has yielded an assemblage that lived about 34–24 ka BP.

Moving now to the Pleistocene NALMAs, we find that two are presently recognized (see Fig. 10), the Irvingtonian, which is the older and much the longer of the two; and the very short Rancholabrean. Most of the Pleistocene anuran records in North America, however, are from the Rancholabrean.

The Irvingtonian land-mammal age was originally defined on the basis of a mammalian fauna from a gravel pit southeast of Irvington, in Alameda County, California. Recent work by Bell (2000) on the dispersal of microtine rodents from Eurasia into North America suggests that the Irvingtonian began about 1.9 Ma BP and lasted until about 150 ka BP.

Three Irvingtonian subunits, I, II, and III, are also based on rodent-dispersal studies. Irvingtonian I is considered to have begun about 1.9 Ma BP (a little before the 1.8 Ma BP date recognized by the international geological community for the beginning of the Pleistocene) and to have lasted until about 850 ka BP. Irvingtonian II is thought to have begun about 850 ka BP and to have lasted to about 400 ka BP. Irvingtonian III is positioned between about 400 and 150 ka BP.

The Rancholabrean land-mammal age was originally defined on the basis of the famous Rancho La Brea faunal assemblage in Los Angeles, California. Bell's (2000) rodent-dispersal study suggests that the Rancholabrean began about 150 ka BP and lasted until about 10 ka BP, when the Pleistocene is considered to have ended. The period from about 10 ka BP until the present is called the Holocene by most of the international geological community. Some workers express doubt that the Holocene is truly a discrete unit of geologic time, believing that we are still in the Pleistocene epoch.

2

Systematic Accounts

This chapter documents the occurrence of fossil anurans in North America in systematic order. In the headings and lists that follow, the symbol # before a taxon indicates that the genus or higher taxon is extinct, and the symbol * indicates an extinct species of an extant genus. Multiple dates associated with authors of frog taxa, some in quotation marks and others in brackets (see p. 39), may be confusing to the layman. For an explanation of these kinds of notations see Sanchiz (1998, p. 11). English names of anuran taxa are now capitalized. This has been done in ornithology for years and has been recently adopted by the Committee on Standard English and Scientific Names, whose members represent the three most prominent herpetological societies in North America (the Society for the Study of Amphibians and Reptiles, the American Society of Ichthyologists and Herpetologists, and the Herpetologists' League) (Crother, 2001). This capitalization rule has also been adapted by the Center for North American Herpetology, a group that has published a separate list of the common and scientific names of amphibians and reptiles (Collins and Taggart, 2002). I have chosen to follow Crother (2001) for the common and scientific names.

ANURAN CLASSIFICATION

Recent proposals to uproot the traditional Linnean system of biological classification and replace it with ones that translate cladistic phylogenies directly into classification and define taxonomic names in terms of clades have created a dither among biologists, unknown since the days of the Darwin debates. As Benton (2000, p. 633) aptly put it, "In practice, phylogenetic nomenclature will be disastrous, promoting confusion and instability, and should be abandoned. It is based on a fundamental misunderstanding of the difference between a phylogeny (which is real) and a classification (which is utilitarian). Under the new view, classifications

are identical to phylogenies, and so the proponents of phylogenetic nomenclature will end up abandoning classifications altogether." I would
modify this by saying that "a phylogeny attempts to depict evolutionary
reality" rather than claiming that the phylogeny is necessarily reality itself.

In their comprehensive review of the paleontology and evolution of
the Amphibia, Heatwole and Carroll (2000, p. 1462) stated the following
about the classification of fossil Amphibia: "Because of major gaps in
knowledge of the fossil record, understanding of the principal groups of
amphibians remains incomplete. Any cladogram would give an erroneous
impression of the comprehension of their phylogenetic history." Thus,
rather than including cladograms that have appeared in the literature (see
Sanchiz, 1998, for some), I follow the classification of Sanchiz (1998) for
anurans known in the fossil record. This classification follows the custom-
ary Linnean taxonomic ranks (Linnaeus, 1758). In the text (except for
the first five genera of uncertain relationships) genera will be placed in
families and species will be placed in genera and subgenera in alphabetical order. Families occurring in the North American fossil record in the
classification below are preceded by **NA**.

Class Amphibia Linnaeus, 1758
  Superorder Salientia Laurenti, 1768
    Order Anura Rafinesque, 1815
      Suborder Discoglossoidea Sokol, 1977
        ?NA Family Leiopelmatidae Mivart, 1869 (New Zealand Frogs)
        NA Family Discoglossidae Günther, 1859 "1858" (Disc-
        tongued Toads)
      Suborder Mesobatrachia Laurent, 1979
        Superfamily Pipoidea Fitzinger, 1843
          NA Family #Palaeobatrachidae Cope, 1865 (Palaeobatrachid
          Frogs)
          Family Pipidae Gray, 1825 (Clawed and Surinam Frogs)
          NA Family Rhinophrynidae Günther, 1859 "1858" (Burrow-
          ing Frogs)
        Superfamily Pelobatoidea Bolkay, 1919
          NA Family Pelobatidae Bonaparte, 1850 (Spadefoot Toads)
          NA Family Pelodytidae Bonaparte, 1850 (Parsley Frogs)
      Suborder Neobatrachia Reig, 1958
        Superfamily Hyloidea Wied, 1856
          NA Family Leptodactylidae Werner, 1896 (1838) (Leptodac-
          tylid Frogs)
          Family Myobatrachidae Schlegel, 1850 (Myobatrachid Frogs)
          NA Family Bufonidae Gray, 1825 (True Toads)
          Family Brachycephalidae Günther, 1859 "1858" (Gold Frogs)
          NA Family Hylidae Gray, 1825 (1815) (True Treefrogs)
        Superfamily Ranoidea Fitzinger, 1826
          NA Family Ranidae Gray, 1825 (True Frogs)
          Family Rhacophoridae Hoffman, 1932 (1859) (Old World
          Treefrogs)
          NA Family Microhylidae Günther, 1859 "1858" (1843)
          (Narrow-mouthed Frogs)

ORDER ANURA RAFINESQUE, 1815

FROGS AND TOADS

Anurans occur on all continents except Antarctica. They are not found in extremely dry deserts or in polar regions. The osteological characters that define the group (see Milner, 1988; Sanchiz, 1998) are as follows. The prefrontal is absent in the skull. A centrum is present in the atlas. Fewer than 11 presacral vertebrae are present. No free ribs are present in the posterior vertebrae of the adults. An elongated, spike-like urostyle (coccyx) is present in the adults. The radius and the ulna are fused (radio-ulna), as are the tibia and the fibula (tibiofibula). The larval feeding complex lacks true teeth. The order Anura includes two stem genera, *Prosalirus* (Early Jurassic) and *Vieraella* (Middle Jurassic), that are of uncertain familial status (familia incertae sedis).

NORTH AMERICAN ANURAN GENERA OF UNCERTAIN FAMILY RELATIONSHIPS

Five North American anuran genera representing Early Jurassic, Late Cretaceous, Early Eocene, and Late Miocene times are not assignable to families (familia incertae sedis), and one of them may not be salientian. They are listed in temporal rather than alphabetical order.

GENUS #*PROSALIRUS* SHUBIN AND JENKINS, 1995

Genotype. *Prosalirus bitis* Shubin and Jenkins, 1995 (incorrectly given as *Prosalirus vitis* in Sanchiz, 1998, p. 98).

Etymology. The generic name is from the Latin *pro*, "in front of, before," and *salire*, "to spring, leap, bound."

Diagnosis. The diagnosis is the same as for the genotype and only known species, *Prosalirus bitis.*

#*PROSALIRUS BITIS* SHUBIN AND JENKINS, 1995
(FIGS. 11, 12)

Holotype. Because the holotype designation made by Shubin and Jenkins (1995) was based on the remains of at least two individuals, Sanchiz (1998) designated a sphenethmoid (Museum of Northern Arizona: MNA V8725) as the holotype.

Locality and Horizon. Gold Springs Quarry, Coconino County, Arizona: Early Jurassic (Pliensbachian). Kayenta Formation.

Other Material. The assemblage of anuran bones (MNA V8725) originally designated as the holotype by Shubin and Jenkins (1995) contains a frontoparietal, stapes, a sphenethmoid (the present holotype designated by Sanchiz [1998] and referred to MNA V8725), a parasphenoid, a maxilla, a premaxilla, an angulosplenial, six presacral vertebrae, ribs, three coracoids, a clavicle, three humeri, a radio-ulna, a carpalia, a metacarpalia, two incomplete ilia, and a tibiofibula. Two other specimens were also referred to this taxon by Shubin and Jenkins (1995): MCZ 9324A (figs. 1C, 3A–3C, 4A, 4B) represents at least two individuals and consists of a parasphenoid, two cervical vertebrae, eight presacral and one sacral vertebrae, the urostyle, a scapula, a cora-coid, a radio-ulna, six carpalia, four meta-carpalia, phalanges, and two ilia; MCZ 9323A (figs. 4C, 4D) consists of a cora-coid, a femur, a tibiofibula, and proximal tarsals (one complete).

Diagnosis. This diagnosis follows San-chiz (1998) after Shubin and Jenkins (1995), with some modifications based on Jenkins and Shubin (1998). The assump-tion here is that only one species is present in the fossil sample. The premaxilla has about 21 pedicellate tooth positions; the maxilla, about 40. The sphenethmoid has a well-developed osseus nasal septum and an ossified orbitonasal canal; this latter el-ement extends posteriorly to include the anterior margin of the optic foramen. The angulosplenial has a slight sinusoidal cur-vature and a sulcus for Meckel's cartilage bordered by a low, longitudinal flange. A bony stapes is present. The atlas has an in-tercotylar notochordal fossa. Notochordal vertebrae (four or five of them) have short ribs: one pair of ribs is ankylosed; the other is free. Uncinate processes are present on two ribs, and another rib is trifurcate dis-tally. A non-condylar articulation is present between the sacrum and the urostyle. The scapula has an elongate glenoid facet with a straplike form. The ilium is elongate and essentially straight. The dorsal ilial promi-nence is represented by an extensive, ovoid, slightly raised, rugose area anterior to the acetabulum. The dorsal ilial crest and supra-acetabular fossa are absent. The dorsal acetabular expansion is short. The

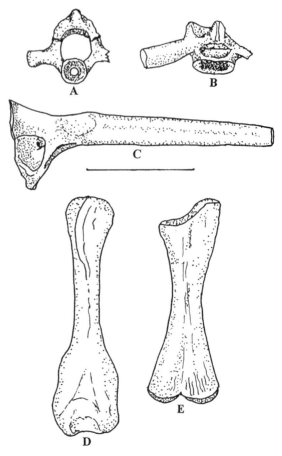

FIGURE 11. *Prosalirus bitis* from the Early Jurassic of Ar-izona, the earliest North American true jumping frog. (A) Anterior presacral vertebra in posterior view. (B) Sacral vertebra in dorsal view. (C) Right ilium in lateral view. (D) Left humerus in ventral view. (E) Left radio-ulna in anterolateral view. Scale bar = 5 mm and applies to all figures.

ilium articulates ventrally with the sacral wings. The tibiale and fibulare are unfused.

Description. This description is modified from Roček (2000). *Prosa-lirus bitis* was a frog with a snout–vent length (SVL) of about 50 mm. In the cranium, the teeth are pedicellate and occur on both the premaxilla and the maxilla. The premaxilla is about one-fourth the length of the maxilla and has 21 tooth positions. The maxilla has about 40 tooth po-sitions and becomes thin and edentulous (toothless) posteriorly. Since there is no articular facet for the quadratojugal, this bone is believed to be absent. The parasphenoid has well-developed posterolateral wings. The frontoparietal and other components of the cranium are smooth dorsally. The frontoparietal has a posterolateral flange that presumably overlapped the otic capsule. The large sphenethmoid extends posteriorly to incor-

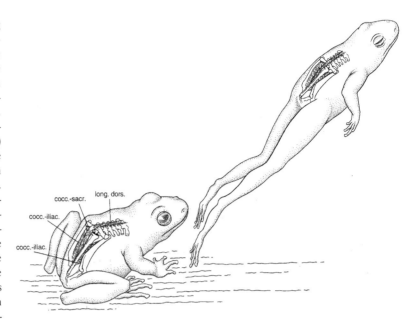

FIGURE 12. Jumping mechanics of *Prosalirus bitis*. During the jump, extension of the trunk occurs through the action of the longissimus dorsi (long. dors.) and coccygeo-sacralis (cocc.-sacr.) muscles, aided by the thrust of the push off of the forelimbs. The activity of the coccygeo-iliacus (cocc.-iliac.) muscles transmits propulsive force from the ilia to the coccyx and thus to the trunk and supports the linkage between the ilia and the sacrum.

porate the anterior border of the optic foramen and anteriorly has an ossified nasal septum. The prearticular is somewhat sinusoidal. The columella is broad and oval proximally and tapers slightly at its distal end. Near the footplate, the shaft of the columella extends as a proximally directed process, which is similar to the situation in many Paleozoic temnospondyls.

Turning to the vertebral column and ribs, we see that the atlas lacks transverse processes, and its centrum has its anterior end broadened and bearing two cotyles slightly separated by a small fossa. A funnel-shaped fossa at the posterior end of the cervical centrum is wider and deeper than one at the anterior end. These fossae do not join to form a continuous notochordal canal. It is evident that the other presacral vertebrae have notochordal canals, but the number of these cannot be determined. The pre- and postzygapophyseal articular facets are nearly horizontal. Neural spines occur merely as low crests on most vertebrae. The anterior and middle presacrals have wide neural canals but relatively constricted notochordal canals. The transverse processes of the middle presacrals have swollen expansions at their distal end, where they articulate with the free ribs. Vertebrae from the posterior part of the column have narrow neural canals but relatively wide notochordal canals. Their transverse processes are gracile.

A single vertebra forms the sacrum; it has a short centrum with a posterior notochordal fossa. The roof of the neural canal extends only over the anterior part of the centrum. The neural spine and prezygapophyses of the sacrum are strong, but the postzygapophyses are missing. The transverse processes of the sacral vertebra are directed posterolaterally and are triangular in cross section. The articulation between the sacrum and the urostyle was cartilaginous, as there is no evidence of a condylar structure. The urostyle itself is short and conical. Four or five pairs of ribs are present in the hypodigm (total assemblage of bones assigned to the species). One pair is ankylosed, but the others are jointed. These single-headed, short ribs extend proximally to form a broad contact with the transverse processes. Two ribs that articulate with vertebrae have anteriorly pointing uncinate processes, and one rib has both anterior and posterior uncinate processes.

Considering the limb girdles and limbs, we find that the clavicle is bowed laterally and straight medially. It has a longitudinal groove that suggests an extensive articulation with the anterior border of the scapula. The scapula, which obviously had an attached suprascapular cartilage, has a shallowly concave glenoid facet that faced posteriorly and somewhat ventrally. The coracoid has an elongate articular facet that corresponds to the glenoid facet of the scapula. The medial end of the coracoid forms an expanded blade, but the middle part of the bone is constricted. The humerus is strongly built and has a well-developed deltopectoral crest on its ventral surface. The head of the humerus is not ossified. The radio-ulna has a shallow, longitudinal groove along the length of the bone that shows where radial and ulnar fusion occurred. The ilia are nearly straight and articulate with one another ventrally to the sacral diapophyses. There is a slight indication of a narrow crest along the dorsal ilial margin, but this is not like the distinct dorsal crests that occur in many modern anurans. The dorsal prominence is a somewhat roughened large muscle scar anterior and dorsal to the acetabulum. The femur has a sigmoid curvature, and the tibiofibula is fused, as in modern anurans. The radio-ulna bears a shallow groove along the length of the bone, and this groove delineates the area of radial and ulnar fusion.

General Remarks. The pelvic girdle of *Prosalirus bitis* has movable sacro-urostylic and iliosacral joints. This, coupled with the fused radio-ulna and tibiofibula, strongly suggests that *Prosalirus* was able to jump. Jenkins and Shubin (1998) explained that the angular look of the back in living frogs at rest reflects the flexure of the sacro-urostylic and iliosacral joints. But during the launch part of the jump (see Fig. 12) the sacrum plus presacral vertebral column extends at these joints because of the contraction of the longissimus dorsi muscles, accompanied by forelimb thrust. During this process the urostyle remains fixed between the ilia by the coccygeo-iliacus muscles, which transmit propulsive thrust from the ilia to the urostyle and along to the sacrum and presacral vertebral column.

As comparisons of *Prosalirus* with other fossil and extant anuran families indicated no close relationships, Shubin and Jenkins (1995) established a new family, Prosaliridae, for the earliest fossil anuran. Sanchiz (1998, p. 98), however, decided to put *Prosalirus* among anurans of uncertain familial status because "the available morphological information was not enough to discuss systematic relationships at the family level properly."

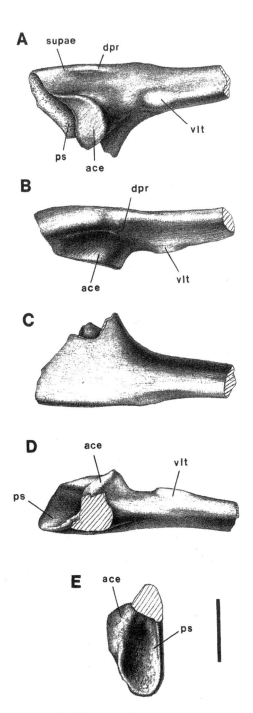

A supae dpr
vlt
ps
ace

B dpr
ace vlt

C

D ace
vlt
ps

E ace
ps

FIGURE 13. Holotype right ilium of *Nezpercius dodsoni* from the Late Cretaceous of Montana. (A) Lateral view. (B) Dorsal view. (C) Medial view. (D) Ventral view. (E) View of surface for articulation with the ischium. Abbreviations: ace, acetabulum; dpr, dorsal prominence; ps, posterior surface; supae, supra-acetabular expansion; vlt, ventrolateral tuberosity. Scale bar = 1 mm and applies to all figures.

GENUS #*NEZPERCIUS* BLOB, CARRANO, ROGERS, FORSTER, AND ESPINOZA, 2001

Genotype. Nezpercius dodsoni Blob, Carrano, Rogers, Forster, and Espinoza, 2001.

Etymology. The generic name honors the Nez Perce tribe. The specific name refers to vertebrate paleontologist Peter Dodson.

Diagnosis. The diagnosis is the same as for the genotype and only known species, *Nezpercius dodsoni*.

#*NEZPERCIUS DODSONI* BLOB, CARRANO, ROGERS, FORSTER, AND ESPINOZA, 2001

(FIGS. 13, 14)

Holotype. A small, well-preserved right ilium (Field Museum of Natural History: FMNH PR 2078).

Locality and Horizon. Judith River Formation type area (University of Chicago: locality UC-913), T23N, R22E (township and range) Baker Monument quadrangle, north-central Montana: Late Cretaceous (Campanian). Judith River Formation.

Other Material. A large right ilium (FMNH PR 2079) missing most of the shaft and a left ilium (FMNH 2080) with the shaft broken anteriorly. FMNH PR 2079 is from the same locality as the holotype. FMNH PR 2080 is from UC-8303, also from the Judith River Formation type area, T23N, R22E (township and range) Baker Monument quadrangle, north-central Montana: Late Cretaceous (Campanian).

Diagnosis. The diagnosis is quoted from Blob et al. (2001, p. 190). "A frog that differs from all other anurans in possessing a pronounced, anteroposteriorly elongate tuberosity on the ventrolateral aspect of the ilial shaft, just anterior to the preacetabular zone, that does not contact the dorsal margin of the ilial shaft but contacts or nearly contacts the ventral margin of the shaft. Relative to other frogs of Judithian age from western North America, it differs from *Palaeobatrachus* in possessing a semicircular (rather than 'bell-shaped') acetabulum, possessing a subacetabular expansion, and lacking interiliac synchondrosis. It differs from *Eopelobates* in possessing a large tuberosity on the ilial shaft and in lacking a spiral groove at the base of the ilial shaft."

Description. This description is an abbreviated version of the description in Blob et al. (2001). The holotype and FMNH PR 2080 are from much smaller frogs than FMNH PR 2079 and could represent juveniles or subadults. The estimated body masses of the three specimens, based on criteria in Esteban et al. (1995), are the following: FMNH PR 2078, 10 g (incorrectly stated as 1 g in Blob et al., 2001, p. 190); FMNH PR 2080, 15 g; and FMNH PR 2079, 29 g.

The ilial shaft is tubular and nearly circular in cross section and lacks a dorsal crest. A ventrolateral tuberosity occurs at the base of the shaft near the acetabulum but is separated from the acetabular margin. In lateral view, the dorsal border of the of the tuberosity lies about midway between the dorsal and ventral edges of the shaft. The tuberosity is anteroposteriorly elongate and forms a smooth, ovoid bump approximately twice as long anteroposteriorly as dorsoventrally. The length of the tuberosity is slightly more than one-half of the anteroposterior length of the acetabulum. The tuberosity is more strongly developed as it approaches the acetabulum, gently sloping away distally as it merges into the shaft.

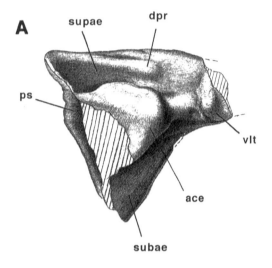

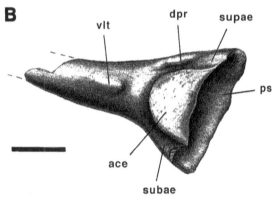

FIGURE 14. Referred material of *Nezpercius dodsoni.* (A) Lateral view of right ilium. (B) Lateral view of left ilium. Abbreviations: subae, subacetabular expansion; others as in Figure 13. Scale bar = 1 mm and applies to both figures.

The dorsal and ventral acetabular expansions are unequal in size and small relative to the acetabular fossa, and their dorsal and ventral flaring is slight. The acetabulum projects strongly from the surface of the surrounding region of the ilium and is slightly shorter anteroposteriorly than dorsoventrally. The fossa is somewhat rugose in all of the specimens, suggesting to Blob et al. (2001) that there was a cartilaginous component to its surface. The dorsal prominence is roughened and slightly raised. There is no distinct tubercle on the dorsal portion of the shaft. The sutural surface of the posterior ilium is strongly concave and rugose, indicating to Blob et al. a cartilaginous attachment to other pelvic elements. I believe, however, that the posterior surface of the "ilium" is an ovoid cotyle that probably formed a movable articulation with some other element of the skeleton.

General Remarks. The assignment of these ilia to the Anura or even the Salientia is highly questionable for the following reasons: (1) The circular (in cross section) ilial shaft, to my knowledge, is not present in either proanurans or anurans. (2) The ventrolateral tuberosity, the most prominent structural feature of the ilial shaft, is unknown in both proanurans and anurans. (3) As far as I am aware the laterally produced acetabular border is not found in any proanurans or anurans. (4) The posterior end of the "ilium" forms an ovoid cotyle that must have articulated with a condyle that would not allow for the anuran type of saltation, which had already evolved in the Early Jurassic (see Shubin and Jenkins, 1995; Jenkins and Shubin, 1998). Blob et al. (2001) mentioned that this posterior "ilial" concavity is rugose (the figures indicate that it is only slightly so) and interpreted this as indicating a cartilaginous attachment to the other pelvic elements. But they also reported that the "acetabulum," which of course is part of the movable joint with the femur, is also rugose. If indeed *Nezpercius* was an anuran and the elements described actually were ilia, *Nezpercius* must have had a different type of locomotion than any other anuran. More *Nezpercius* material is needed.

GENUS *#THEATONIUS* FOX, 1976

Genotype. Theatonius lancensis Fox, 1976.

Etymology. The generic name honors North American vertebrate paleontologist Theodore H. Eaton. The specific name refers to the Lance Formation, Wyoming.

Diagnosis. The diagnosis is the same as for the genotype and only known species, *Theatonius lancensis.*

#THEATONIUS LANCENSIS FOX, 1976
(FIGS. 15, 16)

Holotype. A complete left maxilla (University of Alberta Laboratory for Vertebrate Paleontology: UALVP 12073).

Locality and Horizon. Bushy Tailed Blowout, near Lance Creek, Niobrara County, Wyoming: Late Cretaceous (late Maastrichtian). Lance Formation.

Other Material. Referred material from Gardner (2000), all from the holotype locality, consists of one maxilla (UALVP 12075), five frontoparietals (UALVP 12074, 12076–12078, 40161), and three squamosals (UALVP 12079, 12080, 40162).

Diagnosis. The diagnosis is modified from Fox (1976), Sanchiz (1998), and Gardner (2000). An anuran species differing from other known anurans, especially

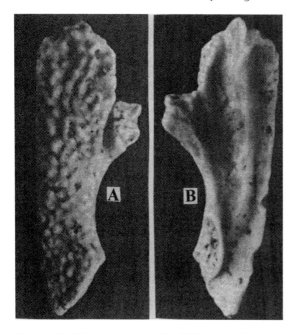

FIGURE 15. Holotype left maxilla of *Theatonius lancensis* from the Late Cretaceous of Wyoming. (A) External view. (B) Internal view. The specimen figured is about 5 mm long in its greatest length.

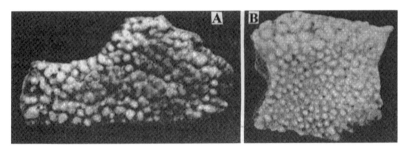

FIGURE 16. *Theatonius lancensis* referred elements. (A) Right frontoparietal in dorsal view. (B) (?) Left squamosal in external view. The greatest length of the element in (A) is about 4.5 mm; of that in (B), about 3.0 mm.

other North American Cretaceous species, in the following unique combination of characters. Body size was small. The maxilla, squamosal, and frontoparietal are ornamented externally with moderate-sized, tightly packed pustules. The maxilla is toothless, and its palatine area forms only an indistinct ridge. The preorbital area of the maxilla is higher than the postorbital area, and its massive palatine process projects more medially than in the postorbital portion. The maxilla has a prominent facet developed dorsally for a sutured contact with the nasal; a deep groove for the nasolacrimal duct that extends labially across the base of the palatine process to the orbital margin; and a pterygoid process that is lingually short and broad and bears a prominent articular facet that wraps posterodorsally onto the dorsal edge of the zygomatico-maxillary process. The posterior process of the maxilla is bluntly pointed. In the squamosal, the lamella alaris is a broad plate that has tiny spines along its posteroventral margin. The frontoparietals are relatively short and broad, paired, and broadly sutured along the midline. The anterior ends of the frontoparietals are transverse and bear ventrally an extensive facet for a broad sutured contact laterally with the squamosal. The supraorbital roof of the frontoparietal is narrow and is developed only along the orbital margin of the lateral process and bears ventrally a thick, ventrally descending flange that parallels the lateral end of the bone.

Description. The original description, modified from Fox (1976), is also useful in the characterization of this species. The maxilla is heavily ossified and has a dorsally ascending facial portion that curves medially anterior to the orbit and has a strongly developed preorbital process on its dorsal edge for articulation with the nasal. The facial portion of the maxilla posterior to the orbit is relatively low, about two-thirds of its preorbital height, and is less than one-half the height of the preorbital process. Internally, adjacent to the dorsal edge of the maxilla and directly behind the suborbital rim, is a small, rough-surfaced depression that probably articulated with the maxillary processes of both the squamosal and the pterygoid. The dental portion of the maxilla is toothless. The palatine process is poorly developed in *Theatonius*; and even anteriorly, where in most frogs this process is strongest, it is but a low, rounded ridge that weakens posteriorly and disappears beneath the suborbital rim.

The frontoparietals are not fused, and they lack crests. The four specimens that are complete anteriorly evince a fully transverse articulation with the nasal and thus indicate an absence of a dorsally open fontanelle. The frontoparietal widens abruptly posteriorly, and ventral notches indicate it extensively overlapped the prootic beneath it. Ventrally, a vertical flange meets with the dorsal edges of the neurocranial walls in specimens that displayed this area. Dorsal to the orbit, this flange joins the lateral edge of the frontoparietal roof, swinging laterally with this roof to a point where the roof finally extends beyond the flange to form a supraorbital shelf. At this point, the external surface of the flange is penetrated by the anterior opening of the occipital canal, which is completely embedded in bone throughout its length.

The questionable squamosals are represented by specimens that are thought to represent the otic processes of this bone. These elements are flat and ornamented and bear short spines along their posterior edges. It appears that in life these otic processes probably abutted against the lateral edge of the frontoparietal, but the contact is not preserved in the putative squamosal specimens.

General Remarks. Theatonius lancensis is the only anuran in the Mesozoic that undoubtedly lacks teeth on the maxilla. Even though the premaxilla of this taxon is unknown, it is probably safe to assume that it was also toothless. In addition to the original description by Fox (1976) this taxon has been commented on by Estes and Sanchiz (1982a), Sanchiz (1988), Gardner (2000), and Roček (2000), and none of these workers offered any suggestions about its familial relationships.

Gardner (2000) additionally noted two features of *Theatonius*, not mentioned by Fox (1976), that may be taxonomically important. First, on the maxilla, the groove occupied by the nasolacrimal duct is a relatively broad, deep channel that wraps around the labial base of the palatine process and opens posteriorly in the anteroventral corner of the orbital margin. Second, on the frontoparietal, the impression that indicates the attachment of the dura mater on the ventral surface of the skull roof (incrassatio frontoparietalis of Špinar, 1976) is undivided and slightly broader posteriorly than anteriorly.

Gardner (2000) also pointed out that the known cranial elements of *Theatonius lancensis* are distinctive, as well as being reliably associated, based on their provenance, their small size, the evidence of sutured contacts between the elements, and the characteristic ornamentation. He pointed out that the ornamentation resembles that of *Scotiophryne pustulosa* in consisting of convex tubercles, but its tubercles are larger and tightly packed, whereas those of *S. pustulosa* are rarely in contact. Gardner also mentioned that the drawings of the holotype maxillary of *T. lancensis* figured by Sanchiz (1998, figs. 134A, 135) do not depict the ornamentation of these fossils accurately.

GENUS #*EORUBETA* HECHT, 1960

Genotype. Eorubeta nevadensis Hecht, 1960.

Etymology. The generic name comprises *Eo*, referring to the Eocene epoch, and *rubeta*, Latin for a species of toad found in bramble. The specific name refers to Nevada, where the specimen was collected.

Diagnosis. The diagnosis is the same as for the genotype and only known species, *Eorubeta nevadensis*.

#*EORUBETA NEVADENSIS* HECHT, 1960

Holotype. A badly crushed frog represented as an organic imprint on two different fragments of a well core (American Museum of Natural History: AMNH 7602) (Hecht, 1960).

Locality and Horizon. Core 8, Standard Oil Company of California line unit 1, White Pine County, Nevada: Early Eocene (Wasatchian NALMA). Member C, Sheep Pass Formation.

Diagnosis. After Hecht (1960) as modified by Lynch (1971) and Sanchiz (1998). Maxilla toothed; with transverse processes of the eight presacral vertebrae long and flattened, and those of the posterior vertebrae as long as the anterior ones and almost as long as the sacral transverse processes; sacral diapophysis expanded distally and oriented at right angles to the sagittal line; ilium long without dorsal prominence and lacking a dorsal crest.

Description. The diagnosis above cannot be much improved by a detailed description of the holotype (see Hecht, 1960, pp. 3, 10; Lynch, 1971, pp. 195–197).

General Remarks. Lynch's analysis of *Eorubeta* (Lynch, 1971, pp. 195–197) shows that this taxon is not a member of the Leptodactylidae (as stated by Hecht, 1960) and that the skeleton of *Eorubeta* does not closely resemble that of any known modern frog genus and for that reason should be assigned to "family incertae sedis, order Salientia (Anura)."

GENUS #*TREGOBATRACHUS* HOLMAN, 1975

Genotype. Tregobatrachus hibbardi Holman, 1975.

Etymology. The generic name refers to Trego County, Kansas (where the genus was collected), and the Greek *batrachos,* "frog." The specific name recognizes North American vertebrate paleontologist Claude W. Hibbard.

Diagnosis. The diagnosis is the same as for the genotype and only known species, *Tregobatrachus hibbardi.*

#*TREGOBATRACHUS HIBBARDI* HOLMAN, 1975

(FIG. 17)

Holotype. A left ilium (Michigan State University, Museum Vertebrate Paleontology Collection: MSUVP 766).

Locality and Horizon. WaKeeney Local Fauna, Trego County, Kansas: Miocene (Clarendonian NALMA). Ogallala Formation.

Diagnosis. The diagnosis is slightly modified from Holman (1975). A moderately large anuran ilium differing from other anuran taxa by the following combination of characters: (1) dorsal acetabular expansion reduced in size, its dorsal border extending almost straight back from the dorsal prominence; (2) no dorsal ilial crest or ilial shaft ridge present; (3) dorsal prominence well developed but lacking a protuberance or tubercle, about twice as long as high, with one-half of its extent anterior to anterior acetabular fossa, with a smoothly rising posterior slope and a more abruptly rising anterior slope, and with tip of prominence deflected medially; (4) ilial shaft compressed, its medial surface with an anterodorsally

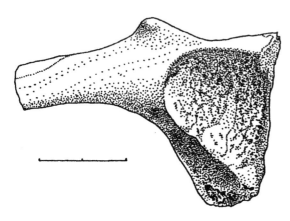

FIGURE 17. Holotype left ilium of *Tregobatrachus hibbardi* from the Late Miocene (Clarendonian) of Kansas. Scale bar = 4 mm.

directed wide groove and entire shaft abruptly constricted anteriorly; (5) large foramen present on shaft just anterior to anterior edge of ventral acetabular expansion; (6) ventral acetabular expansion limited in extent, its ventral border truncated; (7) acetabular fossa very large, its surface pitted.

Description. The description is modified from Holman (1975). The acetabular fossa is well excavated, and its borders are distinct. It is higher than it is long and is subrounded. Its surface is strongly pitted. The dorsal acetabular expansion is quite limited in extent, and its dorsal border extends straight back from the posterior extent of the dorsal prominence. The dorsal prominence is well developed and is about twice as long as it is high. The prominence is smooth in structure, lacking a protuberance or tubercles. The posterior border of the prominence rises gently from the shaft, but the anterior border slopes more precipitously into the shaft. The tip of the prominence is deflected medially and extends anterior to the anterior edge of the acetabular fossa by about one-half the length of this fossa.

There is no dorsal ilial crest or ilial shaft ridge. But the shaft is unique in that it has a distinct, wide groove running anteroventrally from the level of the dorsal prominence toward the ventral border of the shaft. Moreover, the shaft becomes abruptly constricted anteriorly such that the shaft height decreases rapidly in an anterior direction. A very large foramen occurs on the shaft just anterior to and partially hidden by the anterior edge of the acetabular border. The ventral acetabular expansion is limited in extent, and its ventral border is truncated.

Measurements are as follows: anterior height of shaft, 2.1 mm; height of shaft just anterior to dorsal prominence, 3.4 mm; height of shaft through dorsal prominence, 4.6 mm; greatest height of acetabular fossa, 5.8 mm; greatest length of acetabular fossa, 4.5 mm; greatest height through dorsal and ventral acetabular expansion, 8.0 mm.

General Remarks. Sanchiz (1998) stated that the fossil could represent an anomaly in which the ventral ischial and pubic section of the pelvic girdle are coossified with the iliac acetabulum and that in fact a line that could correspond to such a fusion is visible on the acetabulum and ventral acetabular expansion. I disagree with this hypothesis, for the line looks like a postmortem crack in the bone. Moreover, even if the line did represent a natural point of fusion, the diagnostic characters given above would still be present.

I agree that it is difficult to ascertain the familial relationships of *Tregobatrachus*, but it certainly is an anuran with primitive character states. The great encroachment of the acetabular fossa is seen to a degree in the primitive families Leiopelmatidae, Discoglossidae, and Pipidae. More advanced families tend to have smaller acetabular fossae and larger dorsal and ventral acetabular expansions. The lack of an ilial blade or ilial shaft ridge appears typical of primitive anuran families, as many advanced anuran families (e.g., Leptodactylidae, Hylidae, Ranidae) contain taxa that have these structures.

Moreover, several features are unique in *Tregobatrachus*. These include the very large dorsal prominence without protuberances or tubercles and with its tip reflected medially, as well as the widely grooved ilial shaft. Certainly, unexpected morphologies occur in Tertiary reptiles (e.g., an Eocene snake with a natural, dorsoventrally flattened vertebral column, previously unknown in snakes; Holman and Harrison, 2000). Also, one of the most primitive anuran families (in fact the only currently recognized extinct anuran family that has ever existed), with a distinctive ilial structure—the Palaeobatrachidae—survived until the middle Pleistocene in Europe (see Holman, 1998).

SUBORDER DISCOGLOSSOIDEA SOKOL, 1977

The suborder Discoglossoidea contains primitive anuran families. Possibly the Leiopelmatidae and certainly the Discoglossidae occur as fossils in North America.

Osteological definitions of anuran families in the following pages of this volume generally follow those of Duellman and Trueb (1994) unless stated otherwise. Osteological terms used in these definitions follow Duellman and Trueb (1994) and Trueb (1973).

Hyoid Skeleton. The *cricoid ring* is a circular ring in the larynx that may be incomplete dorsally; this incomplete condition occurs mainly in primitive frogs. The *parahyoid* either is a flat bone straddling the midline or occurs as one of a pair of bones on either side of the midline of the cartilaginous body of the hyoid.

Vertebral Column. Seven *presacral* vertebrae usually occur (either six or eight is considered the primitive condition). The vertebra that articulates with the skull is called the *cervical* vertebra. Two *cervical cotyles* of the cervical vertebra articulate with the skull. The *sacral vertebra* or *sacrum* is the last vertebra in the vertebral column and is modified to articulate with the ilia of the pelvic girdle as well as the elongate *urostyle* (referred to as coccyx by Duellman and Trueb, 1994). Lateral extensions of the sacrum, called *diapophyses*, articulate with the anterior ends of the ilia. These diapophyses may be rounded (advanced) or moderately or widely expanded (more primitive). The posterior end of sacrum articulates with the urostyle by way of two *sacral condyles*, but sometimes the sacrum is fused to the urostyle to form a *sacrococcyx*.

The vertebrae may be considered *non-imbricate* (primitive condition) when the vertebrae do not overlap dorsally; or *imbricate* (more advanced condition) when the vertebrae overlap dorsally.

Short *ribs* occur anteriorly (primitive), or ribs are absent (advanced). The vertebral *centra*, which form the body of the vertebrae, may be *ectochordal*, where the centra ossify as cylinders enclosing the notochord (Leiopelmatidae, Rhinophrynidae); *stegochordal*, where the centra are transversely depressed (Pipidae, Discoglossidae, Pelobatidae); or *holochor-*

dal, where the centra are cylindrical and solidly ossified (all other anurans). The *intervertebral cartilage* is a block of cartilage, usually ossified, that lies between successive centra in all frogs.

The vertebral column as a unit may be (1) *amphicoelous* — vertebral column with ectochordal centra, slightly biconcave or flat on both ends, intervertebral cartilages contiguous to and not subdivided between successive presacral vertebrae; (2) *anomocoelous* — vertebral column with stegochordal centra, slightly biconcave or flat on both ends, each intervertebral cartilage subdivided anteriorly and posteriorly producing a free intervertebral body between adjacent centra, the centrum ossifying and remaining free; (3) *opisthocoelous* — vertebral column with ectochordal or stegochordal centra, each intervertebral cartilage anteriorly subdivided, subsequently ossified and fused to the posteriorly adjacent centrum, the mature centra then being convex anteriorly and concave posteriorly; (4) *procoelous* — vertebral column with holochordal centra, each intervertebral cartilage posteriorly subdivided, subsequently ossified and fused to the anteriorly adjacent centrum, the mature centra then being concave anteriorly and convex posteriorly; (5) *diplasiocoelous* — vertebral column with holochordal centra, each intervertebral cartilage posteriorly subdivided in the procoelous manner, with the exception of the cartilage located between the eighth presacral, this cartilage being divided in the opisthocoelous manner, causing the first seven presacrals to be concave anteriorly and convex posteriorly (whereas the eighth is biconcave).

Pectoral Girdle. Two general types of pectoral girdles occur in the anurans, the *arciferal* and *firmisternal* conditions; the firmisternal is derived from the arciferal. In the arciferal type, epicoracoidal cartilaginous horns are posteriorly directed, and these horns articulate with the dorsal surface of the sternum, providing a surface for the insertion of derivatives of the abdominal rectus muscle. In the firmisternal type, epicoracoidal horns are absent, and a fusion of the pectoral arch with the sternum does not occur. The *omosternum* is a somewhat triangular, nonpaired element that may be cartilaginous (or not), and that is the anteriorly terminal element of the pectoral girdle; it may be present or absent in frogs. Whether the *scapula is overlain by the clavicle or not* is a character used in the definition of anuran families.

Phalangeal Formula. The forefoot (or hand) of frogs has four digits that usually consist of 10 phalangeal elements that number (medially to laterally) 2–2–3–3. Sometimes phalanges are lost, but in some frogs a gain occurs, usually producing a 3–3–4–4 number. The distal two phalanges are referred to (proximally to distally) as the *penultimate phalanx* and the *ultimate phalanx*.

FAMILY LEIOPELMATIDAE MIVART, 1869

Here, a single family, Leiopelmatidae Mivart, 1869, is recognized for the modern genera *Ascaphus* Stejneger, 1899, and *Leiopelma* Fitzinger, 1861, as well as for the extinct genus #*Notobatrachus* Reig, 1956 (for comment on the family name see Frost, 1985, p. 233). Another view is taken by Sanchiz (1998), who recognized two families, the Ascaphidae

Plate 1. *Acris crepitans* Northern Cricket Frog
(James H. Harding)

Plate 4. *Bufo americanus* (amplexed) American Toad
(James H. Harding)

Plate 2. *Acris crepitans* Northern Cricket Frog
(James H. Harding)

Plate 5. *Bufo americanus* American Toad
(James H. Harding)

Plate 3. *Acris gryllus* Southern Cricket
Frog (James H. Harding)

Plate 6. *Bufo americanus* American Toad
(James H. Harding)

Plate 7. *Bufo cognatus* Great Plains Toad
(James H. Harding)

Plate 10. *Bufo hemiophrys* Canadian Toad
(Richard D. Bartlett)

Plate 8. *Bufo cognatus* Great Plains Toad
(James H. Harding)

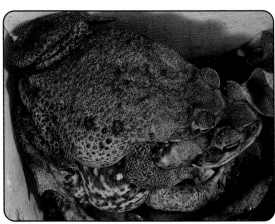

Plate 11. *Bufo marinus* Cane Toad, mating
(Richard D. Bartlett)

Plate 9. *Bufo fowleri* Fowler's Toad
(James H. Harding)

Plate 12. *Bufo punctatus* Red-spotted Toad
(James H. Harding)

Plate 13. *Bufo quercicus* Oak Toad
(Richard D. Bartlett)

Plate 16. *Bufo woodhousii* Woodhouse's Toad
(William Leonard)

Plate 14. *Bufo terrestris* Southern Toad
(James H. Harding)

Plate 17. *Gastrophryne carolinensis* Eastern
Narrowmouthed Toad (James H. Harding)

Plate 15. *Bufo valliceps* Gulf Coast
Toad (Richard D. Bartlett)

Plate 18. *Hyla arenicolor* Canyon Treefrog
(Richard D. Bartlett)

Plate 19. *Hyla cinerea* Green Treefrog
(Richard D. Bartlett)

Plate 22. *Pseudacris crucifer* Spring Peeper
male, calling (James H. Harding)

Plate 20. *Hyla versicolor* Gray Treefrog male, calling
(James H. Harding)

Plate 23. *Pseudacris ornata* Ornate Chorus Frog
(Richard D. Bartlett)

Plate 21. *Pseudacris crucifer* Spring Peeper
(James H. Harding)

Plate 24. *Pseudacris streckeri* Strecker's Chorus Frog
(Wayne Van Devender)

Plate 25. *Pseudacris triseriata* Western Chorus Frog (James H. Harding)

Plate 28. *Pternohyla fodiens* Lowland Burrowing Treefrog (Richard D. Bartlett)

Plate 26. *Pseudacris triseriata* Western Chorus Frog (James H. Harding)

Plate 29. *Rana capito* Gopher Frog (Wayne Van Devender)

Plate 27. *Pseudacris triseriata* Western Chorus Frog (James H. Harding)

Plate 30. *Rana catesbeiana* American Bullfrog (James H. Harding)

Plate 31. *Rana clamitans* Green frog, "blue" and normal color (James H. Harding)

Plate 34. *Rana palustris* Pickerel Frog (James H. Harding)

Plate 32. *Rana clamitans* Green Frog (James H. Harding)

Plate 35. *Rana palustris* Pickerel Frog (James H. Harding)

Plate 33. *Rana grylio* Pig Frog (Richard D. Bartlett)

Plate 36. *Rana pipiens* Northern Leopard Frog (James H. Harding)

Plate 37. *Rana sylvatica* Wood Frog
(James H. Harding)

Plate 40. *Scaphiopus bombifrons* Plains Spadefoot
(Richard D. Bartlett)

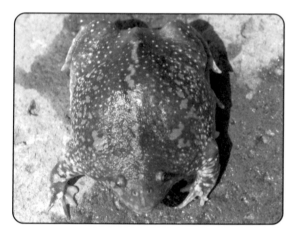

Plate 38. *Rhinophryus dorsalis* Mexican Burrowing Toad
(J. Alan Holman)

Plate 41. *Scaphiopus couchii* Couch's Spadefoot
(James H. Harding)

Plate 39. *Scaphiopus bombifrons* Plains Spadefoot
(Richard D. Bartlett)

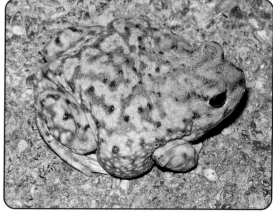

Plate 42. *Scaphiopus couchii* Couch's Spadefoot
(James H. Harding)

Plate 43. *Scaphiopus holbrookii* Eastern Spadefoot (Richard D. Bartlett)

Plate 46. *Ascaphus truei* Western Tailed Frog female (William Leonard)

Plate 44. *Scaphiopus intermontana* Great Basin Spadefoot (William Leonard)

Plate 47. *Ascaphus truei* Western Tailed Frog male (William Leonard)

Plate 45. *Scaphiopus intermontana* Great Basin Spadefoot (William Leonard)

Plate 48. Right ilium of a frog of the *Rana pipiens* complex from the late Tertiary Pipe Creek Sinkhole site, Grant County, Indiana. Length about 14 mm. Scanning electron micrograph created by A. S. Argast.

for *Ascaphus* and the Leiopelmatidae for *Leiopelma* and *Notobatrachus*, each of which occupies its own subfamily. At present, *Ascaphus* occurs in the northwestern United States and southwestern Canada, *Leiopelma* occurs in New Zealand, and *Notobatrachus* occurs in the Jurassic of Argentina.

An osteological definition of the family follows. Nine ectochordal presacral vertebrae are present; these have cartilaginous intervertebral joints and are non-imbricate. Presacrals I and II are not fused, and presacral I has the atlantal cotyles closely juxtaposed. Free ribs are present on presacrals II–IV and sometimes on presacral V in adults. In *Notobatrachus* free ribs are present on presacrals II–V and occasionally on VI. Narrowly dilated diapophyses occur on the sacrum, which has a contiguous cartilaginous connection with the urostyle; the urostyle has transverse processes proximally. The pectoral girdle is arciferal and has the anterior end of the scapula overlain by the clavicle in *Ascaphus*, but not in *Leiopelma* and *Notobatrachus*. A cartilaginous omosternum and sternum are present. Palatine bones are absent, but a parahyoid bone is present and the cricoid ring is complete. Both the maxilla and the premaxilla bear teeth. The astragalus and calcaneum are only fused proximally and distally, there are three tarsalia, and the phalangeal formula is normal.

A possible North American leiopelmatid. A left ilium (Fig. 18C) (University of California Museum of Paleontology: UCMP 55703) from the Late Cretaceous Lance Formation of Wyoming appears to be nearly identical to two ilia of modern *Leiopelma* (Figs. 18A, 18B), and, in fact, Estes (1964) assigned the element to "Family ?Ascaphidae (= Leiopelmatidae) Unidentified genus and species." UC 55703 is simple, as in *Leiopelma*, lacking any distinct protuberances, tubercles, or prominence on either the shaft or the dorsal or ventral acetabular expansions and having an almost identically shaped acetabular cup. The ilial shaft is also gently curved as in *Leiopelma* rather than almost straight as in other groups of primitive anurans. This ilium was not discussed by Sanchiz (1998), who recognized no members of the family Leiopelmatidae or "Ascaphidae" in North America. This ilium was dismissed by Estes and Reig (1973) and Estes and Sanchiz (1982b). Estes and Reig (1973) stated that the questionable ascaphid (= leiopelmatid) cited in Estes (1964, fig. 27) probably was from a small individual of the pelobatid present in the Lance Formation, namely, *Eopelobates* (see Estes, 1970). As far as I can ascertain, Estes never again specifically referred to UC 55703, an element remarkably similar to that in modern *Leiopelma*. I have purposely not made any formal classificatory statement about this element, but merely wish to alert the paleontological community of its existence, as UC 55703 is as

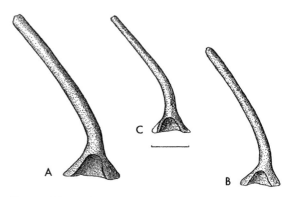

FIGURE 18. Modern leiopelmatid left ilia and a possible Late Cretaceous one. (A) Modern *Leiopelma hamiltoni*. (B) Modern *Leiopelma hochstetteri*. (C) ? Leiopelmatidae from the Late Cretaceous of Wyoming. Scale bar = 4 mm and applies to all figures.

clearly assignable to the Leiopelmatidae as other Mesozoic frogs are to their respective families (e.g., see next generic account).

FAMILY DISCOGLOSSIDAE GÜNTHER, 1859 "1858"

Living taxa of the primitive family Discoglossidae occur in Europe, North Africa, Israel, probably Turkey, the western and eastern portions of the former USSR (disjunctively), China, the Koreas, Vietnam, Borneo, and the Philippines (Frost, 1985). Classification of the family Discoglossidae is a prime example of the current chaos that exists in frog systematics above the generic level.

As pointed out by Gardner (2000), at one extreme Sanchiz (1988) recognized four modern and eight extinct genera in the family; and at the other extreme, Cannatella (1985) and Ford and Cannatella (1993) restricted the name Discoglossidae to two modern genera, *Discoglossus* and *Alytes*, and even suggested dropping the family. Other workers have used the family name in intermediate ways by excluding some extinct genera (e.g., Clarke, 1988; Roček, 1994). Gao and Wang (2001) have suggested four synapomorphies (derived characters) that support a monophyletic Discoglossidae. Reflecting the existing situation, Sanchiz (1998) did not attempt to provide an osteological diagnosis of the Discoglossidae.

Considering the fact that there have probably been more parallel evolutionary tracks in the Anura than in any other large vertebrate group, this type of taxonomic instability is not likely to stabilize — it may become even more chaotic. This is because new anuran specific names are appearing at a rather alarming rate (see Duellman, 1993), because new material is being found in previously unexplored areas or is derived from existing subspecies (e.g., *Bufo woodhousii fowleri* is now recognized as *Bufo fowleri*).

An osteological definition of the Discoglossidae follows. Eight stegochordal, opisthocoelous, imbricate presacral vertebrae are present. Presacrals I and II are unfused. The atlantal cotyles of presacral I are closely juxtaposed. Presacrals II–IV bear free ribs. The sacrum bears expanded diapophyses and has a bicondylar articulation with the urostyle, which has transverse processes proximally. The sacrococcygeal articulation is monocondylar in *Barbourula*. The pectoral girdle is arciferal and has a cartilaginous omosternum and sternum. The omosternum is highly reduced in *Bombina* and *Discoglossus*. The anterior end of the scapula is overlain by the clavicle. Palatines are absent, but single or paired parahyoid bones are present; the cricoid ring is complete. Both the maxilla and the premaxilla bear teeth. The astragalus and the calcaneum are only fused proximally and distally. Three tarsalia are present, and the phalangeal formula is normal.

Here, I am following Sanchiz (1998) in the recognition of three North American fossil genera in the Discoglossidae, realizing of course, that there are dissenting opinions.

GENUS #*ENNEABATRACHUS* EVANS AND MILNER, 1993

Genotype. Enneabatrachus hechti Evans and Milner, 1993.

Etymology. The generic name is derived from the Greek *enneas*, "nine" (after Como Bluff Quarry Nine, Wyoming), and *batrachos*, "frog." The specific name acknowledges anuran paleontologist Max K. Hecht.

Diagnosis. The diagnosis is same as for the genotype and only known species, *Enneabatrachus hechti.*

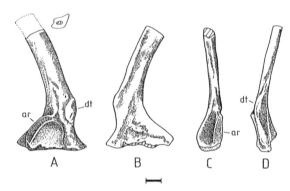

FIGURE 19. *Enneabatrachus hechti* from the Late Jurassic of Wyoming. Holotype left ilium. (A) Lateral. (B) Medial. (C) Ventral. (D) Dorsal. Abbreviations: ar, acetabular rim; dt, dorsal tubercle. Scale bar = 1 mm and applies to all figures.

#*ENNEABATRACHUS HECHTI* EVANS AND MILNER, 1993
(FIG. 19)

Holotype. A left ilium lacking only the dorsal extremity of the ilial shaft (National Museum of Natural History: USNM 460388).

Locality and Horizon. The Como Bluff Quarry Nine locality (bed 24 of Loomis, in Gilmore, 1928), Como Bluff, Albany County, Wyoming: Late Jurassic (probably Tithonian; see Evans and Milner, 1993).

Other Material. No other material could be assuredly assigned to this taxon.

Diagnosis. This diagnosis follows Evans and Milner (1993), with some modification of their terminology. A discoglossid frog having an ilium with the following characteristics: ilial shaft narrow and showing only a slight development of a dorsal crest; well-developed dorsal prominence; distinct but moderately developed dorsal and ventral acetabular expansions; expanded acetabular border. The following characters diagnose this ilium within the Discoglossidae: ilium differs from the ilia of *Bombina, Latonia,* and *Eodiscoglossus* in the absence of a pronounced dorsal prominence; differs from the preceding genera and also *Alytes, Balaeophryne, Wealdenbatrachus,* and *Paradiscoglossus* in the modest size of the dorsal acetabular expansion; differs from *Eodiscoglossus, Paradiscoglossus,* and to a lesser degree, *Wealdenbatrachus,* in having an expanded acetabular border.

Description. This description is modified somewhat from Evans and Milner (1993). The holotype left ilium is well preserved and lacks only the dorsal extremity of the ilial shaft. The acetabular region has a broad base with small dorsal and ventral acetabular expansion areas and a shallow supra-acetabular fossa. The dorsal prominence is large and has its long axis vertically oriented. The acetabulum is extended by an inflated anterior acetabular crest. The medial surface bears rugosities for ligaments, but there is no evidence of a synchondrosis. The facet for the puboischium is complete. The ilial shaft is slender and parallel-sided, with a weak dorsal crest.

General Remarks. These general remarks are modified from the discussion section of Evans and Milner (1993), with some of my comments inserted. The shape of the acetabular region, with the strong dorsal prominence (they refer to this as a dorsal tubercle), weakly developed dorsal crest, and small ventral acetabular expansion, most strongly resembles that of frogs of the family Discoglossidae. *Enneabatrachus* differs from *Eodiscoglossus* from the Middle Jurassic of Britain and the Early Cretaceous of Spain in the shape of the ilial shaft and in possessing an expanded acetabular rim; and from *Wealdenbatrachus* of the Early Cretaceous of Spain in the small size of the of the dorsal acetabular expansion and in having a somewhat greater expansion of the acetabular rim. *Enneabatrachus* differs from *Paradiscoglossus* of the Late Cretaceous Lance Formation of western North America in having a much weaker dorsal crest, a less prominent dorsal acetabular expansion, and a better developed acetabular rim. Among living frogs, *Enneabatrachus* is said to resemble *Bombina* most closely, except that most species of *Bombina* lack a dorsal prominence (Estes and Sanchiz, 1982b).

Author's Comment. The oddly expanded acetabular rim or border is certainly a strong character in *Enneabatrachus*, and the fact that it shares this with *Eodiscoglossus* may be quite significant. Henrici (1998) stated that additional *Enneabatrachus* material may occur at the type locality, and there is a referred partial skeleton from the Late Jurassic Rainbow Park microsite in Utah. Hopefully, this will more firmly establish the generic status and relationships of this frog.

GENUS #*SCOTIOPHRYNE* ESTES, 1969

Genotype. *Scotiophryne pustulosa* Estes, 1969.

Etymology. The generic name is from the Greek *skotios*, "dark" (referring to the darkness of the fossil bones), and *phryne*, "toad." The specific name comes from the Greek *pustulosa*, referring to the distinctive pustular sculpturing of the referred skull elements.

Diagnosis. The diagnosis is the same as for the genotype and only known species, *Scotiophryne pustulosa* Estes, 1969.

#*SCOTIOPHRYNE PUSTULOSA* ESTES, 1969
(FIG. 20)

Holotype. A left ilium (Museum of Comparative Zoology, Harvard University: MCZ 3623).

Locality and Horizon. Bug Creek Anthills, McCone County, Montana: Late Cretaceous (Maastrichtian). Hell Creek Formation.

Other Material. Material reported by Estes (1969) follows: Paratypes from the same locality as the holotype are four left and two right ilia (MCZ 3624), 11 distal ends of humeri (MCZ 3625), and 14 anterior and posterior fragments of maxillae (MCZ 3626). Referred specimens from other localities are a right squamosal (American Museum of Natural History: AMNH 8102), a left maxilla (AMNH 8132), a right ilium (AMNH 8137), and a left ilium (University of California Museum of Paleontology: UCMP 55703), all from the Lance Formation of Wyoming, Late Cretaceous (Campanian). A left ilium (Princeton University Museum, currently deposited at the Yale Peabody Museum, Yale University: YPM [PU] 1703) and three humeri (YPM[PU] 16784, 16827, 16828) are from the Tongue River Formation, Early Paleocene (Torrejonian) of Montana. The Paleocene specimens are considered to be probable records. Estes and Sanchiz (1982b) noted that other records of *Scotiophryne* are known from the Late Cretaceous El Gallo locality at El Rosario, Baja California del Norte. Undescribed skull and postcranial elements have been found there (Clemens et al., 1979). I am not aware if these remains have ever been described. *Scotiophryne* was also reported from the Late Cretaceous (Campanian) Fruitland Formation of New Mexico (Armstrong-Ziegler, 1980; Hunt and Lucas, 1992, 1993) and from the latest Maastrichtian–Early Paleocene (Lancian–Torrejonian) Tullock Formation of Montana (Bryant, 1989).

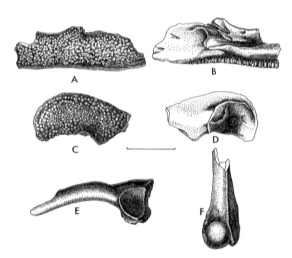

FIGURE 20. *Scotiophryne pustulosa* elements from the Late Cretaceous of Montana and Wyoming. (A) Lateral and (B) medial views of left maxilla. (C) Lateral and (D) medial views of right squamosal. (E) Lateral view of holotype left ilium. (F) Distal part of right humerus. Scale bar = 4 mm and applies to all figures.

Gardner (2000) recently provided more material from the Bug Creek locality in the Hell Creek Formation of Montana. This material includes two ilia (AMNH 26422; University of Alberta Laboratory for Vertebrate Paleontology: UALVP 40155), two maxillae (UALVP 40156, 40157), a nasal (UALVP 40158), a squamosal (UALVP 40159), and a frontoparietal (UALVP 40160). This material will be described in a forthcoming publication.

Diagnosis. The diagnosis is modified from Estes (1969), Sanchiz (1988), and Gardner (2000). An anuran species differing from other known anurans (especially North American Cretaceous forms) in the following unique combination of characters. The body size is moderate. The ilium lacks a dorsal crest, and its dorsal prominence (tubercle) is poorly developed. The ilial preacetabular region is short anteriorly, the dorsal acetabular expansion is low, and the ventral acetabular expansion is moderately well developed, with its depth greater than the height of the dorsal acetabular expansion. The maxilla, nasal, squamosal, and frontoparietal are ornamented externally with fine, beadlike tubercles that are rarely in contact. The maxilla is tooth-bearing and has moderately high postorbital regions that are subequal in height. The palatine and pterygoid processes are moderately well developed, and the palatine area is developed as a well-defined ridge. The squamosal is dorsally elongate, is inclined slightly posteriorly, and has an obtuse bend midway along its length. The base of the squamosal is broadly sutured with the maxilla; its dorsal end is blunt and free. The frontoparietals are paired and appear not to be in broad contact anteriorly with either the nasals or the sphenethmoid.

Description. This description is somewhat modified from Estes (1969). The ilium has a robust shaft with a deeply marked dorsal groove that extends onto the dorsomedial side of the shaft. This groove is well defined in the type specimen but less strongly marked on the others. The acetabular fossa is relatively large and has its anteroventral border strongly produced. No dorsal protuberance is present as such, but a dorsal prominence showing irregularities of muscle attachment is present. The ventral acetabular expansion is large and is markedly set off from the acetabular fossa and directed somewhat medially.

Turning to the referred humeri, the olecranon scar (terminology of Hecht and Estes, 1960) is oblique. A small but deep ventral cubital fossa is present. A well-developed ulnar epicondyle and a small, bituberculate radial epicondyle is present. The distal condyle (humeral ball of Estes, 1969) is well developed and bears a prominent lateral crest leading to the lateral epicondyle and a wider medial crest leading to the ulnar epicondyle. Variable development of the medial crest probably reflects sexual dimorphism in these frogs (see Fig. 20).

The maxilla has a broadly expanded anterior end with a prominent nasal process (terminology of Estes, 1969), and the posterior end is also expanded. A strong pterygoid process occurs medially. Pedicellate teeth are numerous. The external surface is covered with a relatively fine pustular sculpturing. The kidney-shaped squamosal has an expanded tympanic process and has pustular sculpturing as in the maxillae.

General Remarks. The ilium was chosen as the type specimen because this element is more often recovered than the more delicate skull elements (Estes, 1969). The referral of *Scotiophryne* to the family Discoglossidae originally was based on the similarity of the ilium to that of modern Eurasian *Bombina*, and the humeri also show general similarity to *Bombina* in the shape of the distal condyle, epicondyles, and oblique olecranon scar and in the dimorphism of the medial crest development (Estes, 1969). Later, Estes and Sanchiz (1982b) compared character states of *Scotiophryne pustulosa* with those of some other frog groups (Table 2). The assignment of the family Discoglossidae at the present time is uncertain because there are no observable synapomorphies (Sanchiz, 1998). Estes and Sanchiz (1982b, pp. 15–16) established this uncertainty on the basis of an extensive character analysis, yet they decided it was preferable to maintain *Scotiophryne* for the time being in the Discoglossidae, remarking that, in any case, the presence in North America is no longer in doubt (as demonstrated by the following genus, *Paradiscoglossus*). Gardner (2000), however, included *Scotiophryne* in his category of uncertain families.

GENUS #*PARADISCOGLOSSUS* ESTES AND SANCHIZ, 1982

Genotype. Paradiscoglossus americanus Estes and Sanchiz, 1982.

Etymology. The generic name comprises the Greek *para*, "near"; and *Discoglossus*, a living genus in the Discoglossidae. The specific name reflects the American occurrence of the genus.

TABLE 2. Character states of *Scotiophryne pustulosa* compared with those of other anuran groups.

	Alytes Baleaphryne	Discoglossus Eodiscoglossus Latonia	Bombina Barbourula	Eopelobates Pelobates Scaphiopus	Megophrys	Pelodytes
1. Supra-acetabular expansion not very long, but large and with a dorsal prominence (ILIUM)	0	0	+	0	0	+
2. Dorsal protuberance (tubercle) absent (ILIUM)	0	0	+	+	+	+
3. Dorsal crest absent (ILIUM)	+	0	+	+	+	+
4. Subacetabular expansion moderately developed and interiliac tubercle present (ILIUM)	+	+	0	0	+	0
5. Ventral cubital fossa present but not well defined laterally (HUMERUS)	0	0	+	+	0	+

Note: +, present; 0, absent.

Diagnosis. The diagnosis is the same as for the genotype and only known species, *Paradiscoglossus americanus*.

#*PARADISCOGLOSSUS AMERICANUS* ESTES AND SANCHIZ, 1982
(FIG. 21)

Holotype. A left ilium lacking the distal end of the ilial shaft and the dorsal edge along the anterior two-thirds of the ilial crest (University of California Museum of Paleontology: UCMP 125827).

Locality and Horizon. Bushy Tailed Blowout (UCMP locality V-5711), Niobrara County, Wyoming: Late Cretaceous (late Maastrichtian). Lance Formation.

Other Material. All referred material of *Paradiscoglossus americanus* comes from the Late Cretaceous (late Maastrichtian) Lance Formation of Wyoming. These fossils include fragments of a left and a right distal humerus (UCMP 125828, 125829) from Bushy Tailed Blowout (UCMP locality V-5711) and a left ilium (UCMP 125830) from the Lull 2 site (UCMP locality V-5620).

Diagnosis. Gardner (2000) has provided a revised diagnosis as modified from Estes and Sanchiz (1982b) and Sanchiz (1998), but for the purpose of this volume I favor the original diagnosis provided by Estes and Sanchiz (1982b) because it provides comparisons of both living and fossil genera. This latter diagnosis is slightly modified as follows. A discoglossid frog with an ilium that differs from those in the species of the genera *Alytes, Balaeophryne, Bombina,* and *Scotiophryne* in having a well-developed dorsal crest, an elongated dorsal tubercle (prominence), and an interiliac tubercle medially, in the center of the symphyseal articulation; differs from *Barbourula* in the absence of an interiliac synchondrosis; differs from *Eodiscoglossus* and *Discoglossus* in having a more compressed dorsal tubercle (prominence) in dorsal view and a less swollen tubercle — the posterior slope of this tubercle also forms a more open angle with

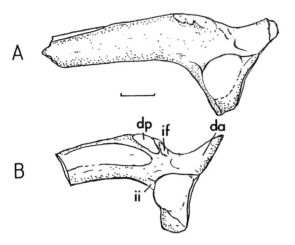

A

B

dp if da

ii

FIGURE 21. (A) Lateral view of left ilium of *Paradiscoglossus americanus* from the Late Cretaceous Lance Formation of Wyoming. (B) Lateral view of left ilium of modern *Discoglossus pictus*. Abbreviations: da, dorsal acetabular expansion; dp, dorsal prominence (origin of gluteus muscle); if, origin of iliofibularis and iliofemoralis muscles; ii, origin of iliacus internus muscle. Scale bar = 2 mm and applies to both figures.

respect to the supra-acetabular expansion (dorsal acetabular expansion), and the sub-acetabular expansion (ventral acetabular expansion) is less developed than in the latter two genera — differs from all anurans in having a very large and deep supra-acetabular fossa.

Description. In this description, slightly modified here, Estes and Sanchiz (1982b) used North American terms, which were followed in parenthesis by the terminology of Vergnaud-Grazzini (1966) and other European authors. For the purpose of this volume, I use this description rather than the "remarks" of Gardner (2000). Both ilia agree in all of the observable features. The holotype is an almost complete proximal half of an ilium, with only the most posterior parts of the supra- and subacetabular expansions missing, as well as the distal parts of the ilial shaft. The supra-acetabular expansion (pars ascendens ilii) is very large, and, although slightly incomplete, it is clear that it terminated in a very long ischiatic process. A very large and deep supra-acetabular fossa is present on its anterior part.

The dorsal tubercle (or prominence) (tuber superius), is well developed, occurs in an anterior position with respect to the acetabulum, and is elongated and directed forward. The dorsal tubercle continues onto a large dorsal crest (crista dorsalis ilii) that is slightly curved inward. There is no deep concavity between the dorsal tubercle and the main ilial shaft. The tubercle slopes gently into the supra-acetabular expansion, with which it forms a very open angle. The dorsal tubercle is not swollen and thus is not clearly separated from the crest. A longitudinal groove is present on the lateral side of the tubercle, deeper in UCMP 125830 than in the holotype. The ventral acetabular expansion (pars descendens ilii) is small and forms an open angle with the ventral border of the ilial shaft. The ilial shaft has a small ridge that extends longitudinally along the middle of the lateral surface. The inner side of the ilium has a distinct interiliac tubercle (tuber interiliacus) and, aside from this tubercle, the internal surface of the ilium forms a concave depression.

Two distinct distal humeri are tentatively referred to this species. Each of them has a large distal condyle (eminentia capitata) and a rather long ulnar epicondyle reaching posteriorly to the distal level of the distal condyle. The ventral cubital fossa is well defined, and there are no well-developed distal crests present. The olecranon scar is elongate and of moderate size. The maximum distal width of the humerus is 5.98 mm, and the anteroposterior diameter of the distal condyle is 3.5 mm. These features are similar to the discoglossid genera *Discoglossus* and *Latonia*, although the ventral cubital fossa appears to be slightly deeper and the ulnar epicondyle longer in the Lance Formation specimens. The disarticulated humeral remains of *Eodiscoglossus* from the Late Cretaceous of Spain (Estes and Sanchiz, 1982a) show a similar development of the ventral cubital fossa, but not in epicondyle size. Moreover, the Spanish specimens are much smaller.

General Remarks. Estes and Sanchiz (1982b) were positive about the assignment of *Paradiscoglossus americanus* to the family Discoglossidae and remarked (p. 11), "It would be possible to refer this frog to the family Discoglossidae based only on the presence in the ilium of a long ischiatic process on the supra-acetabular expansion, which seems a common derived feature since it occurs to such an extent only in discoglossids." Sanchiz made no specific comments about the familial validity of *Paradiscoglossus* but nevertheless included it with the Discoglossidae in his systematic accounts (Sanchiz, 1998). Gardner (2000) included *Paradiscoglossus americanus* in his category of indeterminate family relationships.

Indeterminate "Discoglossidae"

Roček (2000, table 1, p. 1296) indicated that three discoglossids, "Montana discoglossids I, II, and III," occurred in the Late Cretaceous (Campanian) of Montana.

Suborder Mesobatrachia Laurent, 1979

The suborder Mesobatrachia also contains primitive anuran families.

Superfamily Pipoidea Fitzinger, 1843

Families in the superfamily Pipoidea that are represented by North American fossils are the #Palaeobatrachidae and the Rhinophrynidae.

Family #Palaeobatrachidae Cope, 1865

The family Palaeobatrachidae is the only named extinct anuran family that is currently given credibility by all vertebrate paleontologists. Palaeobatrachids are extinct anurans that are thought to be similar to the modern genus *Xenopus* in being obligatorily aquatic. They were up to 120 mm in SVL. Some general osteological characters of the family provided by Špinar (1972) and Sanchiz (1998) are presented here in modified form. In the skull, the orbits are large and in an anterior position on the skull. Both the premaxillae and the maxillae are tooth-bearing. The frontoparietal is undivided into right and left portions. The nasal bones are bent and separated. The squamosals are widened laterally and have a short columella process. The parasphenoid is dagger-shaped and lacks lateral wings. The pterygoid is triradiate, and its medial ramus is wide and long. A mentomandibular is present.

In the vertebral column, nine procoelous, stegochordal, imbricate vertebra are present. Five pairs of free ribs are present and are fused with the transverse processes of vertebrae II–VI. The "atlas" is composed of vertebrae I and II. Also present is a synsacrum, in which at least vertebrae VIII and IX are incorporated. Broadly dilated lateral processes are present on the sacrum. The articulation between the sacrum and the urostyle is facilitated by two cotyles that are partially fused medially.

The shoulder girdle has a short, uncleft scapula, and the coracoid is axe-shaped medially, with a conspicuous proximal process that connects to the clavicle. The humerus lacks a ventral cubital fossa. The metacarpalia are about the same length as the radio-ulna. The phalangeal formula for the hand is 2–2–3–3 (inner to outer). In the pelvic girdle, a small ossified pubis is present. The ilium lacks a dorsal crest but has a well-developed dorsal prominence. The femur is slightly longer than the tibiofibula. The calcaneum and astragalus are separate elements and not fused either proximally or distally. The metatarsals are long and about the same length as the calcaneum and astragalus. The phalangeal formula for the bones of the foot is 2–2–3–4–4 (inner to outer).

It is noteworthy that the Palaeobatrachidae, the only extinct family of anurans that is known, survived until late in geologic time, occurring in Germany until at least the middle Pleistocene (Ice Age) (Holman, 1998). Roček (1995) suggested that the reason these palaeobatrachid frogs were able to live through the glacial onslaughts of earlier times was that their completely aquatic habits would allow them to survive, in contrast to less aquatic anurans that would either emigrate or perish when all aquatic areas were frozen over. In North America, palaeobatrachids did not survive into the Tertiary.

Genus #*Palaeobatrachus* Tschudi, 1838

Genotype. Palaeobatrachus diluvianus (Goldfuss, 1831).

Etymology. The generic name is from the Greek *palaeo*, "ancient," and *batrachos*, "frog."

Diagnosis. This diagnosis follows Sanchiz (1998, p. 36): "Palaeobatrachid frogs with unsculptured frontoparietal; small boomerang-shaped nasals; maxillae with about 12–16 tooth positions and short processus anterior; sphenethmoid without long rostral process; long parasphenoid with a dagger-like cultriform process (rostrum)." A detailed treatment of the genus was given by Špinar (1972), who provided descriptions and illustrations.

#*Palaeobatrachus occidentalis* Estes and Sanchiz, 1982
(Fig. 22)

Holotype. A right ilium missing the posterior part of the dorsal acetabular expansion and the distal end of the ilial shaft (University of California Museum of Paleontology: UCMP 55704).

Locality and Horizon. Lull 2 site (UCMP locality V-5620), Niobrara County, Wyoming: Late Cretaceous (late Maastrichtian). Lance Formation.

Etymology. The specific name is from the Latin *occidentalis*, "from the west."

Other Material. Referred material from Estes and Sanchiz (1982b) is as follows. A left ilium (UCMP 55705) from Bushy Tailed Blowout (UCMP locality V-5711), Niobrara County, Wyoming, Late Cretaceous

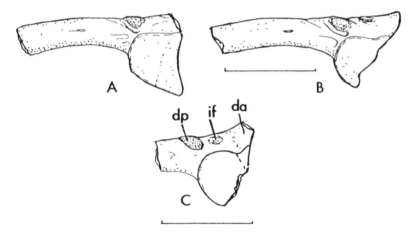

FIGURE 22. Palaeobatrachid ilia in lateral view. (A) Holotype right ilium of *Palaeobatrachus occidentalis* from the Late Cretaceous Lance Formation of Wyoming (image reversed for comparison). (B) Referred left ilium of *P. occidentalis* from the same locality. (C) Left ilium of *Palaeobatrachus* sp. from the Paleocene of France for comparison. Abbreviations: da, dorsal acetabular expansion; dp, dorsal prominence; if, origin of iliofibularis and iliofemoralis muscles. Scale bars = 3 mm; upper one applies to both (A) and (B).

(late Maastrichtian), Lance Formation. A right ilium (MCZ 3653) from the Bug Creek Ant Hills locality, McCone County, Montana, Late Cretaceous (late Maastrichtian), Hell Creek Formation. Gardner (2000) referred a fused atlas and first trunk vertebra (University of Alberta Laboratory for Vertebrate Paleontology: UALVP 40163) from the Bug Creek Ant Hills locality to this species.

Diagnosis. This diagnosis is modified from Estes and Sanchiz (1982b), Sanchiz (1998), and Gardner (2000). A palaeobatrachid species differing from other *Palaeobatrachus* species and from *Pliobatrachus* in the following combination of ilial and atlantal characters. The dorsal margin above the acetabulum is relatively straight in lateral aspect. A sulcus extends around the anteroventral and anterior margin of the dorsal prominence, separating the prominence anteriorly from the ilial shaft. The dorsal prominence is oval in outline and is separated from the iliofibularis–iliofemoralis attachment area, which is low, small, and lenticular in outline. The dorsal acetabular expansion is relatively low. The anterior margin of the acetabulum is squarish in lateral outline. The ventral acetabulum is quite reduced. The ilial synchondrosis on the medial surface is relatively large. The atlas differs from *Pliobatrachus* in having the anterior cotyles paired and not confluent and from *Albionbatrachus* in lacking a median intercotylar process between the anterior cotyles; it differs further from *Pliobatrachus* and from *Palaeobatrachus robustus* in having a smaller implied body.

Description. The ilial characters are described in detail in the diagnosis above. The fused cervical vertebra and first trunk vertebra are described in the following general remarks section.

General Remarks. Gardner (2000) commented that the familial assignment of *Palaeobatrachus occidentalis* is supported by detailed resemblances between the two Lance Formation ilia and those of European palaeobatrachids. These resemblances include (1) the absence of the dorsal crest; (2) the small dorsal prominence, which is separate from the attachment area for the iliofibularis and iliofemoralis muscles; (3) the much reduced ventral acetabulum; (4) the large acetabulum, which is incomplete posteriorly, is expanded laterally, and has its ventral rim projecting anteroventrally to overhang the subacetabular area; and (5) the prominent interiliac synchondrosis.

Gardner (2000) also commented on the fused cervical vertebra and first trunk vertebra he studied. He pointed out that this element compares well with homologous European specimens and is characteristic of the Palaeobatrachidae in (1) being solidly fused; (2) having a foramen for the exit of the first spinal nerve at the point of fusion; (3) having the posterior condyle broader than it is wide; (4) having a narrow, although distally complete, transverse process; (5) having the neural arch low and simple in construction and posteriorly elongate; and (6) having the postzygapophyseal processes unelaborated.

Unfortunately, most European palaeobatrachid taxa are based on articulated skeletons in which ilial and atlas complex characters are obscure (Špinar, 1972; Sanchiz, 1998), thus making the generic and to a lesser degree the specific assignment of *Palaeobatrachus occidentalis* somewhat uncertain. Although both Sanchiz (1998) and Gardner (2000) referred to the above taxon as *Palaeobatrachus? occidentalis*, I will leave out the "?" in this taxonomic name until it is demonstrated that another generic epithet is more realistic.

FAMILY RHINOPHRYNIDAE GÜNTHER, 1859 "1858"

The family Rhinophrynidae is represented in the modern world by a single species, *Rhinophrynus dorsalis*, that occurs from extreme southern Texas and Michoacan, Mexico, south to Costa Rica (Frost, 1985). It is a primitive, burrowing species but has several specializations, including a tongue that may be slowly protruded to gather ants and their larvae.

Henrici (1991, pp. 101, 106) has provided a practical diagnosis of the family, quoted here. "Can be distinguished from all other anuran families by the following combination of characters: supraorbital flange present; maxilla and premaxilla edentate; palatine process of premaxilla absent; palatines absent; parahyoid bone present; eight imbricate, ectochordal, modified opisthocoelous presacral vertebrae; atlantal cotyles closely juxtaposed; atlantal neural arch elongate; presacrals I and II unfused; transverse process of second presacral vertebra hook shaped; free ribs absent; expanded sacral diapophyses; urostyle with bicondylar articulation with sacrum; arciferal pectoral girdle; clavicle overlying anterior end of scapula; humeral shaft strongly angulated; femur bearing lateral and medial crests; tibiale and fibulare fused proximally and distally; and prehallux and distal phalanx of first digit modified as bony digging spades."

A standard osteological definition of the Rhinophrynidae is as follows. Eight ectochordal vertebrae are present. They are modified opisthocoelous vertebrae where the intervertebral body tends to adhere to the anterior end of the centrum but is not fused to it. The vertebrae are also imbricate. Presacrals I and II are not fused, and the atlantal condyles of presacral I are closely juxtaposed. Free ribs are absent. The sacrum has expanded diapophyses and a bicondylar articulation with the urostyle. The urostyle lacks transverse processes. The pectoral girdle is arciferal, and both the omosternum and sternum are absent. The anterior end of the scapula is overlain by the clavicle. Palatine bones are absent, but a parahyoid bone is present. The cricoid ring is incomplete dorsally. Both the premaxilla and maxilla are toothless. The astragalus and calcaneum are only fused proximally and distally. Two tarsalia are present. The phalangeal formula is normal except for the loss of one phalanx on the first toe.

GENUS #*CHELOMOPHRYNUS* HENRICI, 1991

Genotype. Chelomophrynus bayi Henrici, 1991.

Etymology. The generic name is from the Greek *cheloma*, "notch" (in reference to the anterior notch on the maxilla), and *phryne*, "toad." The specific name recognizes Kirby Bay, who discovered the frog type locality.

Diagnosis. The diagnosis is the same as for the genotype and only known species, *Chelomophrynus bayi.*

#*CHELOMOPHRYNUS BAYI* HENRICI, 1991

(FIG. 23)

Holotype. A partially complete, articulated to loosely articulated skeleton of an adult consisting of a partially broken and somewhat disarranged skull, displaced incomplete ?sphenethmoid and parasphenoid, left humerus, vertebral column and sacral vertebra, the urostyle, both ilia, and the left hind limb (Carnegie Museum of Natural History: CM 46712).

Locality and Horizon. North end of Battle Mountain in the Lysite Mountain area, Hot Springs County, Wyoming (specific locality data on file in the Section of Vertebrate Paleontology, Carnegie Museum of Natural History): Middle Eocene (believed to be Uintan NALMA; Henrici, 1991). Wagon Bed Formation.

Other Material. Other elements, considered part of the hypodigm by Henrici (1991) and not listed as either paratype or referred material, include a premaxilla (CM 46714), maxillae (CM 46706, 46707, 46753, 46769), parasphenoids (CM 46753, 46772), vomers (CM 46710, 46754), a pterygoid (CM 44488), a squamosal (CM 44488), a sphenethmoid (CM 46742), exoccipital–prootic complexes (CM 44496, 46701, 46707, 46775, 46776), a dentary (CM 46706), angulosplenials (CM 44496, 46724), a hyoid (CM 46781), a scapula (CM 46716), humeri (CM 44488, 44496, 46754, 46762, 46763), a radio-ulna (CM 44485), cervicals articulated with second vertebra (CM 46722, 46749), free cervicals (CM 44481, 44483, 44496, 46711, 46714, 46722, 46749), presacral vertebrae (CM 46706, 46722), a sacral vertebra (CM 46764), ilia (CM 44486, 46716), ischia (CM 44486, 46716), femora (CM 46714, 46716, 46761), and tarsi (CM 46755, 46756). These bones are often portions of partially associated specimens.

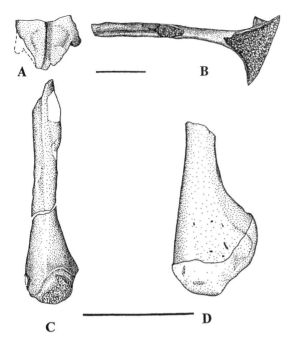

FIGURE 23. Bones of *Chelomophrynus bayi* from the Middle Eocene of Wyoming. (A) Cervical vertebra in dorsal view. (B) Left ilium in lateral view. (C) Left humerus in ventral view. (D) Distal end of right femur in medial view. Scale bars = 5 mm.

Diagnosis. The diagnosis is quoted from Henrici (1991, p. 107): "Differs from all rhinophrynids in unique characters of broad, triangular alary (winglike) process of premaxilla, anterior notch of maxilla, elongate posterior process of vomer, elongate postchoanal ramus of vomer, and laterally oriented transverse processes of last three presacral vertebrae. Differs from *Eorhinophrynus* sp. in having a round humeral ball (distal condyle) and from *E. septentrionalis* in having atlantal (cervical) neural arch that tapers posteriorly. Differs from *Rhinophrynus* in having well-developed lateral crest of humerus, shaft of femur slightly bowed, shaft of femur expanded directly proximal to distal condyle, and distal condyle of femur of equal length and width."

Description. This section is modified from the descriptions of *Chelomophrynus bayi* by Henrici (1991). The shape of the frontoparietal varies greatly, ranging from roughly pentagonal to octagonal. In some frontoparietal bones the sides and angles are apparent, whereas in others the bone is nearly oval. Some of the specimens have more robust frontoparietals than others. The shape of the anterior rostrum in the frontoparietals varies from long and pointed, to long and blunt, to short and blunt. The marked variation displayed in *Chelomophrynus* casts doubts on the usefulness of this bone as a single element upon which to define a frog genus (e.g., Meszoely et al., 1984 [*Albionbatrachus*]).

The sickle-shaped nasal has both rostral and lateral processes; the rostral process is long and slender and directed anteriorly, and the lateral process is shorter, broader, and directed anterolaterally. The posterior margin of the nasal is rounded. The dental portion of the premaxilla is tooth-bearing, and the palatine area lacks a palatine process. The maxilla is also tooth-bearing; its anterior notch is expanded dorsally, forming the facial portion of the bone; anterior and palatine processes are absent.

In the parasphenoid, the cultriform process broadens laterally to form the body of the bone, which lacks lateral wings. The length and shape of the body of the parasphenoid varies from tapering rather abruptly into the cultriform process (rostrum) to tapering gradually into this process; posterior lateral expansions are small, and the posterior margin is notched.

The orientation of the vomer is uncertain. Henrici (1991) pointed out that the posterior part of the vomer lacks teeth, but it is not clear in her description whether the anterior portion is tooth-bearing or not. The palatine was not found in the *Chelomophrynus* hypodigm, and Henrici (1991) suggested that it is absent in the genus as it is in the living genus

Rhinophrynus (Trueb and Cannatella, 1982). A pterygoid has not been positively identified in the material. The squamosal consists of a broad ventral ramus and a small zygomatic ramus, but it cannot be determined whether an otic ramus is present.

The rostral process of the sphenethmoid is rounded, and the right and left lateral processes are located about one-third the length of the bone from the anterior end. At the base of the lateral processes the sphenethmoid narrows medially. The dorsolateral edge of the posterior end of the sphenethmoid is curved, and this curvature may represent the anterior border of the optic–prootic foramen as occurs in *Rhinophrynus dorsalis*.

The exoccipital and prootic are indistinguishably fused as in most anurans. The prootic region is roughly diamond-shaped in dorsal view, bears a lateral rectangular process, and houses the otic capsule. The anteromedial edge of the prootic is concave, and this concavity probably forms the posterior, posterodorsal, and posteroventral walls of the optic–prootic foramen as in *Rhinophrynus dorsalis*. The paired exoccipitals are fused medially and form the posterior portion of the braincase. The occipital condyles are closely juxtaposed, and the jugular foramen lies laterally and slightly ventrally to each condyle. A quadrate was not identified among the material; thus, it is probable that the mandible articulated with the skull by a cartilaginous quadrate.

In the lower jaw, the dentaries are toothless. The left dentary is elongate and tapers at the anterior and posterior ends. Both ends are rounded. Ossified mentomeckelian bones were not identified in the material. The angulosplenial is robust, with a deep posterior end that at its midlength begins to taper dorsoventrally toward the anterior end. The angulosplenial is curved medially, has a troughlike medial surface, and has a coronoid process that is not well developed.

As in some modern frogs, the hyoid contains three bony elements: the parahyoid bone and the paired posteromedial processes. These hyoid elements were identified in only one specimen (CM 46781). The two laterally directed rami of the parahyoid bone are broad at their medial juncture and taper to a point laterally. A third posteriorly projecting process is short and broad. The paired posteromedial processes are somewhat robust, with their anterior and posterior ends expanded and the shaft broad and depressed.

In the postcranial skeleton, the clavicle is strongly curved, which shows that *Chelomophrynus* has an arciferal pectoral girdle. The clavicle has a broad articular surface on the posterior edge of its lateral end, indicating that it probably overlapped the scapular anteriorly.

Since a suprascapula was not identified among the material, it is assumed that it probably was not ossified, a situation that occurs in most anurans. The anterior edge of the scapula is straight, and the posterior edge is concave; the widths of the dorsal and ventral ends are subequal. The ventral end of the anterior margin of the scapula has an articulating surface for the clavicle.

Both ends of the coracoid are expanded, and the medial end is significantly larger than the lateral end. The medial end of the cleithrum is forked and forms anterior and posterior processes. The anterior process is approximately twice as long as the posterior process and, in adults, joins the anterior edge of the main body at an angle of about 130 degrees. This angle is somewhat greater in tadpoles. A ridge extends along the antero-dorsal surface.

The humerus has a weakly developed ventral crest on the ventral surface of the proximal half of the shaft. The distal condyle is bounded by two epicondyles, one each on the medial and lateral sides. The medial epicondyle is the largest, and its own medial surface is triangular. A ridge extends from the ventral crest to the medial epicondyle. A raised, triangular olecranon scar occurs medially on the dorsal surface of the distal end.

The proximal end of the radio-ulna is narrower than the distal end and bears a well-developed olecranon process. A longitudinal sulcus is restricted to the proximal and distal ends in large specimens but extends the length of the shaft in smaller ones. The bones of the manus are disarticulated and scattered; thus, it is impossible to determine the phalangeal formula or even to identify the individual elements.

In the vertebral column, the eight amphicoelous ectochordal vertebrae that are present bear no free ribs. Elongate transverse processes occur on vertebrae II–IV; those of vertebra III are the longest. The distal end of the transverse processes on vertebra II is oriented anteriorly, forming a hook. The transverse processes on vertebrae IV–VIII are short and laterally oriented. The sacral vertebra has two posterior condyles, its sacral diapophyses are expanded anteroposteriorly, and the neural arch is posteriorly notched. The urostyle is relatively robust and is without vestigial transverse processes; a dorsal crest runs from the anterior end to just short of the posterior tip.

In the pelvic girdle, the ilium has its shaft slightly curved and laterally keeled and is without ridges. A well-developed, variably shaped, triangular dorsal prominence arises from the dorsal acetabular expansion. The apex of the dorsal prominence is rounded in six specimens and pointed in two. The anterior margin of the dorsal prominence is longer than the posterior margin in four specimens, is shorter in three, and is subequal in one. It is believed that this variation is not based on growth, as the variation occurs in individuals of similar size. The dorsal prominence, however, is always wider than it is high. The junction of the ilial shaft with the ventral acetabular expansion is more deeply excavated than that with the dorsal acetabular expansion. The ilial portion of the acetabulum is higher than it is wide and projects laterally from the acetabular expansion; the ventral lip of the cup projects laterally more than the dorsal lip. The ilial junction is V-shaped. The paired ischia are medially fused, and the ischial portion of the acetabulum is wider than it is high; the posterodorsal lip projects farther laterally than the ventral lip.

In the hind limb, the shaft of the femur is slightly bowed medially. Its distal end is dorsoventrally flattened and bears lateral and medial condyles. The lateral condyle is larger and posterolaterally oriented, extending from the distal end of the femur to the posterior side of the shaft. The medial condyle is small in large femora and absent in small ones. The femoral shaft bears two crests, a lateral one and a medial one. Since the lateral crest occurs in the same position, it is thought to be homologous to the single crest (crista femoralis) in many other anurans.

The tibiofibula, which is shorter than the femur, is hourglass-shaped, and its proximal and distal ends are subequal in width. The medial surface of the tibiofibula is concave, and its lateral surface is convex; the dorsal edge of the proximal end is rounded, and the ventral edge forms a ridge. The medial surface of the distal end is smooth, but the lateral distal surface is not preserved in any specimens. A distinct longitudinal sulcus extends the length of the bone in small specimens but becomes indistinct and occurs only in the proximal and distal ends in older individuals.

In the tarsus, the proximal series of bone consists of the tibiale and fibulare. The tibiale is about 1.4 mm shorter than the fibulare. The proximal and distal ends of the fibulare are expanded, and its lateral side is strongly concave and longer than the medial side, which is only slightly concave. Both the proximal and the distal ends of the fibulare are expanded; the lateral side is slightly concave and longer than the strongly concave medial side. The proximal and distal ends of the tibiale and fibulare are fused in larger specimens but unfused in smaller ones. In the specimens where they are fused the space between the two bones forms an oval.

In the hind foot, the centrale lies distal to the tibiale and proximal to and slightly between the prehallux and the first metatarsal. Tarsals 2 and 3 are fused and lie distal to the point where the tibiale and fibulare meet. Five metatarsals are present. The prehallux is modified as a digging spade. A smaller spade, closely associated with the first digit, is most likely a modified phalanx of this digit as in *Rhinophrynus dorsalis*. The phalangeal formula cannot be determined because of incomplete material.

General Remarks. Chelomophrynus bayi is represented by the largest sample of any anuran species currently known from any single fossil locality in North America. This allowed the identification of a growth series representing at least six developmental stages, from tadpoles in the beginning stages on through to adults. The presence of two digging spades on each hind limb and the rather robust skeleton of *C. bayi* indicate that this taxon was a burrowing form like its fossil and modern relative *Rhinophrynus*. From a paleoecological standpoint, it has been shown that tadpoles and subadults of *C. bayi* inhabited an Eocene lake but that the adults probably lived in a terrestrial environment (Henrici, 1991; Roček and Rage, 2000b). Henrici and Fiorillo (1993) additionally reported on a large accumulation of bones from the type locality that consisted exclusively of small, postmetamorphic animals.

GENUS *#EORHINOPHRYNUS* HECHT, 1959

Genotype. Eorhinophrynus septentrionalis Hecht, 1959.

Etymology. The generic name comprises the Latin *Eo*, "old"; and *Rhinophrynus*, a fossil and modern genus of the family Rhinophrynidae. The specific name is from the Latin *septentrionalis*, "northern."

Diagnosis. The diagnosis is the same as for the genotype and only known species, *Eorhinophrynus septentrionalis*.

#*Eorhinophrynus septentrionalis* Hecht, 1959
(Fig. 24)

Holotype. A complete atlas (American Museum of Natural History: AMNH 3818).

Locality and Horizon. Tabernacle Butte locality 5, Sublette County, Wyoming: Middle Eocene (Bridgerian NALMA). Bridger Formation.

Other Material. Material referred to *Eorhinophrynus septentrionalis* by Henrici (1991) includes a left fibulare (University of California Museum of Paleontology: UCMP 113252), the right exoccipital–prootic complex (Carnegie Museum of Natural History: CM 46783), the right scapula (CM 46784), the distal half of the left humerus (CM 46782), and the proximal end of the right tibiofibula (CM 46785). All of this material is topotypic.

Diagnosis. The revised diagnosis of Henrici (1991, p. 106) is quoted here. "Differs from all rhinophrynids in having atlantal (cervical) neural arch that is laterally broad with square posterior margin. Differs from *Eorhinophrynus* sp. (*sensu* Estes, 1975) in having round humeral ball (distal condyle). Differs from *Rhinophrynus* in having well-developed lateral crest on humerus."

Description. This description is modified from Hecht (1959) and Henrici (1991). In the detailed description of the holotype atlas, Hecht (1959, pp. 130–131) mentioned that the fossil atlas may be distinguished by its nearly vertical condylar facets and its elongated neural arch and that the elongate neural arch and its posterior overhang appear to be new adaptations to a fossorial habitat.

With regard to the referred topotypic material, Henrici (1991) pointed out that the exoccipital–prootic complex resembles that of *Chelomophrynus bayi* in every aspect except that in *Eorhinophrynus* it is less robust. The scapula is relatively long and is cleft, with an articulating surface for the clavicle on the ventral end of the anterior edge. This condition indicates that the scapula is anteriorly overlain by the clavicle, a feature found in the Leiopelmatidae, Discoglossidae, Rhinophrynidae, and Palaeobatrachidae and in *Pipa* of the Pipidae. Nevertheless, the scapula is not a leiopelmatid, discoglossid, or palaeobatrachid, because these have short and uncleft scapulae. In the Pipidae the scapula is cleft but short as in the Rhinophrynidae.

In the humerus, the distal condyle is round, the olecranon scar is situated medially, and the lateral and medial crests are moderately developed, long, and of subequal length. The shaft of the humerus is strongly angulated. A ventral crest is present, but the degree of its development cannot be determined. A ridge

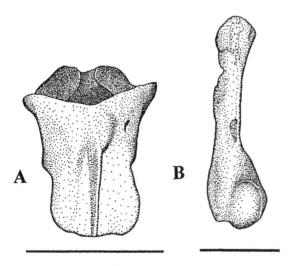

FIGURE 24. Bones of *Eorhinophrynus* from the Middle Eocene of Wyoming. (A) Cervical vertebra of *Eorhinophrynus septentrionalis* in dorsal view. (B) Left humerus of *Eorhinophrynus* sp. in ventral view. Scale bars = 5 mm.

extends from the ventral crest to the medial epicondyle. The strongly angulated shaft, the medially situated olecranon scar, and the ridge running from the ventral crest to the medial epicondyle all indicate assignment to the Rhinophrynidae. Relative to the tibiofibula, the somewhat robust character of this bone suggests rhinophrynid affinities, but its assignment to this group is tentative.

General Remarks. More material of *Eorhinophrynus* is needed so that this genus can be adequately compared with the other North American rhinophrynid genera *Chelomophrynus* and *Rhinophrynus.*

EORHINOPHRYNUS SP. INDET.

Locality and Horizon. Princeton and Fritz quarries, Park County, Wyoming: Late Paleocene (Tiffanian NALMA). Fort Union Formation.

Material. A left humerus (Princeton University Museum: PU 14664), the distal end of the right humerus (PU 14665), a partial sacral vertebra (PU 14669), a right tibiale (PU 21773), and a PU unnumbered humeral fragment. This material currently resides in the Yale Peabody Museum, Yale University.

Description. The description is modified from Estes (1975). The left humerus is well preserved, relatively short, and robust and has a distinctly angular shaft. The distal condyle is of moderate size and has a slightly oblique orientation. The medial condyle is relatively small, and the lateral condyle is minute.

A strong ridge extends from the distal condyle of the humerus near the medial epicondyle obliquely across the humeral shaft and connects with the ventral crest. Medially, the ventral crest has a prominent channel bounded by a ridge. The ventral crest itself is prominent and projects strongly from the shaft. The olecranon scar is prominent, medial, and symmetrical and has its surface slightly raised. Slight channels are present on either side of this scar. The total length of the humerus is 12.6 mm.

The sacral vertebra has two posterior condyles and one anterior cotyle; the cotyle has a prominent notochordal canal. The condyles are subrounded. The remnants of the sacral diapophyses appear as if they had been expanded and extend much more posteriorly than they do medially. A deep notch occurs on the neural arch. The tibiale (mistakenly called a fibulare by Estes, 1975) is robust and short and ends broadly expanded. The total length of the tibiale is 6.8 mm.

GENUS *RHINOPHRYNUS* DUMÉRIL AND BIBRON, 1841

BURROWING TOADS

Modern and fossil *Rhinophrynus* may easily be distinguished from *Chelomophrynus* on the basis of the ilia, which are quite morphologically distinct. In *Rhinophrynus* the entire dorsal acetabular area has been rotated forward. In fact, it is almost perpendicular to the ilial shaft in some specimens (Figs. 25A–25C). The forward rotation of the area also affects the orientation of the elongate dorsal prominence, whose long axis has become almost dorsoventral in position (see Fig. 25C). Moreover, the elongated dorsal prominence of *Rhinophrynus* is laterally produced rather than produced from the dorsal margin of the ilium. Most anurans with an elongate dorsal prominence (e.g., *Palaeobatrachus*) have this prominence horizontally oriented, and in many (e.g., some *Hyla* and a few *Rana*) it is dorsally produced from the margin of the ilium. *Chelomophrynus* has a typical anuran dorsal acetabular area and has a triangular ilial prominence that is produced from the dorsal margin of the ilial shaft (see Fig. 23). Unfortunately, the ilium of *Eorhinophrynus* is unknown.

RHINOPHRYNUS CANADENSIS HOLMAN, 1963
(SEE FIG. 25)

Holotype. A right ilium (Saskatchewan Museum of Natural History: SMNH 1425). As far as I am aware, the holotype, paratypes, and all of the referred material (see Holman, 1963a, 1968, 1972b) can no longer be located.

Etymology. The specific name refers to the fact that all the material is from Canada.

Locality and Horizon. Calf Creek Local Fauna, near East End, Saskatchewan, Canada: Late Eocene (Chadronian NALMA) (incorrectly cited as Lower Oligocene in Sanchiz, 1998). Cypress Hills Formation.

Other Material. All of the material of *Rhinophrynus canadensis* is from a specific lens of the Cypress Hills Formation that lies along the north Branch of Calf Creek, 16 km northwest of East End, Saskatchewan (see Holman, 1972b for a more specific location). The locality was termed the Calf Creek Local Fauna (Holman, 1972b). The fossils were all taken from a matrix of conglomerates of sandstones and sands. The richest matrix included small clay pellets. Paratypes designated by Holman (1963a) include one right ilium (SMNH 1426) and two left ilia (SMNH 1427, 1428). Referred material designated by Holman (1963a) includes two right distal femora (SMNH 1429) and one right distal humerus (SMNH 1430). Additional topotypic referred material was listed by Holman (1968). This material includes 1 vertebra, 10 humeri, 4 radio-ulnae, 30 femora, 12 tibiofibulae, 6 tarsals, and 21 ilia, all under SMNH 1433. Henrici (1991) listed additional material (all numbers prefixed by P661.): 9 tarsals (SMNH 1907–1915), 18 femora (SMNH 1901, 1902, 1916–1931), 23 ilia (SMNH 1905, 1906, 1951–1971), 1 humerus (SMNH 1903), 4 radio-ulnae (SMNH 1947–1950), and 1 vertebra (SMNH 1946).

Note. At this point it should be stated that the figure of *Rhinophrynus canadensis* presented in Henrici (1991, fig. 21E, p. 134) and repeated in Sanchiz (1998)

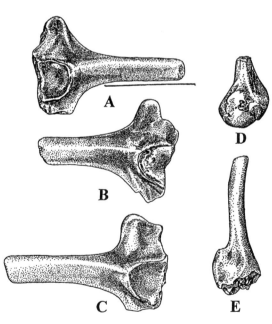

FIGURE 25. Bones of *Rhinophrynus canadensis* from the Late Eocene of Saskatchewan. (A) Holotype right ilium. (B) and (C) Paratype left ilia. (D) Distal end of right humerus in ventral view. (E) Distal end of femur in lateral view. Scale bar = 5 mm and applies to all figures.

and Roček and Rage (2000b) bears little resemblance to the holotype or other ilia from the type material (see Fig. 25, this volume; Holman, 1963a, fig. 1, p. 706). Henrici's specimen (SMNH 1907) is broken and highly eroded and is not at all typical of the species.

New Diagnosis. The new diagnosis differs substantially from that of Holman (1963a). *Rhinophrynus canadensis* differs from *Chelomophrynus bayi* in having the ilium with the entire dorsal acetabular region rotated forward so that it is almost perpendicular to the ilial shaft; differs in having an elongate dorsal prominence almost dorsoventrally located and produced laterally rather than dorsally from the shaft (*C. bayi* has dorsal acetabular area with the normal anuran condition and dorsal prominence triangular and dorsally produced from the shaft). Differs from *Eorhinophrynus septentrionalis* in having distal condyle (humeral ball) round rather than oval. Similar to the living species *Rhinophrynus dorsalis* but differs in being smaller; having an ilium with its dorsal acetabular expansion (supra-acetabular expansion) relatively low and with its apex rounded (*R. dorsalis* has very high dorsal acetabular expansion, with its apex flattened); ilium with its dorsal prominence more extensive and more strongly built and with its ventral portion usually directed inward from the anterior edge of the dorsal acetabular expansion (*R. dorsalis* has a less extensive and less strongly built dorsal prominence, which has its ventral portion usually running congruent with the anterior edge of the dorsal acetabular expansion). Humerus with medial crest more flared than in *R. dorsalis.* Femur with its shaft slightly curved and with its distal portion relatively thick (*R. dorsalis* has femur with its shaft distinctly curved and with its distal portion relatively gracile).

Description of the original material. Since the ilium has been well described in the diagnosis section above, the description of the original material here deals with the referred femora and humerus only. This description is modified from Holman (1963a). The femora (see Fig. 25E), as in *Rhinophrynus dorsalis*, have the distal condyles well ossified and with no line of demarcation from the shaft. The condyles are produced posteriorly, are much compressed, are about twice as high as they are long, are placed obliquely on the shaft, and have the proximal borders of their lateral and ventral surfaces excavated. The shaft is round, moderately thick, sharply keeled medially, and slightly bowed posteriorly. The proximal ends of both femora are missing. The humerus (see Fig. 25D) is represented by only the distal portion. The smooth, rounded distal condyle is slightly wider than it is high. The medial crest is robust and moderately flared. Measurements comparing seven skeletons of adult modern *R. dorsalis* with the ilia, femora, and humerus of *Rhinophrynus canadensis* indicate that the latter species is about half as large as *R. dorsalis.*

Additional Rhinophrynus dorsalis *and* Rhinophrynus canadensis *Material.* In 1968 new material of modern *Rhinophrynus dorsalis*, including 16 adult individuals from a single breeding assemblage in Veracruz, Veracruz, Mexico, collected in August 1965, was compared with the additional *Rhinophrynus canadensis* specimens. In the *R. dorsalis* skeletons, the vertebral centra, other than those of the atlas and sacrum, are hourglass-shaped, biconcave discs, with their notochordal canals large and, in most specimens, completely open. The only evidence of the intervertebral bodies of Walker (1938) lies in the anterior concavities of the centra, which are partially filled or encircled by roughened bone. The posterior concavities of the centra are much better excavated and entirely lack a bony filling. In one specimen (Michigan State University, Museum Herpetological Skeletal Collection: MSUHS 2279) the anterior concavities of two vertebral centra are completely plugged with roughened bone. Two other specimens (MSUHS 2275, 2276) have a single vertebra with this condition. These are the only vertebrae in the 16 specimens (cervical vertebrae and sacra not included) that lack a perforate centrum. The single known fossil *R. canadensis* centrum is a biconcave, completely perforate centrum. But the fossil vertebra is so badly worn that many details of structure are obscured.

The humeri of the Recent and fossil *Rhinophrynus* are quite characteristic. The original fossil humerus discussed (Holman, 1963a) consisted only of the distal end, but the new material is more complete. In both the living and the fossil *Rhinophrynus* the humeri are short and stout and are bowed not only dorsoventrally, but laterally as well. Two distinct ridges occur on the posterior face of the shaft: a long lateral ridge that runs about three-fourths the length of the shaft, and a shorter medial one that runs about one-third the length of the shaft. But the humeri of the fossils are not as short and stout as those of the Recent specimens, and the fossils have less robust processes and ridges and a less robust distal condyle and are less laterally bowed than the Recent specimens. The radio-ulnae of the fossil species are also less robust than those of the modern species.

The ilia of *Rhinophrynus canadensis* are as strong and robust as those of Recent *Rhinophrynus dorsalis*. On the other hand, the femora of most of the fossils are much less robust than in the Recent species, having the distal condyles weaker, as well as the two ridges for muscle attachment on the shaft. Nevertheless, a few of the larger fossils are almost as robust as in Recent *Rhinophrynus*. The tibiofibulae of *R. canadensis* and *R. dorsalis* are quite characteristic: they are exceptionally short and stout and have expanded ends. The fossil tibiofibulae are less robust than in *R. dorsalis*.

General Remarks. Although the skeletal elements of *Rhinophrynus canadensis* have not been found in the articulated position, they generally are much better preserved than those of other taxa of fossil rhinophrynids. The resemblance to the modern species *Rhinophrynus dorsalis* is truly remarkable, considering the many millions of years that separate the fossil from the modern taxon. The highly ossified distal condyle of the femur in fossil and living *Rhinophrynus* is unusual in anurans, and combined with the fact that the proximal portion of the femur is unossified in the many specimens collected would tend to indicate that the condition is associated with burrowing, perhaps to the detriment of the usual anuran type of saltation.

The occurrence of this species in Saskatchewan, Canada, is noteworthy given the probable climate there in the latest part of the Eocene. The only other species in the genus, *Rhinophrynus dorsalis*, which is morphologically similar to *Rhinophrynus canadensis*, occurs from extreme southern Texas to Costa Rica.

Rhinophrynus dorsalis Duméril and Bibron, 1841

Mexican Burrowing Toad

(Fig. 26)

Fossil Locality. **Pleistocene (Rancho-labrean NALMA):** Cueva (Cave) de Abra (various meanings, but most likely "haven" or "shelter") near Antiguo Morelos, Tamaulipas, Mexico—Holman (1970c).

Called the Mexican Burrowing Toad, this toothless species has a rather egg-shaped body, is short-legged, has smooth skin, and has hind feet specialized for digging. The head is not "froglike," as it is rounded rather than flattened and the eyes are small. The front portion of the tongue is not attached to the floor of the mouth as in most anurans; thus, the Mexican Burrowing Toad can protrude the tongue slowly in a mammalian fashion rather than flipping it out like most frogs. Ants and termites are the preferred food. During the breeding season, which is infrequent and usually initiated by cyclones, the males emerge from their burrows and inflate the whole body as a resonating chamber to make their breeding calls, which are loud and unmusical. During these times the males look like small, orange balloons.

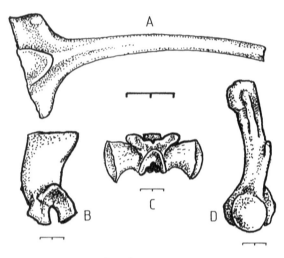

FIGURE 26. Bones of modern *Rhinophrynus dorsalis* from Veracruz, Mexico. (A) Right ilium in lateral view. (B) Left scapula in lateral view. (C) Sacral vertebra in dorsal view. (D) Right humerus in ventral view. Scale bar = 4 mm and applies to all figures.

Excerpts from my field notes from Veracruz, Veracruz, Mexico, August 1964, reflect on the natural history of this species. "I was warned by Hobart Smith that I might not see any *Rhinophrynus* as they hardly ever came out on the surface." "August 16 at 2 a.m., was wakened by a furious rainstorm and opening the window of my hotel room was greeted by a chorus of loud, unmusical, attenuated bleats. Under the streetlight, I could see the little orange balloons that are calling male *Rhinophrynus* everywhere on the wet streets. They puff up to call, call, then collapse into a baggy mess." "August 17, 9 a.m., found an open well that many *Rhinophrynus* had fallen into and drowned. I got 12 by lowering a weighted hook on the end of a casting rod to snag the dead bodies."

Identification of Pleistocene Fossil Remains. Many of the bones of *Rhinophrynus dorsalis* can be immediately separated from those of other Pleistocene and modern anurans. Fortunately, the elements that often turn up in Pleistocene deposits, such as the ilium, the scapula, and the humerus (see Fig. 26 for examples), all easily distinguish *R. dorsalis* from other anurans.

?RHINOPHRYNIDAE

The following genus has ectochordal vertebrae in which the centrum is an ossified cylinder that encloses a persistent notochord. This suggests that it is a member of the Rhinophrynidae according to Henrici (1998).

If this is true, then it is the earliest member of the family known, as it is from the Late Jurassic.

GENUS #*RHADINOSTEUS* HENRICI, 1998

Genotype. Rhadinosteus parvus Henrici, 1998.

Etymology. The generic name is from the Greek *rhadinos,* "slender," and *osteon,* "bone" (in reference to the slim limb bones of the genus). The specific name is from the Latin *parvus,* "small" (in reference to the small size of the taxon).

Diagnosis. The diagnosis is the same as for the genotype and only known species, *Rhadinosteus parvus.*

#*RHADINOSTEUS PARVUS* HENRICI, 1998
(FIG. 27)

Holotype. A partial skeleton that is articulated to loosely articulated (Dinosaur National Monument: DINO 14693).

Locality and Horizon. Rainbow Park microsite 96, Dinosaur National Monument, Uintah County, Utah: Late Jurassic (Kimmeridgian). Brushy Basin Member, Morrison Formation.

Other Material. Specific referred material listed by Henrici (1998) comprises (1) an individual that consists of a part–counterpart specimen, including an associated vertebral column, urostyle, and left scapula in a slab (DINO 13143A [part], 13143B [counterpart]); (2) partial part–counterpart skeletons and closely associated but disarticulated bones that occurred on a large slab (DINO 13105A [part], 13105B [counterpart]) — these partial skeletons are numbered DINO 14694–14698, and the closely associated but disarticulated bones are given a single number, DINO 14700; of these, the following elements were given individual numbers: two parasphenoids (DINO 14701, 14702), a right cleithrum (DINO 14703), right scapulae (DINO 14704), two right scapulae (DINO 14721,

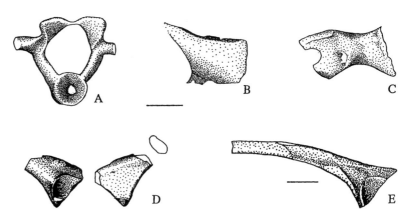

FIGURE 27. Bones of *Rhadinosteus parvus* from the Late Jurassic Morrison Formation of Utah. (A) Fifth vertebra in posterior view. (B) Right cleithrum in dorsal view. (C) Right scapula in lateral view. (D) Left ilium in lateral, medial, and transverse views. (E) Left ilium in lateral view. Scale bars = 1 mm; upper one applies to (A)–(D).

14722), two right cleithra (DINO 14723, 14724), and fourth and fifth presacral vertebrae (DINO 14725)—and (3) a left partial ilium recovered from screen washing (DINO 14726).

Diagnosis. The diagnosis of the hypodigm is quoted from Henrici (1998, p. 322): "A primitive pipoid [superfamily Pipoidea Fitzinger, 1843, recognized by Sanchiz (1998), that contains the families Palaeobatrachidae, Pipidae, and Rhinophrynidae] anuran on the basis of azygous frontoparietal and parasphenoid that lacks lateral alae (wings). Differs from all currently known pipoids in lack of burrowing and swimming specializations known to occur in other pipoids. Possession of ectochordal vertebrae suggest it is a rhinophrynid. This small anuran differs from other known rhinophrynids by the following unique combination of primitive characters: 1) bicuspid, pedicellate teeth present; 2) hind limb bones slender; and 3) femur has sigmoid curve."

Description. This description is modified from Henrici (1998). In the skull, the frontoparietal is azygous (unpaired); it is broad and then narrows slightly anteriorly, ending in a blunt apex at the anterior end. The premaxillae are tooth-bearing, and both specimens lack the tip of the rectangular alary process. The surface of the facial portion of the maxilla is smooth and unridged. The facial portion of the maxilla is deep at the anterior end and abruptly decreases in depth just past the midpoint of the preserved length of the bone.

In lingual aspect in the maxilla, a palatine portion and a broad tooth-bearing dental portion are present. In most cases only the tooth pedicels are preserved; these pedicels exhibit the pleurodont (attached to the inside of the jaw) condition. Several tooth crowns detached from the pedicels are present, and at least one of these teeth is bicuspid. There is no evidence that the maxilla overlapped the premaxilla. The anterior end of the maxilla is deep and lacks the narrow, toothless process of the Pipidae. On the other hand, there is no shelf of bone on the labial surface of the premaxilla.

The left angulosplenial is gently bowed lingually and well ossified at the rather deep posterior end, but it becomes poorly ossified at the narrower anterior end. The dorsal margin of the angulosplenial is irregular because of bone loss or deformation. Two isolated parasphenoids have a diamond-shaped base that lacks lateral alae and processes. The elongate cultriform process of the parasphenoid is narrower than the base and tapers anteriorly to about the middle of its length.

In the postcranial skeleton, the vertebral columns show varying degrees of ossification, and in most examples the centra are incompletely ossified. Ectochordal development is evident and best exhibited in the vertebral column in slab DINO 13143A (Henrici, 1998, fig. 4, p. 326). The atlas in this vertebral column has the most completely ossified centrum of the series and lacks a posterior condyle. Although the neural arch is not exposed, the ventral edges of the cervical cotyles are exposed and closely juxtaposed. In the second presacral vertebra, ossification occurs down both lateral walls, just reaching the ventral surface. Ossification in the third and fourth presacral centra is almost complete, as a small seam on the ventral midline indicates the meeting but not the fusion of the two centers of ossification. In the fifth presacral, the two ossification centers barely meet along the midline, and in the sixth they are separate. The seventh presacral has the same degree of ossification as the second; both the eighth presacral and the sacrum are damaged.

Additional evidence that vertebrae of *Rhadinosteus* are ectochordal and not procoelous comes from two well-ossified vertebrae (DINO 14725) of uncertain association with other skeletal remains in the slab. These vertebrae are in probable association with each other, as the posterior ends of both vertebrae face upward and the centrum of one is placed inside the neural arch of the other. These vertebra are believed to represent the fourth and fifth presacrals, as indicated by the long transverse processes on one (fourth) and the short, laterally directed transverse processes on the other (fifth). In both centra, the posterior ends have a circular outline and are concave with an open notochordal canal, indicating they are ectochordal and not procoelous. Unfortunately, it cannot be determined whether they were opisthocoelous or amphicoelous, as the anterior ends are not visible. In both, a neural spine is lacking and a notch occurs between the postzygapophyses.

In other vertebral material, slight overlapping of neural arches suggests they are imbricate. The transverse processes of the anterior presacrals are long and cylindrical. Small, slender pieces of bone that could possibly represent free ribs are found in several specimens. Nevertheless, some forelimb and shoulder girdle bones near the second through fourth transverse processes could actually be poorly preserved phalanges, clavicles, or coracoids rather than free ribs. A urostyle (DINO 14693) bears two cotyles, is proximally broad, and tapers posteriorly. Its exact length may not be determined, but it is at least as long as five presacral vertebrae.

Turning to the shoulder girdle, we find that the scapula (DINO 14704) is rectangular and relatively long. A distinct cleft forms a broad U. A suture scar for the articulation of the clavicle runs along the anterior edge of the acromial process and forms a narrow, elongate U that opens ventrally. Both the anterior and the posterior edges of the scapula are concave, but the concavity of the posterior edge is greater. In the only specimen where the scapula and clavicle are in association, the scapula is slightly shorter than the clavicle. The clavicle itself is thin and slightly bowed, and it tapers posteriorly.

The only complete humerus of *Rhadinosteus* is DINO 14693. The shaft appears to be straight, but this could be the result of crushing. A straight humeral shaft immediately separates it from *Rhinophrynus*, with its strongly bowed humeral shaft. A thin, high ventral crest extends to the narrowest point of the shaft near its middle. The shaft broadens from this point to its distal end. The distal condyle is small and oval, and its long axis is oriented mediolaterally. The medial epicondyle is slightly larger than the lateral epicondyle.

In the radio-ulna, the olecranon process is not well developed, the sulcus appears to run the length of the bone, and the distal ends are free. It is believed that carpal bones were probably cartilaginous in this frog, as carpal bones were not found in any of the material. Some bones of the manus were found, but there were not enough to determine a phalangeal formula or to significantly add to the systematic value of the specimen.

Although the ilia of *Rhadinosteus* are poorly preserved, they are quite different from those of *Rhinophrynus*. Ilia were found in DINO 14693 and 14695, and an isolated piece of distal ilial shaft (DINO 14726) was referred to *Rhadinosteus*. Most of the following description is based on DINO 14693 and 14726. The ilium has a slightly bowed shaft that is oval in cross section and lacks a dorsal crest. The shaft is narrow anteriorly and expanded in the acetabular region. A preacetabular zone may be identified. Located slightly anterior to the acetabular fossa is an obsolete, oval dorsal prominence ("protuberance" of Henrici, 1998), which in the two specimens figured by Henrici (see Fig. 27, this volume) is not produced above the ilial margin.

The acetabular fossa is symmetrical, and its height (measured along the base of the ilium) is nearly twice its length (distance between the base and the anterior rim). The dorsal acetabular expansion is larger than the ventral acetabular expansion in lateral view. In lateral view, the ventral rim of the acetabular fossa does not hide the ventral acetabular expansion from view. Medially, the interiliac tubercle is absent and the acetabular rim is not visible.

In the hind limb, the bones of *Rhadinosteus* are gracile. The femur has a gentle sigmoid curve; its proximal and distal ends are subequal in width and are only slightly wider than the narrowest width of the shaft. The distal condyle is unossified, thus differing strongly from *Rhinophrynus*. Crests on the femoral shaft appear to be absent, but bone damage obscures the shaft to such an extent that it is impossible to establish the lack of such crests with assurance.

The length of the femur is exceeded by the length of the tibiofibula. In the tibiofibula, the sulcus that divides the two components is restricted to the proximal and distal ends, which are only slightly wider than the narrowest width of the shaft. The slender tibiale and fibulare are only slightly expanded at their free proximal and distal ends. In both of these elements the width of the distal end is greater than that of the proximal end, more so in the tibiale. Metatarsals and phalanges are incompletely represented. The left limb of DINO 14693 has five metatarsals; the third and the fifth metatarsals have two phalanges each. The metatarsals are not highly elongate and are shorter than both the tibiale and the fibulare. Digging spades formed by the modification of the prehallux and the distal phalanx of the first digit were not identified in the *Rhadinosteus* material.

General Remarks. Henrici (1998) made a phylogenetic analysis that placed *Rhadinosteus* in the Rhinophrynidae as a sister group to the *Chelomophrynus–Rhinophrynus* clade (Fig. 28). The close relationship between *Chelomophrynus* and *Rhinophrynus* is said to be indicated by at least three synapomorphies—(1) teeth absent; (2) femoral shaft bowed; and (3) tibiale and fibulare stout and expanded at their proximal and distal ends—and four other characters that are unknown for *Rhadinosteus*: (1) atlantal neural arch elongate; (2) modified opisthocoelous vertebrae present; (3) distal condyle of femur divided into a large lateral condyle and a smaller medial condyle; and (4) prehallux and distal phalanx of first digit modified into spades. That *Rhadinosteus* is a sister taxon to the *Chelomophrynus–Rhinophrynus* clade and is placed in the Rhinophrynidae

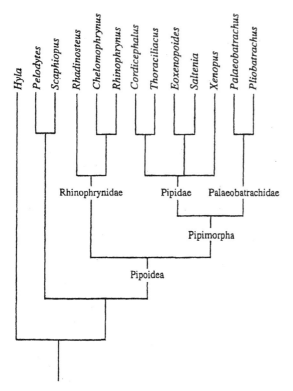

FIGURE 28. Consensus tree showing the relationships of *Rhadinosteus* to other pipoid anurans. After Henrici (1998).

is supported by only one synapomorphy, and that is the presence of ectochordal vertebrae.

As Henrici (1998) pointed out, it would be informative to know if *Rhadinosteus* had the derived character of modified opisthocoelous vertebrae and if ectochordal vertebrae persisted into the adult stages as in the other rhinophrynids. I would agree with Henrici that for the present it is best to consider *Rhadinosteus* as a pipoid that is not closely related to either the Pipidae or Palaeobatrachidae but may possibly be the most primitive rhinophrynid known and, I would add, by far the earliest.

SUPERFAMILY PELOBATOIDEA BOLKAY, 1919

Families in the superfamily Pelobatoidea that occur in the North American fossil record are the Pelobatidae and Pelodytidae.

That the superfamily Pelobatoidea is a monophyletic group (group of taxa descended from a single ancestral taxon) has been established by Cannatella (1985) and confirmed by Ford and Cannatella (1993) on the basis of an ossified sternum, a palatine process on the premaxilla, and an adductor longus muscle. This primitive superfamily is now considered to contain only two families, the Pelobatidae and the Pelodytidae (Henrici, 1994; Sanchiz, 1998; Heatwole and Carroll, 2000). The Megophryidae of Cannatella (1985) is currently considered a subfamily of the Pelobatidae. Henrici (1994) provided a consensus of the phylogenies of the two recognized families in the Pelobatoidea (Fig. 29).

FAMILY PELOBATIDAE BONAPARTE, 1850

At present, pelobatid species are widely distributed, occurring from Pakistan east to the Indo-Australian archipelago and the Philippines; in China, Europe, western Asia, and northern Africa; and in southwestern Canada and the eastern United States down to southern Mexico (Frost, 1985). The family Pelobatidae is considered by Henrici (1994) to be monophyletic and to share four non-homoplasious characters (derived characters considered not to be the result of convergence or parallelism) and four homoplasious derived characters. The non-homoplasious characters are as follows: (1) postchoanal ramus of vomer absent, (2) parahyoid bone absent, (3) scapula long, and (4) scapula not anteriorly overlain by clav-

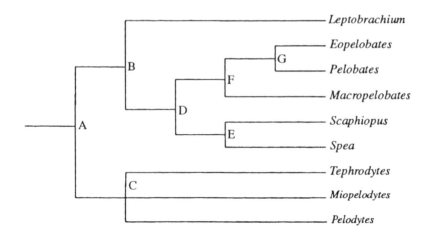

FIGURE 29. Consensus tree of the Pelobatoidea, A. The two major phylogenies are depicted by B and C.

icle. Of these, the "scapula long" character is the only one that is known for all members of the Pelobatidae.

An osteological definition of the family follows. Eight stegochordal presacral vertebrae with the imbricate condition are present. Ossified intervertebral discs fuse with the centrum to effect a procoelous condition in adults. The discs, however, remain free in the Megophryinae. Presacrals I and II are not fused, and the atlantal cotyles of presacral I are closely juxtaposed. The sacrum has widely expanded diapophyses and is fused with the urostyle except in most megophryines, where a monocondylar articulation with the urostyle occurs.

A bony web is present between the urostyle and the sacral diapophyses in some pelobatids. The pectoral girdle is of the arciferal type, and a cartilaginous omosternum occurs. The sternum is osseus in most taxa. The anterior end is not overlain by the clavicle. Palatines may be present, absent, or fused with adjacent elements. The parahyoid is absent, and the cricoid ring is incomplete dorsally. Both the premaxilla and the maxilla are tooth-bearing. The astragalus and calcaneum are fused only proximally and distally, and there are two tarsalia. The phalangeal formula is normal.

I recognize here that only one genus, *Scaphiopus* (with subgenera *Scaphiopus* and *Spea*), represents the Pelobatidae in the fossil record of North America, as in Frost (1985) and Sanchiz (1998). Henrici (1994) recognized *Scaphiopus* and *Spea* as full genera. The status of North American pelobatoid taxa previously designated "*Eopelobates*" by Sanchiz (1998) will be discussed after the section on *Scaphiopus*. Fossil species in *Scaphiopus* are presented in alphabetical order of subgenera and species of each subgenus. Two extant and five extinct species of the subgenus *Scaphiopus* and four extant and two extinct species of the subgenus *Spea* are currently recognized in North America.

GENUS *SCAPHIOPUS* HOLBROOK, 1836

NORTH AMERICAN SPADEFOOTS

Modern species of *Scaphiopus* occur in south-central Canada, the United States, and Mexico south to the southern edge of the Mexican Plateau.

The *Scaphiopus–Spea* clade recognized by Henrici (1994) is defined on the basis of six derived characters: (1) quadratojugal absent, (2) post-choanal ramus of vomer elongate, (3) medial end of coracoid not expanded, (4) sternum cartilaginous, (5) ischium not extended posteriorly, and (6) sartorius muscle and associated tendon concealed, in part, by gracilis major.

SCAPHIOPUS (*SCAPHIOPUS*) *ALEXANDERI* ZWEIFEL, 1956
(FIG. 30)

Holotype. An incomplete, articulated skeleton and the posterior part of the skull (University of California Museum of Paleontology: UCMP 45030).

Locality and Horizon. Fish Lake (UCMP locality V-2804), Nevada: ?Late Miocene (?Hemphillian NALMA). Esmeralda Formation.

Other Material. Fossils assigned to *Scaphiopus* cf. *Scaphiopus alexanderi* have been reported from two Middle Miocene (medial Barstovian NALMA) sites. The first of these is the Norden Bridge Local Fauna of Brown County, Nebraska. Here, two distal portions of ilia, two fragmentary sacrococcyges (sacrum fused to urostyle), and other elements not diagnostic but tentatively referred to this species on the basis of size were assigned to University of Nebraska State Museum (UNSM) 61016 by Estes and Tihen (1964). This material is later referred to *Scaphiopus* (*Scaphiopus*) *hardeni* here. The second locality is the Kleinfelder Farm locality near Rockglen, Saskatchewan, Canada. Here, two left and one right ilia and two sacrococcyges were assigned to Royal Ontario Museum (ROM) 7700 by Holman (1970a).

Revised Diagnosis. The diagnosis of Sanchiz (1998, p. 57)—"*Scaphiopus* species with their ilia similar to those of modern *S. couchii* differing by the absence of a preacetabular foramen"—is replaced with the following revised diagnosis. A *Scaphiopus* of the subgenus *Scaphiopus* that closely resembles *Scaphiopus* (*Scaphiopus*) *couchii* but differs in lacking a preacetabular foramen; in having the ventral acetabular expansion of the ilium rotated backward and not at a right angle to the shaft as in *S. couchii*; and in lacking the sharply produced acetabular rim of *S. couchii*.

Description. The type specimen is three-dimensional and loosely articulated. Important diagnostic elements include a sacrococcyx (sacrourostyle of Sanchiz, 1998) fused to the eighth vertebra, a humerus, two ilia, and two fragmental ischia. The sacral vertebra has expanded transverse processes. Rather extensive webbing occurs between the transverse process and the proximal part of the urostyle on the right side of the bone, but this webbing is broken off on the left side. The sacral vertebra is firmly fused with the eighth vertebra. The humerus lacks the articular part of the proximal end, but the remaining part of the proximal end is relatively compressed. The humeral shaft is moderately bowed, and the distal condyle is round.

The ilium has both the dorsal and the

FIGURE 30. Left ilium of *Scaphiopus* (*Scaphiopus*) *alexanderi* from the Late Miocene (?Hemphillian) of Nevada. Scale bar = 2 mm.

ventral acetabular areas well developed, with the dorsal acetabular expansion much more extensive than the ventral acetabular expansion. The dorsal prominence is obvious, roughened, and somewhat rounded. It is produced laterally, rather than dorsally, from the shaft and is positioned well behind the anterior border of the acetabulum. The ilial portion of the acetabulum is large and well excavated and roughly triangular in lateral view. A preacetabular foramen is absent. The shaft is long and slender and slightly curved throughout its extent. The ischium is strong and has its posterior margin moderately convex in lateral view. The ischial portion of the acetabulum is well excavated, and its posterior acetabular rim is well produced laterally, indicting the acetabular fossa is deeper posteriorly than anteriorly.

General Remarks. As far as I can determine, the type material of *Scaphiopus alexanderi* and the material from the Middle Miocene of Saskatchewan referred to as *Scaphiopus* cf. *Scaphiopus alexanderi* are both, at most, weakly distinguishable from modern *Scaphiopus couchii*, a living species discussed below. Estes and Tihen (1964) incorrectly assigned *S. alexanderi* to the subgenus *Spea*. Both *Scaphiopus alexanderi* and its living representative *Scaphiopus couchii* are members of the subgenus *Scaphiopus* (e.g., Frost, 1985).

SCAPHIOPUS (SCAPHIOPUS) COUCHII BAIRD, 1854
COUCH'S SPADEFOOT
(FIG. 31)

Fossil Localities. All remains in these localities have been referred to *Scaphiopus couchii*, although in a few cases they were originally referred to as *Scaphiopus* cf. *Scaphiopus couchii* (see Holman, 1995b). **Pleistocene (Rancholabrean NALMA):** Dark Canyon, Eddy County, New Mexico—Applegarth (1980), Sanchiz (1998). Deadman Cave, Pima County, Arizona—Mead et al. (1984), Mead and Bell (1994), Holman (1995b), Sanchiz (1998). Howell's Ridge, Grant County, New Mexico—Van Devender and Worthington (1977), Holman (1995b), Sanchiz (1998). Rancho la Brisca, Sonora, Mexico—Van Devender et al. (1985), Sanchiz (1998). Schulze Cave Fauna, Edwards County, Texas—Parmley (1986), Holman (1969b, 1995b), Sanchiz (1998). Shelter Cave, Dona Ana County, New Mexico—Van Devender and Mead (1978), Holman (1995b), Sanchiz (1998).

Scaphiopus couchii, also known as Couch's Spadefoot, derives its specific and vernacular names from D. N. Couch, a professional soldier who collected vertebrate specimens in Mexico while on leave from the U.S. Army (Conant and Collins, 1998). A color pattern of mottled yellows and green makes *S. couchii* one of the most attractive members of the family. This species lacks a rounded boss between the eyes, and its digging spade is long and said to be often sickle-shaped (Conant and Collins, 1998). The SVL in this taxon ranges from about 55 to 57 mm. Couch's Spadefoot occurs from southeastern California to southwestern Oklahoma and south to northern Nayarit, Zacatecas, San Luis Potosi, and northern Veracruz, Mexico (Frost, 1985). This species occupies arid and semiarid habitats with short-grass prairies and mesquite savannahs.

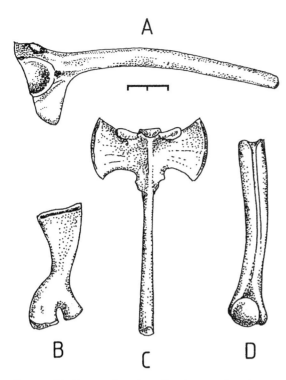

FIGURE 31. Bones of modern *Scaphiopus (Scaphiopus)* *couchii* from San Luis Potosi, Mexico. (A) Right ilium in lateral view. (B) Right scapula in medial view. (C) Sacrococcyx in dorsal view. (D) Right humerus in ventral view. Scale bar = 4 mm and applies to all figures.

Identification of Pleistocene Fossils. Skeletal elements from larger individuals of *Scaphiopus couchii* are considered to be distinctive from those of *Scaphiopus bombifrons* and *Scaphiopus hammondii*, according to Van Devender and Worthington (1977), who were able to distinguish frontoparietals, premaxillae, maxillae, vertebrae, humeri, radio-ulnae, sacrococcyges, tibiofibulae, fibulare, and ilia from the other two species in their study of Howell's Ridge Cave in New Mexico. Distinguishing characters, however, were not given in their discussion.

General Remarks. Sanchiz (1998, p. 57) indicated that *Scaphiopus couchii* has a preacetabular foramen on the ilium. This is not the case, thus indicating that *S. couchii* is even more closely related to *Scaphiopus alexanderi* than previously thought. Both species, however, have a small, well-marked preacetabular depression or fossa.

SCAPHIOPUS (SCAPHIOPUS) GUTHRIEI (ESTES, 1970)
(FIG. 32)

Holotype. A partial skull and associated right scapula (Museum of Comparative Zoology, Harvard University: MCZ 3493).

Etymology. The specific name refers to Daniel Guthrie, who collected the specimen in 1962.

Locality and Horizon. Unnamed locality, Fremont County, Wyoming: Early Eocene (Wasatchian NALMA). Lysite Member, Wind River Formation.

Revised Diagnosis. This diagnosis is based on characters in Henrici (2000). A *Scaphiopus* of the subgenus *Scaphiopus*, based on (1) the implied presence of an elongate postchoanal ramus of the vomer that articulates with the palatine process of the maxilla, (2) presence of sculptured cranial bones, and (3) presence of a long, low, arcuate ventral flange on the pterygoid. Differs from other *Scaphiopus (Scaphiopus)* in having (1) frontoparietal narrowest just posterior to the supraorbital flange, and (2) long and thin otic ramus of squamosal.

Note. Henrici (2000) recognized *Scaphiopus* and *Spea* as full genera and referred the above taxon to cf. *Scaphiopus guthriei* (Estes, 1970), as she found no evidence MCZ 3493 represented a new genus (or subgenus) because it has diagnostic characters of the subgenus *Scaphiopus* and because it has derived characters that distinguish it from other species of the subgenus. I see no reason for this tentative identification and have dropped the *cf.* here.

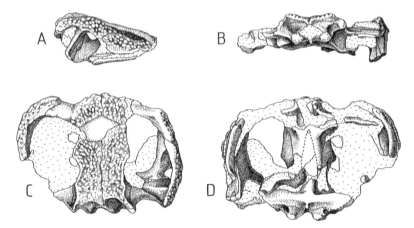

Figure 32. Holotype skull of *Scaphiopus guthriei* from the Early Eocene of Wyoming. (A) Right lateral aspect. (B) Occipital region. (C) Dorsal view. (D) Ventral view. The greatest width of the skull in dorsal view in the actual specimen was 38 mm; the other bones in the figure are proportionally the same. Stippling indicates matrix adhering to the fossil skull.

Description. This description is slightly modified from Estes (1970). The holotype specimen consists of a portion of the skull, a portion of the prearticular region of the jaws, and an associated fragment of the left scapula. The skull is well preserved on the right side, but on the left the temporal region is missing. Also, the premaxillae, the anterior portion of the nasals, and the anterior parts of both maxillae are absent. Although the skull is slightly flattened, distortion is limited for the most part to the peripheral tooth-bearing and temporal bones.

The skull indicates a somewhat broad-headed anuran with subequal dorsal temporal excavations and orbits that are separated by postorbital processes. The dorsal skull region is flattened and somewhat concave medially and bounded by weak crests. The skull as a whole is covered by a well-developed dermal sculpture. Posteriorly, the nasals meet on the midline, diverge at their posterior borders to expose the ethmoid, barely meet the frontoparietals, and then extend laterally to meet the maxillae. The nasals are weakly crested in the area that is continuous with the lateral borders of the frontoparietals and slope toward the midline between the crests. The nasals are sculptured over their entire surface.

Occurring between the frontoparietals and the nasals is a smooth, somewhat diamond-shaped portion of the ethmoid, which is the center of a depression bordered anteriorly by the nasal crests mentioned above and posteriorly by the lateral borders of the frontoparietals. This depression extends to the posterior border of the skull. Henrici (2000) pointed out that this depression, noted by Estes (1970) as being characteristic of *Eopelobates*, was probably the result of postmortem dorsoventral compression during the fossilization process.

The paired frontoparietals are subrectangular and prominently sculptured. The postorbital processes are situated anteriorly about two-thirds the frontoparietal length from the apex of the foramen magnum. The anterior tip of the left frontoparietal is missing, increasing the apparent depth of the ethmoid depression. The anterior tip of the right frontoparietal touches the nasal at its lateral border. The undistorted occiput, the lateral crests of the frontoparietals, and the symmetry of the cranial roof indicate that the midline depression of the frontoparietal, ethmoid, and nasals is natural. The postorbital processes are the widest points on the frontoparietals except for the posterior tips, which extend onto the paired projections of the paroccipital processes on the occiput dorsal to the condyles. Posteriorly, the frontoparietal reaches the apex of the foramen magnum, from which point lambdoidal crests form concave curves, extending toward the paired projections mentioned above.

In occipital view, the median skull roof is depressed, and the highest points on it are on its lateral borders. The occipital surface of the skull is well preserved, and there is little breakage except for the missing left temporal region. The most prominent bones are the otoccipitals (exoccipitals), which meet above and below the triangular foramen magnum. The large, circular foramen for the ninth and tenth cranial nerves is recessed at the base of the prominent, hemispherical occipital condyles.

The stapes is proximally forked and is closely appressed to the ventral surface of the lateral extension of the otoccipital noted above. As in Recent Spadefoots, a large opercular space is present, and since the delicate stapes is preserved in place, a calcified operculum was probably absent. The right squamosal is displaced dorsally at the point of its posterior articulation with the otoccipital and has rotated somewhat (along with the pterygoid and maxilla) on the lateral tip of the otoccipital, so that the greatest dorsal displacement is at the medial end of the squamosal; and the quadrate process of the squamosal has been rotated mediad so as to carry with it the remains of the lower jaw. The quadrate is represented by a small sliver of bone that lies between the squamosal and the and pterygoid. The posterior end of the lower jaw is missing, as are the tip of the quadrate and the posterior border of the maxilla. Apparently the quadratojugal and posterior processes of the maxilla, if such processes were present, were broken off during the dislocation of the temporal region.

Turning to the skull in the ventral view, we find that posterior part of the parasphenoid is well preserved, but the cultriform process is obscured by the right scapula and then is broken at the ethmoid border. The parasphenoid extends anteriorly from the border of the foramen magnum to the posterior border of the ethmoid. The lateral arms of the parasphenoid form the floor of the fenestra ovalis region. Well-developed nuchal, pterygoid, and retractor bulbi muscle scars highlight a trapezoidal, flattened area midway between the lateral arms of the parasphenoid. The otoccipitals extend posteriorly somewhat beyond the posterior borders of the parasphenoid, completing the fenestra ovalis region ventrally.

A large opening in the posterior region of the braincase is bounded anteriorly by the ethmoid, ventrally by the parasphenoid, posteriorly by the otoccipital, and dorsally by the frontoparietal. The ethmoid is well exposed between the parasphenoid and the vomers and is clasped by the ventral processes of the frontoparietals laterodorsally. The ethmoid sends broad, crested processes laterally toward the maxillary arcades; posterodorsally to each arcade, one finds foramina for the orbital extensions of branches of the occipital arteries. Anterior to each of these crested ethmoid processes is a depression where bone is missing as a result of erosion and breakage. Anterior to this area, the curved vomers are present in a natural position. Over the left anterior part of the ethmoid is a raised area that probably represents the left vomerine tooth plate, but the other parts of the vomer are absent. An irregular piece of bone lateral to this probably represents the palatine. The pterygoid occurs as a complete element only on the right side, where it is strongly curved and bends broadly toward the quadrate region on one hand and toward the otoccipital and maxilla on the other.

In lateral view, the relationships of the maxilla, squamosal, quadrate, pterygoid, and prearticular are normal on the right side. On the left side, only the middle part of the maxilla is present; the temporal region and the premaxilla are missing. Pedicellate teeth are present on the maxillae, which are heavily sculptured in the same way as in the frontoparietals. The posterodorsal corner of the maxilla joins the squamosal in a broad suture on the right side. Also on the right, the T-shaped squamosal is well preserved, and the crossbar of the T is sculptured in the same pattern as in the maxilla. The squamosal is much broader anteriorly than it is posteriorly.

The posterior process of the squamosal curves posteriorly over the tympanic cavity, then, expanding slightly at its border, forms an acute angle with the descending process of the squamosal. The posterior process is flattened anteroposteriorly and bears a sharp crest that separates the tympanic cavity from the lower temporal excavation. The descending process of the squamosal is close to the posterolateral border of the pterygoid and is separated from it ventrally by the sliver of the quadrate previously noted.

Unfortunately, the only part of the postcranial skeleton of *Scaphiopus guthriei* that is available is a crushed and fragmentary left scapula that has been rotated during postmortem events 180 degrees and now lies on the right side. The posterior border of the scapula is broken, and little if any lamina appears to have been present.

General Remarks. Estes (1970) originally assigned *Scaphiopus (Scaphiopus) guthriei* to the mainly Old World fossil genus *Eopelobates*. Henrici (2000) then tentatively assigned it to the genus *Scaphiopus* (= subgenus *Scaphiopus* here). If the assignment of Henrici (2000) and my present interpretation are correct, the origin of the *Scaphiopus (Scaphiopus)* monophyly is extended back from the Early Oligocene to the Early Eocene.

* Scaphiopus (Scaphiopus) hardeni Holman, 1975
(Fig. 33)

Holotype. A right ilium (Michigan State University, Museum Vertebrate Paleontology Collection: MSUVP 753).

Etymology. The specific name recognizes Warren L. Hardin for his contributions to the paleontological history of Trego County, Kansas.

Locality and Horizon. WaKeeney Local Fauna, Trego County, Kansas: Late Miocene (Clarendonian NALMA). Ogallala Formation.

Other Material. Paratypes include six right and six left ilia, all under

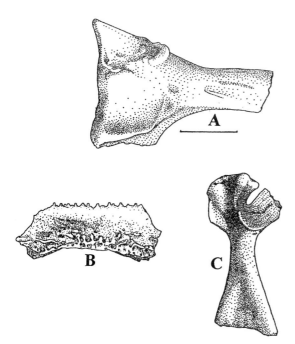

FIGURE 33. *Scaphiopus (Scaphiopus) hardeni* from the Late Miocene (Clarendonian) of Kansas. (A) Holotype right ilium in lateral view. (B) Partial maxilla. (C) Right scapula in lateral view. Scale bar = 2.5 mm and applies only to (A). Other scales not available.

MSUVP 754, collected at the same locality. Referred material includes 1 sphenethmoid, 10 fragmentary maxillae, 1 right and 1 left fragmentary frontoparietal, 1 right scapula, and 6 fragmentary sacrococcyges, all under MSUVP 755, collected at the same locality.

Scaphiopus hardeni has been mistakenly identified as four other species in the subgenus *Scaphiopus*. Wilson (1968) collected material from the type locality of *Scaphiopus hardeni* that he identified as *Scaphiopus couchii*. Holman (1975) restudied this material and assigned it to *S. hardeni*. This material consists of three partial maxillae (University of Michigan, Museum Vertebrate Paleontology Collection: UMMPV 55390, 55391, 55395), six frontoparietal fragments (UMMPV 55399–55404), two sacrococcyges (UMMPV 55397–55398), a nearly complete otic capsule (UMMPV 55396), and nine left and six right ilia (included in UMMPV 55392–55394).

Chantell (1971) identified fossil material as *Scaphiopus* cf. *Scaphiopus holbrookii* from the Middle Miocene (medial Barstovian) Egelhoff Local Fauna of Keya Paha County, Nebraska. This material is assigned here to *Scaphiopus hardeni* because these ilia (UMMPV 59948, 59951) resemble those of *S. hardeni* and differ from those of *Scaphiopus holbrookii* in having a rounded rather than an ovoid acetabular border and in having a narrower and more posteriorly directed ventral acetabular area (see Fig. 33A, this volume; Chantell, 1971, plate 1, figs. 5, 6). Moreover, the two maxillae (UMMPV 56549) assigned to *Scaphiopus* cf. *S. holbrookii* by Chantell (1971) from this site resemble *S. hardeni* and differ from *S. holbrookii* in having the orbital border of the maxillae shallowly convex and without an extruded orbital rim. This border is deeply convex and has an extruded orbital rim in *S. holbrookii* (see Fig. 33B, this volume; Chantell, 1971, plate 1, figs. 3, 4).

Holman (1976) mistakenly assigned Chantell's Egelhoff *Scaphiopus* material to *Scaphiopus wardorum*, an extinct species that will be discussed later. Voorhies et al. (1987) also mistakenly assigned three *Scaphiopus hardeni* ilia from the Middle Miocene (medial Barstovian NALMA) Hottell Ranch rhino quarries, Banner County, Nebraska, to *S. wardorum*. Also, I have recently restudied four additional left ilia (MSUVP 1102) from the Middle Miocene (medial Barstovian) Egelhoff Local Fauna of Nebraska that were mistakenly assigned to *S. wardorum* (Holman, 1987). These ilia clearly represent *S. hardeni* in the structure of the dorsal prominence and in the backswept ventral acetabular expansion that extends beyond the level of the anterior border of the acetabulum.

Finally, Estes and Tihen (1964) tentatively identified material that is clearly *Scaphiopus hardeni* from the Middle Miocene (medial Barstovian NALMA) Norden Bridge Local Fauna as *Scaphiopus* cf. *Scaphiopus alexanderi*.

Revised Diagnosis. A *Scaphiopus* (*Scaphiopus*) species with an SVL of about 70 mm, based on a comparison with a modern *Scaphiopus holbrookii*. Sphenethmoid with a strong posterior median tubercle and with lateral processes at right angles to the longitudinal axis. Ilia most similar to those of extinct *Scaphiopus wardorum* and *Scaphiopus alexanderi* and living *S. holbrookii* and *Scaphiopus couchii*; differ from *S. wardorum* in being smaller and in having a well-developed rather than an obsolete ventral acetabular expansion. Ilia differ from those of *S. alexanderi*, *S. holbrookii*, and *S. couchii* in having a much better developed and more rugged dorsal prominence. Ilia differ from those of *S. alexanderi* in having dorsal part of the ventral acetabular area much less extensive. Ilia differ from those of *S. holbrookii* in having the acetabulum rounded rather than ovoid and in having the ventral acetabular expansion narrower and posteriorly deflected rather than at about right angles to the shaft. Maxilla differs from that of *S. holbrookii* in having its orbital border shallowly concave rather than deep and in lacking an extruded orbital rim.

Description of the Holotype. The description of the holotype mainly follows Holman (1975). The holotype represents an individual with an SVL of about 65 mm, whereas *Scaphiopus* (*Scaphiopus*) *wardorum* represents an individual with an SVL of about 85–90 mm. In lateral view, the holotype of *Scaphiopus hardeni* has a well-developed dorsal prominence that is large and rugged. This prominence is produced dorsally above the dorsal border of the dorsal acetabular expansion and is also well produced laterally. The prominence is swollen anteriorly and has a ridge-like posterior portion, so the entire prominence is elongate. The acetabular fossa is well developed and rounded. Anterior to the acetabular fossa, the ventral acetabular expansion is well developed, rather than obsolete as in *Scaphiopus wardorum*, but is narrower in lateral view than in *Scaphiopus holbrookii*. The ventral acetabular expansion is also reflected posteriorly, in contrast to the expansion in *S. holbrookii*, which makes almost a right angle with the shaft. Just above the ventral acetabular expansion and just anterior to the middle part of the anterior rim of the acetabular fossa is a moderately deep pit, the preacetabular fossa. The ilial shaft is slightly curved and lacks a dorsal crest. Measurements are as follows: greatest height of acetabulum, 3.3 mm; height of ventral border of acetabulum through dorsal acetabular expansion, 5.4 mm; height of shaft anterior to acetabulum, 1.8 mm.

Other Material. In the paratype ilia from Kansas there is little variation in trenchant characters, as all of them represent medium-sized specimens that have the ventral acetabular expansion extending well anterior to the anterior edge of the acetabular fossa and have well-developed and rugged dorsal protuberances. Measurements of these ilia are as follows: greatest height of acetabulum, 2.7–3.2 mm (mean, 3.0 mm; $n = 6$); height of acetabulum from ventral border through dorsal acetabular expansion, 4.3–5.2 mm (mean, 4.83 mm; $n = 3$); height of shaft anterior to acetabulum, 1.4–1.8 mm (mean, 1.60 mm; $n = 6$).

In the referred maxillae the external surface is ornamented with dermal sculpturing. The orbital margin is relatively shallowly concave and lacks an extruded orbital rim. The referred sphenethmoid represents a *Scaphiopus* with an SVL of about 70 mm. *Scaphiopus hardeni* is more similar to modern *Scaphiopus couchii* than to modern *Scaphiopus holbrookii* of about the same size: in dorsal view, the base of the fossil's anterior medial process is wider than in *S. couchii*, but the *S. hardeni* sphenethmoid differs from both of these forms in having (1) a stronger posterior median tubercle; (2) the lateral process at about right angles to the long axis of the bone (it's directed anteriad in *S. holbrookii* and *S. couchii*); (3) the posterior part of sphenethmoid much more highly sloping; and (4) the lateral process about twice as high as it is wide (the lateral process is about twice as high as it is wide in *S. holbrookii* and *S. couchii*). The sphenethmoid of *Scaphiopus wardorum* is unknown. Measurements of the *S. hardeni* sphenethmoid are as follows: greatest posterior height, 4.8 mm; greatest posterior width, 4.5 mm. Characters were few on the partial frontoparietal of *S. hardeni* other than the fact that ornamentation occurred, indicating the subgenus *Scaphiopus* rather than *Spea*.

The right scapula of *Scaphiopus hardeni* resembles that of *Scaphiopus holbrookii* more than that of *Scaphiopus couchii*: the lateral border of the clavicular articular process is truncated in *S. hardeni* and *S. holbrookii*, whereas this process slopes gently into the shaft in *S. couchii*; and the glenoid opening between the clavicular articular process and the coracoid articular process is more constricted in *S. hardeni* and *S. holbrookii* than in *S. couchii*. The sacrococcyges of *S. hardeni* have the slight webbing between the transverse processes and the urostyle that indicates the subgenus *Scaphiopus* rather than *Spea*.

General Remarks. It seems probable that *Scaphiopus hardeni* is in the same clade and is directly ancestral to *Scaphiopus holbrookii* in the same way that *Scaphiopus alexanderi* is thought to be directly ancestral to *Scaphiopus couchii*.

SCAPHIOPUS (SCAPHIOPUS) HOLBROOKII (HARLAN, 1835)
EASTERN SPADEFOOT
(FIG. 34)

Fossil Localities. **Pleistocene (Rancholabrean NALMA):** Arredondo site, Alachua County, Florida—Holman (1962a, 1995b), Lynch (1965), Sanchiz (1998). Bootlegger Sink, York County, Pennsylvania—Guilday et al. (1966), Holman (1995b), Sanchiz (1998). Clark's Cave, Bath County, Virginia—Holman (1986b, 1995b), Fay (1988), Sanchiz (1998). Cutler Hammock Local Fauna, Dade County, Florida—Emslie and Morgan (1995), Hulbert (2001). Devil's Den, Levy County, Florida—Holman (1978b, 1995b), Sanchiz (1998). Kingston Saltpeter Cave, Bartow County, Georgia—Fay (1988), Holman (1995b), Sanchiz (1998). Natural Chimneys site, Augusta County, Virginia—Guilday (1962), Gehlbach (1965), Holman (1986b, 1995b), Fay (1988), Sanchiz (1998). Orange Lake site, Marion County, Florida—Holman (1959b, 1995b), Sanchiz (1998). Peccary Cave, Newton County, Arkansas—Davis (1973) (recorded as *Scaphiopus* cf. *Scaphiopus hol-brookii*); Holman (1995b), Sanchiz (1998). Reddick I site, Marion County, Florida—Gut and Ray (1963), Holman (1995b), Sanchiz (1998). Sabertooth Cave, Citrus County, Florida—Holman (1958, 1995b), Sanchiz (1998). Vero Beach strata 2 and 3, Indian River County, Florida—Weigel (1962), Holman (1995b), Sanchiz (1998). Williston IIIA site, Levy County, Florida—Holman (1959a, 1995b, 1996c), Sanchiz (1998).

Scaphiopus holbrookii, the Eastern Spadefoot, is the only member of the genus *Scaphiopus* whose range extends east of the Mississippi. This species usually has two rather distinct yellowish lines, each originating at the eye and running down the back; the digging spade is long and sickle-shaped; and there is no rounded boss between the eyes. This Spadefoot usually occurs in areas with sandy soils, where it spends most of the time in burrows. It is normally seen on the surface only when it comes out of its burrow at night to feed. The Eastern Spadefoot occurs from Massachusetts to Florida and west to eastern Louisiana; and from eastern Oklahoma and western Louisiana to extreme southern Texas (Frost, 1985).

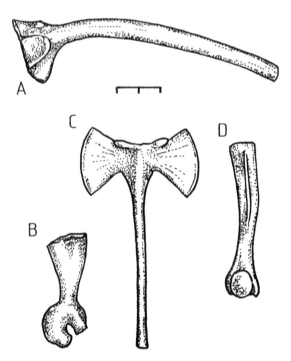

FIGURE 34. Bones of modern *Scaphiopus* (*Scaphiopus*) *holbrookii* from Levy County, Florida. (A) Right ilium in lateral view. (B) Right scapula in medial view. (C) Sacrococcyx in dorsal view. (D) Right humerus in ventral view. Scale bar = 4 mm and applies to all figures.

Identification of Pleistocene Fossils. In Pleistocene sites east of the Mississippi *Scaphiopus holbrookii* may be distinguished easily from those of other anurans by direct comparison of such elements as frontoparietals, scapulae, humeri, ilia, and especially sacrococcyges. Osteological characters of *S. holbrookii* are mentioned above in the section on *Scaphiopus hardeni.*

General Remarks. Fossil *Scaphiopus holbrookii* is known only from the Rancholabrean of North America. A *Scaphiopus* from the Miocene (Hemingfordian NALMA) Thomas Farm site in Gilchrist County, Florida, was referred to *Scaphiopus* cf. *Scaphiopus holbrookii* by Auffenberg (1956). The ilia attributed to this taxon may lack a dorsal prominence, but when it exists, it occurs as a small, rounded protuberance (Auffenberg, 1956). This is characteristic of the subgenus *Spea* rather than of *Scaphiopus holbrookii* or other members of the subgenus *Scaphiopus* that have a well-developed dorsal ilial prominence. Moreover, considering the other western and prairie vertebrate species that lived in Florida during the Oligocene and the Miocene (see Holman, 1999; Holman and Harrison, 2001; Hulbert, 2001), it would not be totally surprising that Spadefoots of the *Spea* clade were there as well. Material previously recorded as *Scaphiopus* cf. *S. holbrookii* from the Middle Miocene (medial Barstovian NALMA) Egelhoff Local Fauna, Keya Paha County, Nebraska, by Chantell (1971) has been reassigned to *Scaphiopus hardeni* in the account of that species.

SCAPHIOPUS (SCAPHIOPUS) SKINNERI ESTES, 1970
(FIG. 35)

Holotype. A three-dimensional skull, a vertebral column, and pectoral girdle elements (including a left scapula, right coracoid, and left? thyroid ossification) associated in an almost natural position (American Museum of Natural History: AMNH 42920 [formerly Frick Laboratory, American Museum of Natural History: FAM 42920; Estes, 1970]).

Etymology. The specific name recognizes North American vertebrate paleontologist Morris F. Skinner.

Locality and Horizon. Leo Fitterer Ranch, Stark County, North Dakota: Early Oligocene (Orellan NALMA). White River Group.

Other Material. Referred specimens from the type locality identified

FIGURE 35. *Scaphiopus (Scaphiopus) skinneri* from the Early Oligocene (Orellan) of North Dakota. (A)–(E) Views of the skull: (A) dorsal; (B) right lateral; (C) ventral; (D) anterior; (E) occipital (posterior). (F), (G) Views of the vertebral column: (F) dorsal; (G) ventral. (H) Coracoid (top); left scapula (middle); ossified thyroid process (bottom). (I) Right frontoparietal in dorsal view with its anterior end broken. The skull in the actual specimen had a greatest width of 29 mm; the other views in the figure are proportional to this.

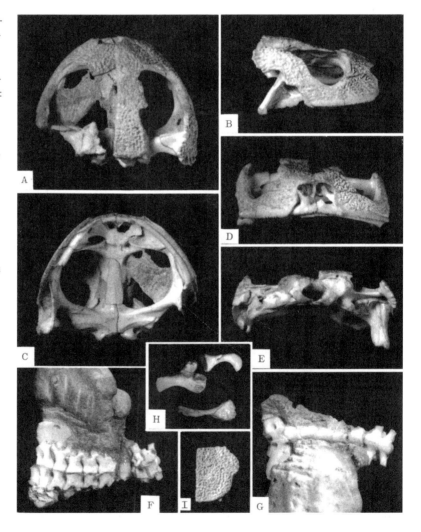

by Estes (1970) include a left and a right fragmentary frontoparietal and a partial vertebral column with an adherent tibiofibular fragment (AMNH [formerly FAM] 42921). *Scaphiopus skinneri*, identified by Richard Estes for J. A. Holman, was also reported from the Late Eocene (Chadronian NALMA) Calf Creek Local Fauna of Saskatchewan, Canada, by Holman (1972b). This material consists of 1 frontoparietal, 16 maxillary fragments, 24 vertebrae, 7 humeri, 8 radio-ulnae, 3 sacrococcyges, 71 ilia, and 7 tibiofibulae, all recorded under Saskatchewan Museum of Natural History SMNH 1434. This material was originally identified as *Scaphiopus* sp. by Holman (1968).

Diagnosis. The diagnosis is slightly modified from Sanchiz (1998) following Estes (1970). A large *Scaphiopus* (subgenus *Scaphiopus*) with extensive skull ornamentation; absence of an open frontoparietal fontanelle; with a large, rounded tympanic process of the squamosal; presence of a squamosal–maxillary contact; presence of a pterygoid process of the maxilla; widely emarginated prootic foramen; length of sacral processes at their ends about equal to the length of three presacral vertebrae.

Description. This description is modified from Estes (1970). The skull, which was mainly undistorted and uncrushed, was separated from the vertebral column for study. In posterior view the skull roof is essentially flat but slightly depressed medially. The occipital canal opens just medial to the prominent paroccipital process. The foramen magnum is a depressed oval, with its apex directed dorsally. The occipital region is well preserved except that the left frontoparietal, left stapes, and lateral edges of the otoccipital are absent. The otoccipitals extend laterally to form the border of the fenestra ovalis. They articulate dorsally with the frontoparietal and ventrally with the parasphenoid, which is excluded from the fenestra ovalis.

The foramen for the ninth and tenth cranial nerves opens prominently just lateral to the large, rounded occipital condyles. The paroccipital process has a prominent boss on its lateral tip, just lateral to the frontoparietal and occipital canal. The prootic is notched laterally and forms the medial border of the foramen for the maxillomandibular branch of the trigeminal nerve. The stapes is just posterior and dorsal to this foramen and has a forked head that fits into the anterodorsal part of the fenestra ovalis. A large opercular space remains, but if a calcified operculum was present, it has been lost. The fact that such delicate structures as tooth crowns, septomaxillae, and stapes remained on the skull indicate that the operculum was not likely to have been present in life. A prominent descending suspensorium is formed by the pterygoid medially and the squamosal laterally, which hold between them the well-developed quadrate.

Turning to the dermal roofing bones of the skull of *Scaphiopus skinneri*, we find that the premaxillae are unsculptured; the right one is well preserved, but the nasal process of the left one is missing. The nasals are prominently sculptured and are complete except for their pointed anterior processes above the nasal openings. The nasals articulate on the midline (where they form a slight depression) and also laterally with the maxillae. No open groove or unsculptured area exists in the nasomaxillary suture. The frontoparietals are also sculptured and have a prominent postorbital projection. Anteriorly, the frontoparietals articulate with the nasals but leave a small trapezoidal area of the ethmoid open on the midline. Posteriorly, the frontoparietal borders are rounded, and these borders curve into the postorbital projection. A tiny, pointed, and unsculptured process of the frontoparietal extends onto the paroccipital process.

The maxillae and squamosals are also completely covered by dermal sculpture. The maxillae and squamosals firmly articulate with one another, but no squamosal projection to or toward the frontoparietals occurs. The tympanic process of the squamosal is prominent and rounded, and a broad prootic process overlies the tip of the otoccipital. Laterally, the maxillae are deep and sculptured over all of their surface except for a narrow band that occurs immediately dorsal to the teeth. The teeth are pedicellate, and most of the narrow, spatulate crowns are present. The rounded tympanic process of the squamosal extends almost to the occipital condyles posteriorly and is notched ventrally for the tympanic membrane. The quadratojugal is not present, but on the lateral surface of the quadrate, a small projection occurs that may represent the quadratojugal.

In palatal view, the vomers have strong processes anterior to the choanae, and small tooth patches occur medially. The vomers do not meet at the midline; however, slim lateral processes to the palatines almost reach the pterygoids. The palatines are completely fused to the maxillae. The ethmoid has robust lateral processes. Well-developed concavities behind the vomers indicate a prominent turbinal-like fold. The anterior tip of the ethmoid is broken away. The pterygoids have a long suture with the maxillae and end in small spaces that separate them from the vomers.

The wings of the parasphenoids clasp the pterygoids laterally. Anteriorly, the cultriform process lies smoothly on the ethmoid without developing a channel, and posteriorly there are well-defined crests for nuchal and retractor bulbi muscles. The prootic foramen is long and open anteriorly. The oculomotor and optic foramina cannot be discerned. The mandibles are broken off posteriorly, but anteriorly mentomeckelian bones are present, separated from the prearticulars by spaces and clasped by the dentaries.

In anterior view, the premaxillae are well preserved, but they are loosely attached. On the right side, the ascending process contacts the small septomaxilla. The anterior process of the ethmoid is broken off, but its remnant indicates that it was not thickened. A well-defined capsular process with a prominent turbinal-like fold is visible.

Turning to the vertebral column, we find that the atlas is missing, as well as the neural arch of the fourth vertebra. The vertebrae are procoelous, and the ninth vertebra (sacral) has well-defined, hatchet-shaped diapophyses, with considerable posterior webbing. The urostyle is broken off, but the narrowness of the portion that remains, as well as the presence of two pairs of postsacral foramina, indicates that it obviously was fused with the sacrum. According to Estes (1970) the scapula, coracoid, and ossified thyroid cartilages (see Fig. 35H, this volume) are all robust and show no unusual characters, and the disarticulated vertebral column (AMNH [FAM] 4291c) is similar to that of the type specimen.

General Remarks. Skull characters of *Scaphiopus skinneri* that place it within the subgenus *Scaphiopus* are given by Kluge (1966) and Estes (1970) as follows: (1) presence of squamoso-maxillary contact, (2) widely emarginate prootic foramen, (3) absence of frontoparietal fontanelle, (4) extensive dermal skull ornamentation, (5) probable absence of calcified operculum, (6) presence of pterygoid process on the maxilla, (7) presence of a palatine bone, and (8) its large size.

It appears that *Scaphiopus skinneri* is the basal member of the subgenus *Scaphiopus* clade, and it is reasoned that, like modern *Scaphiopus holbrookii, S. skinneri* was associated with deciduous forest communities and essentially warm, humid climates (Estes, 1970). The development of the subgenus *Spea* complex, on the other hand, was probably associated with the woodland scrub and grasslands that were developing in the late Paleogene in North America.

SCAPHIOPUS (SCAPHIOPUS) WARDORUM ESTES AND TIHEN, 1964
(FIG. 36)

Holotype. The posterior 15 mm of a left ilium (University of Nebraska State Museum: UNSM 61014).

Etymology. The specific name recognizes the Allen Ward family of Valentine, Nebraska.

Locality and Horizon. Norden Bridge Local Fauna, Brown County, Nebraska: Middle Miocene (medial Barstovian NALMA). Valentine Formation.

Other Material. Material from the type locality includes the posterior portions of seven ilia, one fragment of a maxilla, several fragments of cranial roofing bones, one fragmental ethmoid, and one trunk vertebra, all included under UNSM 61015. Parmley (1992) assigned a left ilium (UNSM 56945) from the Late Miocene (Hemphillian NALMA) of the Santee site, Knox County, Nebraska, to *Scaphiopus wardorum*. Fossil material mistakenly referred to *S. wardorum* has been discussed under previous headings in the subgenus *Scaphiopus* section.

Diagnosis. This diagnosis is modified from that of Sanchiz (1998). A *Scaphiopus* of the subgenus *Scaphiopus* that differs from other members of the subgenus on the basis of its larger size (SVL estimated at 85–100 mm); maxilla ornamented by numerous separate denticles; ilium with obsolete ventral acetabular expansion that ends posterior to the anterior border of the acetabulum; dorsal protuberance ridgelike and lacking a rounded portion.

Description. The description here is modified from Estes and Tihen (1964). The holotype left ilium probably represents an individual with an SVL of about 85–90 mm. The ventral portion of the ventral acetabular expansion is missing. The portion that remains is much reduced and ends posterior to the anterior border of the acetabulum. The dorsal acetabular expan-

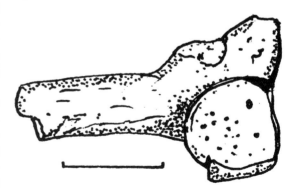

FIGURE 36. Holotype left ilium of *Scaphiopus (Scaphiopus) wardorum* of the Middle Miocene (medial Barstovian) of Nebraska. Scale bar = 5 mm.

sion bears a prominent somewhat ridgelike dorsal prominence. This prominence extends perceptibly dorsad to the dorsal border of the expansion proper at about the level of the anterior border of the acetabular fossa and runs posteriorly and somewhat ventrally for a short distance from that point. One of the referred ilia is much larger than the holotype and represents an individual of at least 100 mm in SVL. A second referred ilium is about the same size as the holotype, and the remaining ones are slightly to very much smaller. In most of the referred ilia, portions of the acetabular area are missing; nevertheless, all are complete enough to show that the ventral acetabular region is very narrow anteroposteriorly as in the holotype. In the larger ilia, the ridgelike dorsal prominence is prominent and oriented as in the holotype. In the smaller ilia this ridge is obscure and poorly defined, and when it is present it is not produced above the dorsal border of the dorsal acetabular expansion.

The most important cranial element recovered is a maxillary fragment from the left side below the anterior part of the orbit. The total length of this element is 10.0 mm, and 9.5 mm of the tooth row is present. About 16 tooth spaces are present. The distance from the ventral maxillary border to the ventral border of the orbit at the narrowest point on the fragment is 7.7 mm. This probably would not be the narrowest portion if more of the posterior part of the bone were preserved; nevertheless, it appears that the maxilla would have been deeper below the orbit than in living forms.

The ornamentation of all of the cranial elements differs from that seen in any modern taxon. In the living species, the ornamentation tends to be vermiculate (inlaid in such a way that it resembles worm tracks) (also vermiculate in *Scaphiopus hardeni*), and if separate denticles are present they are large and few, in contrast to the numerous small denticles in *Scaphiopus wardorum*. The presence of prominent ornamentation clearly eliminates the possibility of assigning *S. wardorum* to the subgenus *Spea*. Estes and Tihen (1964) also pointed out in their description that the intrinsic characteristics in these ornamented skull fragments preclude the possibility of their being of leptodactylid, or perhaps even hylid origin.

General Remarks. At present, *Scaphiopus wardorum* stands as the most individually unique species recognized in the subgenus *Scaphiopus*. It is known only from a single Middle Miocene (medial Barstovian NALMA) site and another single Late Miocene (Hemphillian NALMA) site, both in Nebraska. It seems possible to me that this is a dead-end sister group of the main subgenus *Scaphiopus* clade and that it may not have any living relatives.

A Tentative Phylogeny of the Subgenus *Scaphiopus*

As new fossils are found, even phylogenies that are tenaciously championed by their advocates tend to crumble away. Nevertheless, the tentative phylogeny of the subgenus *Scaphiopus* that follows appears to be parsimonious, at least for the time being. The subgenus *Scaphiopus* is considered here to be a monophyletic group. The earliest member of the group is *Scaphiopus guthriei* of the Early Eocene of Wyoming. Two clades, an eastern one and a western one, apparently originated from this taxon.

The basal member of the eastern clade is considered to be *Scaphiopus skinneri*, first known from the Late Eocene of Saskatchewan and the Early Oligocene of North Dakota. Three species are thought to be derived from *S. skinneri*: *Scaphiopus hardeni*, *Scaphiopus wardorum*, and *Scaphiopus holbrookii*. *Scaphiopus hardeni* of the Middle Miocene (Barstovian) of Nebraska and the Late Miocene (Clarendonian NALMA) of Kansas is thought to be the direct ancestor of the Pleistocene and modern species *S. holbrookii*. *Scaphiopus holbrookii* occurs mainly in the eastern United States, in temperate regions east of the Mississippi River. *Scaphiopus wardorum*, a large species that is known only from the Middle Miocene (medial Barstovian NALMA) and Late Miocene (Hemphillian NALMA) of Nebraska, is thought to be a sister group to the *S. hardeni–S. holbrookii* clade and is thought to have become extinct without replacement.

The basal member of the western clade is thought to be *Scaphiopus alexanderi* of the Middle Miocene (Barstovian) of Nebraska and Saskatchewan. *Scaphiopus couchii* of the Pleistocene and modern fauna of the southwestern United States and Mexico is believed to be the direct descendent of *S. alexanderi*. It may be of interest that the basal members of each major clade had their origins in the high plains of the United States and Canada.

Scaphiopus (Spea) bombifrons (Cope, 1863)
Plains Spadefoot
(Fig. 37)

Fossil Localities. Most of the Miocene and Pliocene records here were originally listed as *Scaphiopus* cf. *S. bombifrons*, but as criteria used for the identification of these "cf." remains were basically the same as those used for the identification of *Scaphiopus bombifrons* in North American Pleistocene localities, I consider the Miocene and Pliocene records here to represent *S. bombifrons*. **Middle Miocene (medial Barstovian NALMA):** Achilles Quarry, Brown County, Nebraska—Voorhies (1990). Egelhoff Local Fauna, Keya Paha County, Nebraska—Chantell (1971), Holman (1976, 1987), Wellstead (1981), Voorhies (1990), Sanchiz (1998). Hottell Ranch rhino quarries, Banner County, Nebraska—Voorhies et al. (1987). Norden Bridge Local Fauna, Brown County, Nebraska—Holman (1976), Wellstead (1981), Voorhies (1990), Sanchiz (1998). Welke locality, Brown County, Nebraska—

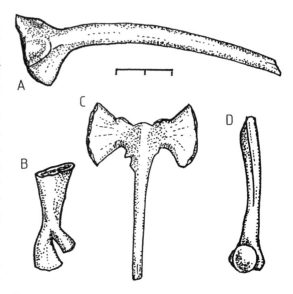

FIGURE 37. Bones of modern *Scaphiopus (Spea) bombifrons* from Trego County, Kansas. (A) Right ilium in lateral view. (B) Right scapula in medial view. (C) Sacrococcyx in ventral view. (D) Right humerus in ventral view. Scale bar = 4 mm and applies to all figures.

Voorhies (1990). **Middle Miocene (late Barstovian NALMA):** Annies Geese Cross Local Fauna, Knox County, Nebraska—Holman (1995a), Sanchiz (1998). **Late Pliocene (Blancan NALMA):** Sand Draw Local Fauna, Brown County, Nebraska—Holman (1972a), Holman and Schloeder (1991), Sanchiz (1998). **Pleistocene (Irvingtonian NALMA):** Albert Ahrens Local Fauna, Nuckolls County, Nebraska—Ford (1992), Holman (1995b), Sanchiz (1998). Hansen Bluff Chronofauna, Alamosa County, County, Colorado—Rogers et al. (1985), Rogers (1987), Holman (1995b), Sanchiz (1998). Nash Local Fauna, Meade County, Kansas—Holman (1979), Sanchiz (1998); the Nash Local Fauna was inadvertently left out of Holman (1995b). **Pleistocene (Rancholabrean NALMA):** Cragin Quarry Fauna, Meade County, Kansas—Tihen (1962b), Sanchiz (1998); the Cragin Quarry record by Tihen was missed by Holman (1995b). Dry Cave Fauna, Eddy County, New Mexico—Holman (1970b, 1995b), Harris (1987), Sanchiz (1998). Jinglebob Local Fauna, Meade County, Kansas—Tihen (1954), Holman (1995b), Sanchiz (1998).

Remains identified as *Scaphiopus bombifrons* or *Scaphiopus hammondii* were recorded from the Howell's Ridge Cave site in Grant County, New Mexico (Rancholabrean portion) by Van Devender and Worthington (1977) and Holman (1995b) and from the Lubbock Lake site, Lubbock County, Texas (Rancholabrean portion) by Johnson (1987) and Holman (1995b). Because Canadian records of fossil amphibians and reptiles are so rare I cite a record of *Scaphiopus bombifrons* that is based on a large number of skeletons covered by postglacial sands near Killam, Alberta, Canada (Bayrock, 1964; Harington, 1978; Holman et al., 1997). These remains are estimated to date from about 7–4 ka BP.

Scaphiopus bombifrons, the Plains Spadefoot, may be immediately recognized by the rounded boss or eminence between the eyes. This species is usually grayish or brownish, sometimes with a vague greenish tinge mixed in. The digging spade is short and rounded and usually somewhat wedge-shaped. Sometimes four faint lines are present on the back.

Plains Spadefoots inhabit regions of low rainfall in open areas and are seldom seen in riparian (river or stream bank) habitats. At present, the species occurs from southern Alberta and southwestern Saskatchewan, Canada; southward into Montana and North Dakota and south to eastern Arizona; and down to Chihuahua and Tamaulipas, Mexico (Frost, 1985).

Identification of Fossils. Scaphiopus (Spea) bombifrons has been identified in fossil deposits from the Middle Miocene to the Pleistocene, mainly on the basis of ilial remains. Plains Spadefoot ilia have two major characteristics: (1) they lack a preacetabular fossa; and (2) they either lack or have an obsolete dorsal prominence, as in Figure 37. Other important characters on *S. bombifrons* ilia include a moderately wide ventral acetabular expansion and a rather poorly defined dorsal rim of the acetabular fossa. Skull bones that define the interorbital boss are other important fossil elements in the identification of this species.

General Remarks. Scaphiopus (Spea) bombifrons occurs much earlier in the fossil record (Middle and Late Miocene) than any other living species of the genus. The species in the next section, *Scaphiopus (Spea) hammondii*, is the second oldest known modern *Scaphiopus* species, as it has been reported from the Late Pliocene. All of the other living species of *Scaphiopus* that have been recorded as fossils are known only from the Pleistocene.

A species from the Late Miocene (Hemphillian NALMA) of the Rhino Hill locality in Logan County, Kansas, which Taylor (1938) named *Scaphiopus studeri*, is indistinguishable as far as I can ascertain from modern and fossil skeletons of *Scaphiopus (Spea) bombifrons*, and I tentatively assign it to that species here. Resemblances include having a bony boss between the orbits (Sanchiz, 1998, fig. 91) and other diagnostic features (see Taylor, 1938; Zweifel, 1956; Tihen, 1960a; Kluge, 1966; Sanchiz, 1998).

Scaphiopus (Spea) hammondii Baird, 1859 "1857"
Western Spadefoot
(Fig. 38)

Fossil Localities. **Late Pliocene (Blancan NALMA):** Hagerman Local Fauna, Twin Falls County, Idaho—Chantell (1970), Mead et al. (1998) (recorded as *Scaphiopus* cf. *S. hammondii* by both). **Pleistocene (Rancholabrean NALMA):** Deadman Cave, Pima County, Arizona—Mead et al. (1984), Holman (1995b), Sanchiz (1998) (recorded as *Scaphiopus* cf. *S. hammondii* by all authors). Dry Cave, Eddy County, New Mexico—Holman (1970b, 1995b), Applegarth (1980), Sanchiz (1998). Smith Creek Cave, White Pine County, Nevada—Brattstrom (1976) (recorded as *Scaphiopus* cf. *S. hammondii*); Holman (1995b), Sanchiz (1998).

Remains identified as *Scaphiopus hammondii* or *Scaphiopus couchii* were listed from the Friesenhahn Cave site, Bexar County, Texas (Rancholabrean NALMA) by Holman (1995b) and Sanchiz (1998). Fossils identified as *S. hammondii* or *Scaphiopus bombifrons* were recorded from the Howell's Ridge Cave site in Grant County, New Mexico (Rancholabrean portion) by Van Devender and Worthington (1977), Holman (1995b), and Sanchiz (1998); and from the Lubbock Lake site, Lubbock County, Texas (Rancholabrean portion) by Johnson (1987), Holman (1995b), and Sanchiz (1998).

The Western Spadefoot, *Scaphiopus (Spea) hammondii*, unlike the preceding species, lacks a rounded boss between the eyes. It is grayish or greenish above and may have four irregular light stripes on the back. The warts on the skin are capped with a reddish or orangish coloration. This species is found mainly in lowlands, but sometimes it may be found in foothills and mountains. Its preferred habitat is areas of

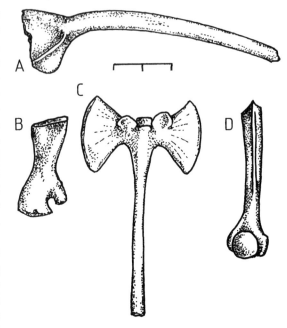

Figure 38. Bones of modern *Scaphiopus (Spea) hammondii* from Jalisco, Mexico. (A) Right ilium in lateral view. (B) Right scapula in medial view. (C) Sacrococcyx in dorsal view. (D) Right humerus in ventral view. Scale bar = 4 mm and applies to all figures.

grasses and sparse vegetation where the soil is loose or sandy. Western Spadefoots occur today only in western California and in northwestern Baja California, Mexico (Frost, 1985).

Identification of Fossils. Chantell (1970, pp. 654–655) characterized the scapula, ilium, and humerus of this taxon as follows. Scapula: "Prominent ball-like projection on terminal, lateral edge of clavicular process; anterior surface of clavicular process drawn out into flat web forming shelflike edge extending basally onto body of scapula below rim of glenoid fossa." Ilia: "No distinct dorsal protuberance (dorsal prominence), but rather an irregularly shaped, enlarged dorsolateral area that merges into the supra-acetabular (dorsal acetabular) expansion (most prominent on [UMMPV] V51524); no dorsal rim on acetabulum (weakly present on V50671); no preacetabular fossa; acetabular outline angular; subacetabular (ventral acetabular) expansion wide (V51524), or appears so on broken specimens, forming relatively sharp angle with ilial shaft." Sacrococcyx: "Cotyles oval; diapophyses partially broken off on V57674 but appear well-expanded with no postsacral webbing, completely broken off on V49840 with moderate postsacral webbing present; two pairs of nerve foramina; shaft laterally compressed; no dorsal crest."

SCAPHIOPUS (SPEA) INTERMONTANA (COPE, 1883)
GREAT BASIN SPADEFOOT

Fossil Localities. **Pleistocene (Rancholabrean NALMA):** Bechan Cave, Colorado Plateau, Utah—Mead and Bell (1994), Sanchiz (1998). Hidden Cave, Lahontan Basin, near Reno, Nevada—Mead and Bell (1994), Holman (1995b), Sanchiz (1998). Smith Creek Cave, White Pine County, Nevada—Brattstrom (1976), Mead et al. (1982), Mead and Bell (1994), Holman (1995b), Sanchiz (1998).

The Great Basin Spadefoot, *Scaphiopus (Spea) intermontana*, resembles the Western Spadefoot, *Scaphiopus hammondii*, but has a non-bony, glandular boss between the eye that distinguishes it from that species. Also, a brown spot usually occurs on each upper eyelid. This toad, as its name implies, occurs in the Great Basin, where it may be found in habitats ranging from flat sagebrush areas, through shrublands and pinon–juniper woodlands, to the spruce–fir zone. The specific range is the Great Basin of western North America, north to southern Idaho and to British Columbia, Canada; south to eastern California, northern Arizona, and northeastern New Mexico; and east to western Colorado and southwestern Wyoming (Frost, 1985).

Identification of Fossils. Scaphiopus intermontana and *Scaphiopus hammondii* are closely related, and it appears that the skeletons of the two species are nearly identical. Identifications of such forms as Pleistocene fossils are usually based on their modern geographic ranges.

SCAPHIOPUS (SPEA) MULTIPLICATA (COPE, 1863)
MEXICAN SPADEFOOT

Fossil Locality. **Pleistocene (Rancholabrean NALMA):** Dark Canyon, Eddy County, New Mexico—Applegarth (1980), Sanchiz (1998).

Scaphiopus (Spea) multiplicata closely resembles *Scaphiopus ham-*

mondii and was formerly considered a subspecies of *Scaphiopus hammondii*. But *Scaphiopus multiplicata* usually has a more brownish back and a longer digging spade than *S. hammondii*. This species occupies a wide variety of habitats, ranging from brushy deserts and short-grass plains to pinon–juniper woodlands and open pine forests. Sandy soils are preferred. *Scaphiopus multiplicata* occurs in the southwestern United States, excluding California, to the southern edge of the Mexican Plateau (Frost, 1985).

Identification of Fossils. The skeleton of *Scaphiopus multiplicata* is essentially similar to those of the related species *Scaphiopus hammondii* and *Scaphiopus intermontana.*

** SCAPHIOPUS (SPEA) NEUTER* KLUGE, 1966
(FIG. 39)

Holotype. A nearly complete articulated skeleton, with the exception of the distal parts of the limbs (Los Angeles County Museum of Natural History: LACM 9209).

Etymology. The specific name is from the Latin *neuter,* "toward neither side."

Locality and Horizon. Wounded Knee area, Shannon County, South Dakota (LACM locality 1982 [= South Dakota School of Mines locality V5360]): Late Oligocene (early Arikareean NALMA). Sharps Formation. Incorrectly listed as Miocene by Sanchiz (1998).

Other Material. Holman (1981) identified a right ilium (Michigan State University, Museum Vertebrate Paleontology Collection: MSUVP 1027) from the Early Miocene (late Arikareean NALMA), Mouth of McCann's Canyon site (Harrison Formation), Cherry County, Nebraska.

Diagnosis. This diagnosis is modified from Sanchiz (1998). A *Scaphiopus* tentatively assigned to the subgenus *Spea* but osteologically somewhat intermediate to both *Spea* and *Scaphiopus.* Cranial roofing bones sparsely ornamented; large frontoparietal fontanelle present; pterygoid process of maxilla absent; quadratojugal absent; transverse processes of fifth through eighth presacral vertebrae not strongly oriented anteriorly; urostyle fused to sacrum; maxilla and squamosal widely separated; ilium with rounded dorsal prominence placed above acetabulum.

Description. This description is a slightly shortened version of Kluge (1966). All skull bones are present except for the premaxillae, which appear to have been eroded away during fossilization. The skull is broad and deep. There is a light encrustation of dermal bone on the frontoparietals and squamosals and a heavy encrustation on the posterior portion of the maxillaries and the posterolateral part of the remaining nasal. The frontoparietals are thin and ragged (emarginate) medially, thus producing a relatively large frontoparietal fontanelle. The frontoparietals are narrow and show no evidence of a lateral winglike extension

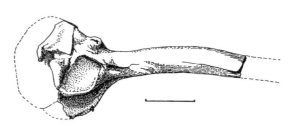

FIGURE 39. Right ilium of *Scaphiopus (Spea) neuter* from the Late Oligocene (early Arikareean) of South Dakota. Scale bar = 5 mm.

of encrusting dermal bone. There is no evidence of a frontoparietal boss. Both the anterior and the posterior openings of the frontoparietal canal can be seen from a dorsal view resulting from the absence of winglike extensions.

The dorsal surface of the prootics is deeply concave. A large operculum is present. The prootic foramen for the passage of the trigeminal nerve is completely encased in bone. The dorsal end of the squamosal extends anteroventrally and is widely separated from the maxilla. The quadratojugal is not present. The maxillae are toothed, but vomerine teeth cannot be identified because the vomers are either eroded or covered with matrix. The palatine is not present. The right nasal is in broad contact with the maxilla. The maxilla lacks a pterygoid process. The mandible is toothless.

Nine presacral vertebrae, all of which appear to be procoelous, are present. Most of the neural arches and the dorsal spines of the vertebrae have been eroded away. The second, third, and fourth vertebrae have long diapophyses; those of the second project anterolaterally, the third laterally, and the fourth slightly posterolaterally. The diapophyses of vertebrae V–IX are mainly destroyed, but there is an indication that they projected laterally, with the possible exception of the ninth, which could have been directed slightly anteriorly. The sacral vertebra is completely fused to the urostyle. Although the sacral diapophyses are mainly missing, the size of their proximal remnants and the position of the intersacral–coccygeal foramen indicate that they were widely dilated. The length of the lateral remnants and the presence of two postsacral coccygeal foramina suggest that the postsacral webbing was extensive.

Only the proximal part of the urostyle remains, and that portion has a relatively low, rounded dorsal crest.

The pectoral girdle appears to be arciferal, with the remaining left clavicle robust and strongly arched anteriorly. The sternal style was not located in the matrix, thus indicating that this element may have been cartilaginous. Both the scapula and the suprascapula are robust. The exposed distal portion of the left humerus has a well-developed medial epicondyle and a prominent, anteriorly directed crest. There is a moderately deep depression on the inner surface of the humerus, proximal to the medial epicondyle.

The pelvis is represented by both ilia (with the proximal end of both ilial shafts missing) and the anterodorsal portions of the ischia. The ilial shaft is robust. A dorsal crest is absent. Anterior to the acetabular fossa on the ilial shaft is a shallow preacetabular fossa. The rounded dorsal prominence is large and is produced dorsally above the dorsal acetabular expansion. The acetabular fossa is relatively shallow, and the dorsal and anterior parts of the acetabular ridge are very low. The ventral part of the acetabular ridge is moderately well developed in association with the extremely concave ventral acetabular expansion. The posterior part of the acetabular ridge, situated on the ischium, appears to be well developed. The posterior ischial depression is quite deep. The exposed proximal portion of the left femur shows a well-developed, sharp, posteriorly directed trochanteric ridge. The more distal limb elements are not preserved.

Measurements of the holotype are as follows: total head and body length, 85.0 mm; length of skull from tip of snout to occipital condyles, 24 mm; width of skull between posterior extremes of maxillae, 27 mm; depth of skull from level of frontoparietals to distal tip of quadrate, 12 mm; depth of deepest part of the body of the ilium, 9 mm. Kluge (1966) cautioned that the above measurements are only approximations because of the incompleteness of many bones.

The right ilium (MSUVP 1027) from the Early Miocene (late Arikareean NALMA) of Cherry County, Nebraska, assigned to *Scaphiopus neuter* by Holman (1981), is similar in size to that of *S. neuter*, having a maximum height of the acetabulum of 4.0 mm. The dorsal prominence is well developed and produced dorsally from the dorsal acetabular expansion, as in *S. neuter*, but the ilial shaft is somewhat more curved and the ventral acetabular expansion somewhat wider anterior to the anterior border of the anterior rim of the acetabulum than in *S. neuter*. The water-worn state of MSUVP 1027 precludes the identification of other ilial characters.

General Remarks. Although *Scaphiopus neuter* shows resemblances to both the subgenera *Spea* and *Scaphiopus*, it is considered here to be more likely basal to the *Spea* than to the *Scaphiopus* lineage because it has a moderately large frontoparietal and only a light degree of sculpturing in the dermal roofing bones of the skull.

A Tentative Phylogeny of the Subgenus *Spea*

The subgenus *Spea* is tentatively considered here to be a monophyletic group, provided that *Scaphiopus neuter* of the Late Oligocene of South Dakota and the Early Miocene of Nebraska is actually the basal member of the subgenus. It appears that there is also an eastern clade and a western clade in *Spea*. The eastern clade consists of *Scaphiopus bombifrons* of the Middle Miocene to the Late Pliocene of Nebraska; the species later spread to the other plains states in the Pleistocene and Holocene. The western clade is composed of the closely related species *Scaphiopus (Spea) hammondii*, *Scaphiopus intermontana*, and *Scaphiopus multiplicata*. *Scaphiopus hammondii* is first known (tentatively) from the Late Pliocene of Idaho on into the Pleistocene and Holocene, and the other two species are known only in the Pleistocene (both within their present range) and Holocene. One might question the Late Pliocene Idaho record (Chantell, 1970) of *Scaphiopus* cf. *S. hammondii* on zoogeographic grounds, as, at present, its closest occurrence to Idaho is in northern California (Stebbins, 1985, map 25). The only Spadefoot that currently occurs in Idaho is *S. intermontana*, which occurs in approximately the eastern half of the state (Stebbins, 1985, map 23).

Scaphiopus sp. indet.

Several Pleistocene records of *Scaphiopus* sp. indet. are found in the literature, but these do little to elucidate the morphological, zoogeographic, or evolutionary aspects of the genus, which is represented by a variety of taxa in the North American fossil record. Sanchiz (1998, p. 57) provided access to the Pleistocene and Holocene localities. I shall list the pre-Pleistocene sites here, as they may be of further interest to students of fossil anurans who wish to restudy these incompletely identified specimens. **Early Oligocene (Whitneyan NALMA):** I-75 Local Fauna, Alachua County, Florida—Patton (1969). **Middle Miocene (late Barstovian NALMA):** Railway Quarry A, Cherry County, Nebraska—Holman and Sullivan (1981). **Late Miocene (Hemphillian NALMA):** Edson Quarry, Sherman County, Kansas—Sanchiz (1998). Mailbox Prospect, Antelope County, Nebraska—Parmley (1992). Santee site, Knox County, Nebraska—Parmley (1992). White Cone Local Fauna, Navajo County, Arizona—Parmley and Peck (2002) [recorded as *Scaphiopus (Scaphiopus)* sp. indet.] **Pliocene (Blancan NALMA):** Beck Ranch, Scurry County, Texas—Rogers (1976). Fox Canyon site, Meade County, Kansas—Sanchiz (1998, p. 57). Rexroad Local Fauna, Meade County, Kansas—Tihen (1960a, fig. 1a, p. 91).

THE *"EOPELOBATES"* PROBLEM

The genus *Eopelobates*, a member of the Pelobatidae, is known in Europe from the Early Eocene to the Pliocene–Pleistocene boundary and contains three well-documented species (see Sanchiz, 1998, pp. 52–54). The two North American species originally referred to *Eopelobates*, however, have been questioned. It has been suggested that *Eopelobates guthriei* Estes, 1970 of the Early Eocene of Wyoming actually represents the living genus *Scaphiopus* (Henrici, 2000), and I have referred to that species as *Scaphiopus guthriei* (Estes, 1970) in the *Scaphiopus* section above.

Eopelobates grandis Zweifel, 1956 was described from the Late Eocene (Chadronian NALMA) of the Indian Creek locality, Pennington County, South Dakota. The age of this locality (see Prothero and Emry, 1996) was incorrectly said to be Lower Oligocene by Sanchiz (1998) and Roček and Rage (2000b). An incomplete sacral vertebra from the Middle Eocene (late Bridgerian NALMA) of the Tabernacle Butte locality, Sublette County, Wyoming, compares favorably with *"Eopelobates" grandis* (Hecht, 1959).

Sanchiz (1998) referred to this taxon as *"Eopelobates" grandis*. Roček and Rage (2000b), however, continued to refer to this species as *Eopelobates grandis*, pointing out that "despite Zweifel's original errors in anatomical interpretation, most features suggest his original taxonomic assignment was correct." This taxon is being redescribed and reassessed by A. Henrici (see Henrici, 2000, p. 146).

To one day help solve the problem of the presence or absence of European-like pelobatids (i.e., the living genus *Pelobates* and the extinct genus *Eopelobates*) in North America is a complete specimen of a *Pelobates–Eopelobates*-like frog (Roček and Rage, 2000b, fig. 22, p. 1362) "that has not yet been formally described" but is thoroughly discussed (Roček and Rage, 2000b, pp. 1361–1364). This specimen is from the Middle Eocene Green River Formation (Lancy Member) near Farson, Wyoming. The principle character that suggests to Roček and Rage (2000b, p. 1363) that this fossil is closely related to the principally European genera *Eopelobates* and *Pelobates* is as follows: "the frontoparietal obviously develops from three elements (two lateral ones homologous with proper frontoparietals, plus a single posteromedian ossification)."

Finally, Sanchiz (1998, p. 51) provided access to records that had been ascribed to what he termed "*Eopelobates*" sp. by several authors; the 11 localities involved ranged in age from the Late Cretaceous to the latter part of the Eocene. This type of taxonomic uncertainty, relative to what I have called the "*Eopelobates*" problem above, is not at all unusual in vertebrate paleontology. The reader will be relieved to know that, oddly enough, such problems are usually worked out in the long run, especially when more complete fossil specimens are unearthed.

FAMILY PELODYTIDAE BONAPARTE, 1850

At present, only one living genus, *Pelodytes*, with two species, represents the small family Pelodytidae. *Pelodytes caucasicus* occurs in the northwestern Caucasus and western Transcaucasus in the former Soviet Union and in adjacent Turkey; *Pelodytes punctatus* occurs in Belgium through France to Spain and northwestern Italy (Frost, 1985). The family Pelodytidae is united by four derived characters (Henrici, 1994): (1) anterior ramus of pterygoid elongate, (2) scapula with anterior tubercle, (3) posterior presacral vertebrae strongly oriented anteriorly, and (4) tibiale (calcaneum) and fibulare (astragalus) completely fused. Two extinct monotypic genera represent the Pelodytidae in North America.

An osteological definition of the Pelodytidae follows. Eight procoelous, stegochordal presacral vertebrae occur, and these have the imbricate condition. Presacrals I and II have their neural arches fused, and the cervical cotyles of presacral I are closely juxtaposed. Free ribs are absent. Broadly dilated diapophyses are present on the sacrum, which has a bicondylar articulation with the urostyle. The urostyle has a transverse process anteriorly. The arciferal pectoral girdle has a cartilaginous omosternum and an osseus sternum. The anterior end of the scapula is not overlain by the clavicle. Palatines are present, as is a parahyoid bone. The cricoid ring is incomplete dorsally. Both the premaxilla and the maxilla are tooth-bearing. The astragalus and the calcaneum are completely fused. Three tarsalia occur, and the phalangeal formula is normal.

GENUS #*MIOPELODYTES* TAYLOR, 1941

Genotype. *Miopelodytes gilmorei* Taylor, 1941.

Etymology. The generic name comprises *Mio*, referring to the Miocene locality where the type specimen was collected; and the generic name of the modern Parsley Frog, *Pelodytes*. The specific name recognizes the North American vertebrate paleontologist Charles W. Gilmore.

Diagnosis. The diagnosis is the same as for the genotype and only known species, *Miopelodytes gilmorei*.

#*MIOPELODYTES GILMOREI* TAYLOR, 1941

Holotype. An articulated specimen (National Museum of Natural History: USNM 12356).

Locality and Horizon. Elko Shales, near Elko, Elko County, Nevada: Middle Miocene (Barstovian or Clarendonian NALMA). Elko Formation.

Diagnosis. The diagnosis is modified from Sanchiz (1998) after Henrici (1994). A pelodytid with fused tibiale–fibulare complex somewhat shorter than in *Tephrodytes*, with squamosal bearing sculpture on the zygomatic and otic rami (cranial sculpture apparently eroded from some dermal roofing bones); frontoparietal paired; frontoparietal fontanelle present; otic ramus of squamosal in lateral view well developed; transverse processes of last two presacral vertebrae strongly directed anteriorly; sacral diapophyses expanded, but less than in *Pelodytes*; scapula short; ischium does not extend posteriorly beyond the dorsal acetabular expansion of the ilium.

Description. The skull is somewhat broader than it is long, Other than the frontoparietals, many dermal roofing bones are eroded. Eight presacral vertebrae are present. Free ribs are absent. The first three post-cervical vertebrae have robust diapophyses that are curved somewhat posteriorly. The scapula is short. The ilia are complete but show little detailed structure in the specimen other than moderate curvature of the ilial shafts. The sacrococcygeal area appears obscure in figures of the specimens, but it is reported that the sacral vertebra has greatly widened diapophyses and is presumably free from the coccyx (Roček and Rage, 2000b). The preserved urostyle is rather strong and is slightly shorter than the ilia. The femora are robust, slightly curved, and about the same length as the tibiofibulae, which are also slightly curved and show evidence of tibiofibular division at both the proximal and the distal ends. The distal ends of the relatively short, robust, and completely fused tibiofibulae are wider than the distal ends of both the femora and the tibiofibulae. The prehallux is absent. The hind feet are short, but the individual elements making up the feet are relatively robust.

General Remarks. The type specimen is not figured here because it is extremely difficult to made out any diagnostic characters from the figures of this taxon that are available. It is probable that *Miopelodytes* is a member of the Pelodytidae, as indicated by Henrici (1994), but if the urostyle is free from the sacral vertebra (see Roček and Rage, 2000b, p. 1370), this would cast doubt on its pelodytid affinities. More remarks about the relationship of *Miopelodytes* to the North American pelodytid genus *Tephrodytes* are presented in the following section.

GENUS #*TEPHRODYTES* HENRICI, 1994

Genotype. Tephrodytes brassicarvalis Henrici, 1994.

Etymology. The generic name is from the Greek *tephra*, "ashes," and *dytes*, "diver, enterer" (in reference to the tuffaceous sediments from which the frog was collected and in which it may have aestivated). The specific name comes from the Latin *brassica*, "cabbage," and *arvalis*, "of a cultivated field" (referring to the Cabbage Patch beds).

Diagnosis. The diagnosis is the same as for the genotype and only known species, *Tephrodytes brassicarvalis*.

#*Tephrodytes brassicarvalis* Henrici, 1994

(Fig. 40)

Holotype. A partially complete, loosely articulated skeleton (Museum of Natural History, University of Kansas: KU 19928).

Locality and Horizon. Cabbage Patch site 10 (Museum of Natural History, University of Kansas locality KU-MT-25), Flint Creek Basin, Powell County, Montana: Late Oligocene to Early Miocene (Arikareean NALMA). Cabbage Patch beds.

Other Material. Referred material from the type locality includes a partial left frontoparietal, left exoccipital–prootic complex, and vertebral column (KU 19221); right scapula and clavicle, sacral vertebra, urostyle, ilia, and hind limbs (KU 23489); right maxilla (KU 18191); left maxilla (KU 19940); right ilium (KU 18195); left ilium (KU 19917); right ilium (KU 19918); and right ilium (KU 19919).

Material from other Cabbage Patch localities includes the following. (1) From KU-MT-12, Cabbage Patch locality 4, Flint Creek Basin, Granite County, Montana, middle Cabbage Patch beds, Arikareean: proximal half of right humerus (KU 18266); proximal end of fused tibiale–fibulare (KU 18270); proximal end of fused tibiale–fibulare (KU 18273). (2) From KU-MT-8, Pikes Peak locality 1, Flint Creek Basin, Powell County, Montana, upper Cabbage Patch beds, Arikareean: sacrum (KU 20654); complete right humerus and proximal end of left humerus from different individuals (KU 20659).

Diagnosis. The diagnosis is directly from Henrici (1994, pp. 159–160): "*Tephrodytes* differs from all other pelodytids by having frontoparietals that meet medially to conceal the frontoparietal fontanelle. It differs from *Pelodytes* in the following unique combination of characters that are not known for *Miopelodytes*: 1) posterior tip of frontoparietal present, 2) otic plate of squamosal present, 3) presacral neural arches elongate, and 4) anterior lamina of scapula absent. It is distinguished from *Miopelodytes* by possession of reduced otic ramus of squamosal and expanded sacral diapophyses."

Description. The description here is abbreviated from Henrici (1994). The holotype is a three-dimensionally preserved, nearly complete skele-

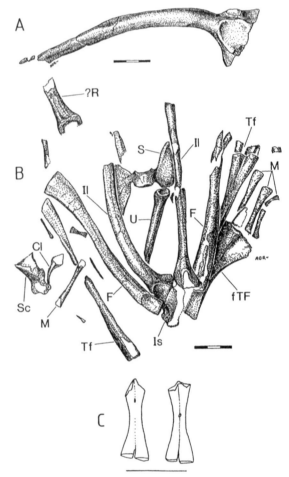

FIGURE 40. *Tephrodytes brassicarvalis* from the Late Oligocene–Early Miocene (Arikareean) of Montana. (A) Left ilium in lateral view. Scale bar = 3 mm. (B) Stippled mass of bones under the top scale bar is most of the pelvic girdle and hind limbs. Scale bar = 3 mm. (C) Unstippled, fused tibiale and fibulare. Scale bar = 5 mm. Abbreviations: Cl, clavicle; F, femur; fTF, fused tibiale and fibulare; Il, ilium; Is, ischium; M, metatarsals; R, radio-ulna; S, sacrum; Sc, scapula; Tf, tibiofibula; U, urostyle (coccyx).

ton. The length from the snout tip to the distal end of the urostyle is approximately 40 mm. The well-ossified bones suggest that it was an adult. The following description is based on the holotype unless stated otherwise.

In the skull, both halves of the subrectangular frontoparietals are present; the right half is slightly crushed. The frontoparietal is widest near the posterior end and from this point narrows to the midpoint of its length, where it slightly flares laterally to form the postorbital process. From there to the anterior end the bone is slightly narrower. The medial edge of the left frontoparietal is straight where it is exposed. The posterolateral corner of the frontoparietal is a drawn out tip, which is oriented posteriorly. Light sculpturing composed of pits and grooves occurs on the somewhat flattened dorsal surfaces of the frontoparietals; this sculpturing is most pronounced on the posterolateral corner and is absent along the medial edge. In KU 19221, a larger individual than the holotype, the posterior half of one frontoparietal is exposed. Sculpturing occurs over most of its dorsal surface, and the pits and grooves are larger. The frontoparietal fontanelle is not exposed, whereas it is in other pelodytids.

Nasal bones occur only in the holotype; the right nasal is undistorted, but the left one is incomplete. There is no sculpturing on the dorsal surface of the nasals. The two nasals meet at the midline for a fairly long distance. The width of the right nasal is greater than its length, its anterior process is a small nubbin, and the concave anterolateral margin forms the posteromedial border of the external narial opening. Laterally, the nasal enters a maxillary process that is directed posterolaterally.

Turning to the premaxillae, we find that the right and the left ones are present in the holotype and bear teeth. The lateral ends of both premaxillae are overlain by the medial ends of the maxillae. An elongate alary process occurs in the left premaxilla but is incomplete in the right one. Both maxillae are present in the holotype, and two isolated maxillae (KU 18191, 19940) have also been collected. In the holotype, the facial portion of the maxilla is slightly crushed at the anterior end, where it reaches its greatest height. This portion is constricted at the level of the orbit and increases in height at the posterior end to form the zygomatic process. In KU 18191 the dorsal surface of the zygomatic process bears a suture scar, which shows that the zygomatic ramus of the squamosal articulated with it.

The posteroventral end of the maxilla forms a point, the posterior process, which extends posteriorly beyond the level of the zygomatic process. Small teeth occur on the facial portion of the maxilla for about three-fourths the length of the bone from its anterior end. A tooth crown removed from the holotype appears fanglike and is recurved medially. Maxillary sculpturing is light in the holotype and occurs as pits and grooves that become larger at the posterior ends of the bone. In KU 19940, the largest maxilla, only a few pits are present on the posterior portion of the element. In the holotype, a sliver of bone located adjacent to the posteroventral edge of the right maxilla is thought to be the quadratojugal, but even if this tiny bone is not the quadratojugal, the presence of this bone is suggested by the presence of a posterior process in the holotype (see Estes, 1970).

Both squamosals occur in the holotype; the left one is complete, and the right one is missing the otic plate. The zygomatic ramus of the squamosal is relatively long and thin and reaches its greatest thickness near the middle of its length. In lateral aspect, the otic ramus appears as an obsolete nubbin. Projecting medially from the otic ramus is the otic plate, which has its medial end rounded. Sculpturing on the squamosals consists of relatively large pits and some tiny pits that are mainly confined to the area where the rami meet and on the zygomatic ramus itself.

Only the right exoccipital–prootic complex is present in the holotype, but a left one occurs in KU 19221 where the medial wall of the auditory capsule is exposed, showing the anterior and posterior acoustic foramina and the superior and inferior perilymphatic foramina. A small, T-shaped bone lies ventral to the right exoccipital–prootic complex and medial to the posterior end of the right squamosal in the holotype. This T-shaped element is believed to be the medial end of the right columella. A small bone ventral to the columella was removed; it shows a dome-like shape with a smooth dorsal surface and a rough ventral one. It was suggested that this element represents a calcified operculum, but it was pointed out that it was equally possible that it was merely a fragment of eroded bone.

The right pterygoid, which is exposed in ventral view, has an elongate anterior ramus, the anterior end of which lies between the upper and lower jaws. This anterior ramus is at least twice the length of the short medial ramus, which has a blunt medial end. The triangular posterior ramus is the shortest of the three. A part of the vomer exposed in the holotype shows several tooth sockets, but no other details are visible. Unfortunately, both lower jaws are pushed dorsally inside the mouth, and not much of either dentary could be exposed by preparation. Moreover, only a part of the angulosplenial could be prepared without damaging the skull. Noteworthy features were not present in either of these bones.

Turning now to the postcranial skeleton, we find that the vertebral column occurs as a part–counterpart in the holotype. The eight presacral vertebrae have imbricate neural arches, as indicated by the elongate neural spine of the only completely preserved vertebra (third presacral).

The vertebrae are procoelous. The cervical condyles lie close together. The dorsal part of the neural arch of the cervical vertebra is smooth, with its posterior end extended to form the neural spine. The transverse processes of vertebrae II–IV are long, and all are laterally directed except the fourth, which is slightly directed posteriorly. Transverse processes of vertebrae VI–VIII are shorter and thinner, and those of the fifth are directed laterally; those of the sixth, moderately directed anteriorly; and those of the seventh and eighth, strongly directed anteriorly. Free ribs are presumably absent.

The sacral vertebra is present in the holotype and in KU 23489, and an isolated sacrum (KU 20654) is preserved. The sacral diapophyses are widely expanded and are longer than they are wide.

In the holotype the length of the sacral diapophyses is roughly the same as the length of the last four presacrals. The shape of the sacral condyle is variable. In the holotype the sacral condyle is neither monocondylar nor bicondylar but forms a depressed oval that is slightly grooved in the middle. The sacral condyle is monocondylar in KU 23489 and forms a depressed oval that is distinctly bicondylar in KU 20654. In the holotype the urostyle occurs as a part–counterpart; in KU 23489 only the anterior half of this bone is exposed. In the holotype the urostyle is at least as long as the last five presacral vertebrae. Transverse processes on the urostyle have not been found in either specimen.

The only bones of the shoulder girdle present are a right scapula and a clavicle exposed in KU 23489. The scapula is short and robust and bears both an acromial process and a glenoid process on its ventral end; the processes themselves are narrowly separated. An anterior tubercle occurs on the anterior edge of the scapula near the acromial process. In the clavicle, the scapular articular process is broken off, but enough of the clavicle is bowed to suggest that the shoulder girdle is arciferal. It is estimated that the incomplete clavicle would have been longer than the scapula.

In the skeleton of the front limb, the right humerus and radio-ulna are preserved in articulation in the holotype. Three isolated humeri also occur (one humerus, KU 18266; and two humeri, KU 20659). In complete humerus KU 20659, the shaft is straight and bears a ventral crest on its proximal half. Well-developed medial and lateral epicondyles lie adjacent to the round distal condyle. The medial epicondyle is produced medially, is triangular in medial aspect, and is joined to the shaft by a thin crest. The smaller lateral epicondyle is crestlike. In lateral view, a groove lies between the roughened bone of the lateral epicondyle and the distal condyle. A triangular olecranon scar is situated medially. In the radio-ulna, the olecranon process is rounded and a sulcus is not visible on the anterior half of the shaft.

Both ilia are present in the holotype; the left ilium is complete, and the right one is without most of its shaft. The best-preserved ilium, however, is an isolated ilium (KU 19917), which is described here. The shaft of KU 19917 lacks a dorsal crest, is bowed ventrally, and is oval in cross section. The dorsal acetabular expansion bears an oval dorsolaterally projecting dorsal prominence that is said to be roughened by "unfinished" bone (Henrici, 1994) but is shown to be rather smoothly rounded in Henrici (1994, fig. 4d, p. 163). This prominence is situated at about the middle of the dorsal border of the acetabular fossa. The prominence is larger in KU 19917 than in the holotype, which, in fact, represents a smaller individual. In KU 19917, a groove runs from the shaft to the dorsal acetabular expansion. The surface of the dorsal acetabular expansion itself is somewhat excavated; its dorsal margin is slightly concave posterior to the dorsal prominence. The anterior margin of the ventral acetabular expansion is slightly concave and is generally at about right angles to the shaft. The acetabulum is large, roughly bell-shaped, and relatively well excavated. Its dorsal border projects slightly from the lateral surface of the dorsal acetabular expansion, whereas the ventral border projects strongly from the lateral surface of the ventral acetabular expansion. The dorsal ends of both ischia are present in KU 23489, and they appear to be medially fused. The ischia do not extend far posteriorly as occurs in *Eopelobates* and *Megophrys*.

Turning to the hind limb, we find that both femora and both tibiofibulae are present in UK 23489, but only partial tibiofibulae occur in the holotype. The femur is slightly sigmoidal, with its distal end wider than the proximal end. Presence of a femoral crest cannot be determined. In KU 23489 the tibiofibula is slightly longer than the femur. The longitudinal sulci of the tibiofibula are restricted to the proximal and distal ends.

Complete fusion of the tibiale and fibulare occurs in *Tephrodytes*. In the holotype the left fused tibiale–fibulare lie under the sacrum. In KU 23489 the left fused tibiale–fibulare are partly covered by other bones. Two isolated fused tibiale–fibulare (KU 18270, 18273) are also present. In the holotype and KU 23489 the longitudinal sulci are restricted to the proximal and distal ends. In KU 18270 and 18273, which are considerably smaller and missing their extreme distal ends, the longitudinal sulci extend most of the length of the bones. In all four examples the fused tibiale–fibulare are hourglass-shaped and have a small foramen located near the midpoint of the shaft.

General Remarks. Henrici (1994) felt that the assignment of *Tephrodytes brassicarvalis* to the Pelodytidae is certain, based on the following criteria. *Tephrodytes* has fused tibiale–fibulare, a condition that as far as is known occurs only in the anuran families Pelodytidae and Centrolenidae. Three of the numerous characters that distinguish the pelodytids from the centrolenids (see Duellman and Trueb, 1994) occur in *Tephrodytes*: (1) closely juxtaposed atlantal (cervical) condyles, (2) imbricate vertebral neural arches, and (3) widely expanded sacral diapophyses. On the other hand, *Tephrodytes* has no characters that occur in the centrolenids but not in the pelodytids.

I am not disputing the family allocation of Henrici (1994) here, who incidentally stated that relationships within the Pelodytidae are unresolved and that more information on the extinct members is needed. But I would like to point out that the ilia of *Tephrodytes*, at least as depicted by Henrici (see Fig. 40, this volume; Henrici, 1994, fig. 4d, p. 163), are completely different from those of species of modern and fossil *Pelodytes*, the only modern genus of the group, and also completely different from the ilia of modern and fossil *Pelobates* and *Scaphiopus* (subgenera *Scaphiopus* and *Spea*). These differences mainly concern the structures on the proximal end in the vicinity of the acetabulum and dorsal and ventral acetabular expansion and the shape and position of the dorsal prominence as shown in Henrici (1994, fig. 4d, p. 163). Such strong ilial structural differences usually do not occur among genera of many modern anuran families.

SUBORDER NEOBATRACHIA REIG, 1958

The suborder Neobatrachia consists of the more advanced anuran families.

SUPERFAMILY HYLOIDEA WIED, 1856

The superfamily Hyloidea contains the North American families Leptodactylidae, Bufonidae, and Hylidae.

Family Leptodactylidae Werner, 1896 (1838)

The huge, New World family Leptodactylidae occurs in the southern United States and the Antilles south to southern South America (Frost, 1985). Four living subfamilies, 54 genera, and more than 900 species are currently recognized. The leptodactylids form a morphologically diverse group, as may be seen in the variation that occurs in the definitive osteological characters of the family presented by Duellman and Trueb (1994, p. 528): "There are eight presacral vertebra with holochordal (stegochordal in some telmatobiines), procoelous centra. The neural arches are imbricate in ceratophryines and some telmatobiines and leptodactylines, and nonimbricate in hylodines, some telmatobiines, and most leptodactylines. Presacrals I and II are not fused (except in *Telmatobufo*); the atlantal condyles of Presacral I are closely juxtaposed in ceratophryines and some telmatobiines, and widely separated in hylodines, leptodactylines, and other telmatobiines. Ribs are absent. The sacrum has rounded diapophyses (weakly dilated in ceratophryines and some telmatobiines) and has a bicondylar articulation with the coccyx, which lacks transverse processes proximally (except in some telmatobiines). The pectoral girdle is arciferal (pseudofirmisternal in *Insuetophrynus*, *Sminthillus*, and *Phrynopus peruvianus*). A cartilaginous omosternum (absent in *Lepidobatrachus*, *Macrogenioglottus*, *Odontophrynus*, *Proceratophrys*, and some *Eleutherodactylus*) and cartilaginous sternum are present (also bony postzonal elements in leptodactylines). The anterior end of the scapula is not overlain by the clavicle. Palatines are present; a parahyoid is absent, and the cricoid ring is complete. The maxillae and premaxillae are dentate (edentate in *Batrachophrynus*, *Lynchophrys*, *Sminthillus*, and some *Physalaemus* and *Telmatobius*). The astragalus and calcaneum are fused only proximally and distally (completely fused in *Geobatrachus*); there are two tarsalia, and the phalangeal formula is normal (reduced in *Euparkerella* and *Phyllonastes*)."

Fossil leptodactylids are represented in North America by only two living genera, *Eleutherodactylus* and *Leptodactylus*, both of which are known only from the Pleistocene.

Genus *Eleutherodactylus* Duméril and Bibron, 1841

Rain Frogs

Living species of the genus *Eleutherodactylus* occur indigenously in the West Indies and from the southwestern United States, through Mexico, to Argentina and Brazil. They were introduced into Florida and Louisiana (Frost, 1985). *Eleutherodactylus* contains many more species than any other anuran genus (see Duellman, 1993, pp. 2–10). As of 1992, 512 species were recognized. Only 405 species were recognized in 1985, so it is likely that many more species have been named since 1992. Lynch (1971) provided a detailed diagnostic definition of this genus that included many osteological characters. I include a modified version of this definition here, as it might be helpful to the anuran paleontologist if skulls or skull parts are available. I caution the reader that the content of *Eleutherodactylus* is not the same as it was in 1971!

Omosternum usually present; maxillary arch toothed; alary process of maxillae relatively broad at the base; palatal shelf of premaxilla usually deeply dissected; facial lobe of maxilla deep and not exostosed (sculptured with excess dermal bone); palatal shelf of maxilla broad and usually with a prominent pterygoid process; nasals large and in broad medial contact (narrowly separated in some species); nasals not in contact with pterygoids, sometimes in contact with maxillae; frontoparietal fontanelle absent in adults except for *Eleutherodactylus palmeri*; frontoparietals not ornamented in most species groups, but bearing lateral crests in *Eleutherodactylus biporcatus*, *Eleutherodactylus cornutus*, *Eleutherodactylus galdi*, and *Eleutherodactylus unistrigatus* complexes; frontoparietals fused in most species in West Indies and northern Andes, free in other groups; epiotic eminences variably prominent to obsolete; cristae paroticae short and thick to long and narrow; zygomatic ramus short to long, sometimes knobbed, in contact with maxilla only in *Eleutherodactylus ruthae*; otic ramus of squamosal short to long, usually forming a small otic plate, ornamented in a few species groups; squamosal–maxillary angle, 44–67%, mostly 50–60%; columella present; prevomers usually toothed, entire, narrowly to broadly separated medially; palatines long, usually expanded laterally, widely separated medially, no odontoid ridges; sphenethmoid entire and extending anteriorly beneath the nasals for varying distances; cultriform process of parasphenoid narrow to broad, relatively long, nearly reaching prevomers, unkeeled medially; parasphenoidal alae in two patterns — alae I: alae deflected posteriorly, short, not overlapped laterally by median rami of pterygoid (West Indian species and some Andean species); and alae II: alae oriented at right angles to anterior ramus, rarely deflected posteriorly, long, broadly overlapped by medial rami of pterygoids (Central American and lowland South American species) — pterygoids slender to robust, lacking ventral flange, anterior rami short, not reaching palatines, median rami short to long, straight or bent; occipital condyles fairly small, stalked or unstalked, widely separated medially; terminal phalanges nearly always T-shaped, inner phalanges usually knobbed, terminal phalanges of toes more T-shaped than those of fingers.

Note: *Eleutherodactylus* sp. has been reported from the ?late Pleistocene Cutler Hammock Local Fauna, Dade County, Florida (Emslie and Morgan, 1995), but further evidence is needed to confirm this taxon in the Florida fossil record (Hulbert, 2001).

Eleutherodactylus augusti (Duges, 1879)
Barking Frog
(Fig. 41)

Fossil Localities. **Pleistocene (Rancholabrean NALMA):** Dark Canyon site, Eddy County, New Mexico—Applegarth (1980), Sanchiz (1998). Friesenhahn Cave, Bexar County, Texas—Mecham (1959), Holman (1969c, 1995b), Sanchiz (1998). Rancho la Brisca locality, Sonora, Mexico—Van Devender et al. (1985), Sanchiz (1998). Schulze Cave

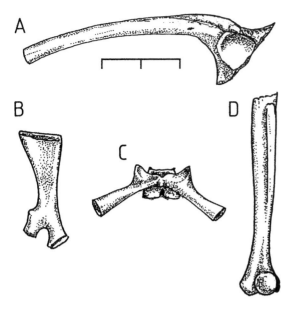

FIGURE 41. Bones of modern *Eleutherodactylus augusti* from Uvalde County, Texas. (A) Left ilium in lateral view. (B) Left scapula in medial view. (C) Sacrum in dorsal view. (D) Left humerus in ventral view. Scale bar = 4 mm and applies to all figures.

Fauna, Edwards County, Texas—Parmley (1986), Holman (1995b), Sanchiz (1998).

The Barking Frog is a large *Eleutherodactylus* species, with an SVL of about 64–75 mm (Conant and Collins, 1998). Some describe this species as looking like a toad. But *Eleutherodactylus augusti* lacks warts and has a dorsolateral fold that runs down each side of the back. The coloration is tan or reddish brown or sometimes somewhat greenish. This species is especially adapted to limestone caves and ledges and seldom ventures forth from this habitat. The present range of *E. augusti* is southeastern Arizona and southeastern New Mexico, south through central and western Mexico, to the Isthmus of Tehuantepec, Oaxaca, Mexico (Frost, 1985). This species had the generic name *Hylactophryne* from 1968 until recently.

Identification of Fossils. I have not been able to find reference to a discussion of specific characters on elements used to identify this large species of *Eleutherodactylus.* I am here including the description of an ilium of a young individual of *Eleutherodactylus augusti* from Durango, Mexico (see Fig. 41). The dorsal acetabular expansion is well developed and rotated posteriorly and has a pointed apex. The acetabular cup is well excavated. The rim of the acetabular cup is sharply defined. The ventral acetabular expansion extends anterior to the anterior edge of the acetabular cup, and the anterior border of the expansion is roughly at a right angle to the long axis of the shaft. The ilial shaft itself is slightly curved ventrally. A low dorsal crest arises at the level of about the anterior third of the acetabular cup and extends to about the middle third of the shaft, where it slopes gently into the dorsal margin of the shaft. This dorsal crest is excavated both laterally and medially. A dorsal prominence is not discernible in this specimen.

ELEUTHERODACTYLUS CF. *ELEUTHERODACTYLUS CYSTIGNATHOIDES* (COPE, 1878 "1877"))

RIO GRANDE CHIRPING FROG

Fossil Locality. **Pleistocene (Rancholabrean NALMA):** Cueva de Abra, near Antiguo Morelos, Tamaulipas, Mexico—Holman (1970c), Sanchiz (1998).

The tiny Rio Grande Chirping Frog, usually less than an inch long (15–25 mm SVL; Conant and Collins, 1998), is about as plain-looking as an anuran can be, being brownish gray or brownish green, with a little darker bar from the eye to the end of the nose. Irregular flecks occur on the posterior part of the body. This is a running frog that can dash under cover to escape predators. This little animal can adapt to a large variety of habitats within its range (as long as some moisture exists), and it readily adapts to civilization. Today it occurs in low to moderate elevations from the Rio Grande embayment, Texas, to central Nuevo Leon, Tamaulipas, eastern San Luis Potosi, and central Veracruz, Mexico (Frost, 1985). This animal formerly had the generic name *Syrrhophus*.

Identification of Fossils. As far as I can ascertain, osteological characters for the identification of this little frog have not been given.

ELEUTHERODACTYLUS MARNOCKII (COPE, 1878)
CLIFF CHIRPING FROG
(FIG. 42)

Fossil Localities. **Pleistocene (Irvingtonian NALMA):** Gilliland Fauna, Knox County, Texas—Tihen (1960b), Gehlbach (1965), Holman (1995b), Sanchiz (1998). Vera Local Fauna, Knox County, Texas—Parmley (1988a), Holman (1995b), Sanchiz (1998) (recorded as *Eleutherodactylus* cf. *Eleutherodactylus marnockii* by all authors). **Pleistocene (Rancholabrean NALMA):** Easley Ranch Local Fauna, Foard County, Texas—Lynch (1964), Holman (1995b), Sanchiz (1998). Schulze Cave Fauna, Edwards County, Texas—Parmley (1986), Holman (1995b), Sanchiz (1998).

Eleutherodactylus marnockii is a small, flattened frog (19–38 mm SVL; Conant and Collins, 1998) with a large head. It has a greenish brown coloration, with dark flecks over the body. This nocturnal anuran lives in limestone crevices and cracks in the cliff country of the Edwards Plateau and the extreme edge of the Stockton Plateau of south-central Texas, which constitutes its known range (Frost, 1985). This frog can run if it needs to. This species also bore the generic name *Syrrhophus*.

Identification of Fossils. I include a description of the ilium of *Eleutherodactylus marnockii* here as a possible aid to the identification of this species as a fossil. This description is based on a specimen (see Fig. 42) from Uvalde County, Texas. The ilium is about 13 mm long. The dorsal acetabular expansion is well developed and has a pointed apex. The portion of the acetabulum present indicates an ovoid condition for the whole unit. The ventral acetabular expansion is slightly larger than the dorsal one and has its apex narrowly rounded. The shaft is high anteriorly but becomes much lower and longer about one-fourth of the distance anteriorly down

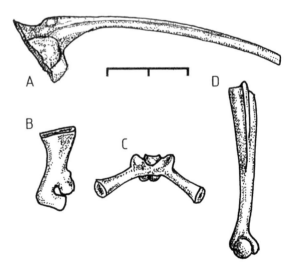

FIGURE 42. Bones of modern *Eleutherodactylus marnockii* from Uvalde County, Texas. (A) Right ilium in lateral view. (B) Right scapula in medial view. (C) Sacrum in dorsal view. (D) Humerus in ventral view. Scale bar = 4 mm and applies to all figures.

the bone. A narrow ilial shaft ridge is present along the dorsal border of the ilium. The shaft is gently curved ventrally. A somewhat ovoid dorsal prominence occurs above the anterior portion of the acetabular cup.

General Remarks. The ilia of *Eleutherodactylus augusti* and *Eleutherodactylus marnockii* appear quite distinct from one another (compare Figs. 41 and 42).

Genus *Leptodactylus* Fitzinger, 1826

Neotropical Grass Frogs

Living species of the genus *Leptodactylus* are indigenous to southern North America, South America, and the West Indies (Frost, 1985). As of 1992, 50 species were recognized, only 1 more than were recognized in 1985 (Duellman, 1993). Lynch (1971) provided a detailed diagnostic definition of the genus that included many osteological characters. These are repeated here in a slightly modified version.

Sternum bearing an elongate osseus style; transverse processes of posterior presacral vertebrae moderately elongate; cervical cotyles widely spaced; omosternum cartilaginous, large, and elongate, with large manubrium; sacral diapophyses rounded; maxillary arch toothed with teeth often pointed; alary processes of premaxillae directed dorsally or posterodorsally with a broad base; palatal shelf of premaxilla moderately broad; facial lobe of maxilla relatively shallow and maintaining the same depth from the anterior portion to the end of the tooth row; palatal shelf of maxillary relatively narrow and without a pterygoid process; maxillary arch complete; nasals large and narrowly separated medially; nasals usually not in contact with maxillae and never in contact with the pterygoids; nasals with elongate maxillary processes in most species and not in contact with the frontoparietals; frontoparietal fontanelle lacking; frontoparietals with some ornamentation posteriorly in the older adults of some taxa; epiotic eminences well defined posteriorly; parotic crests moderately long and somewhat robust; zygomatic ramus of squamosal relatively short and somewhat expanded; otic ramus of squamosal slightly longer than zygomatic ramus and expanded into narrow otic plate that usually rests tenuously on the parotic crests; squamosal–maxillary angle less than 45 degrees; columella present; prevomers large, entirely toothed, and narrowly separated medially; palatines broad and narrowly separated medially and sometimes bearing an odontoid ridge; sphenethmoid extending anteriorly to the middle of the nasals in *Leptodactylus melanonotus, Leptodactylus ocellatus,* and *Leptodactylus pentadactylus* groups and extending anteriorly to a point anterior to the nasals and usually anterior to the premaxillae in *Leptodactylus fuscus* group; anterior ramus of parasphenoid narrow, not keeled medially, and reaching to the palatines; parasphenoid alae deflected posteriorly, narrowly overlapped laterally by median rami of pterygoids in *L. fuscus, L. melanonotus,* and *L. pentadactylus* groups; pterygoids slender with anterior rami reaching to middle of orbit; occipital condyles of moderate size, not stalked, with moderate to wide median separation; terminal phalanges knobbed; alary processes of hyoid on narrow stalks in *L. fuscus, L. melanonotus, L. ocellatus,* and *L. pentadactylus* groups.

LEPTODACTYLUS CF. *LEPTODACTYLUS LABIALIS* (COPE, 1878 "1877")

MEXICAN WHITE-LIPPED FROG

Fossil Locality. **Pleistocene (Rancholabrean NALMA):** Cueva de Abra, near Antiguo Morelos, Tamaulipas, Mexico—Holman (1970c), Sanchiz (1998).

Leptodactylus labialis is a small frog (35–51 mm SVL; Conant and Collins, 1998) with a cream-colored or white line on the upper lip and a very large, round disc (ventral disc) on the abdomen. The color of this species varies from brown to gray, and dorsal spots of a variable size and number are present. The species occupies a wide variety of moist habitats. The distribution of this species is from extreme southern Texas, through eastern and southern Mexico and Central America, to northern Colombia and Venezuela (Frost, 1985).

Identification of Fossils. Definitive osteological characters for the identification of this species as a fossil, to my knowledge, have not been given.

LEPTODACTYLUS MELANONOTUS (HALLOWELL, 1861 "1860")

SABINAL FROG

Fossil Locality. **Pleistocene (Rancholabrean NALMA):** Rancho la Brisca locality, Sonora, Mexico—Van Devender et al. (1985), Sanchiz (1998).

Some external characters that in combination are helpful in identifying this species are as follows: (1) interorbital area with a dark triangle outlined in white, a light and dark bar, or a light triangle reaching the tip of the snout anteriorly and bordered by a dark triangle posteriorly; (2) back brown to gray with darker indistinct spots, blotches, bands, stripes, or without pattern; (3) belly never distinctly spotted; (4) upper limbs barred or uniform in color; (5) posterior thigh mottled; (5) toe tips never expanded into discs (Heyer, 1970). This frog does not occur in the United States, but it ranges from western (from Sonora south) and southern Mexico through Central America to central Ecuador (Frost, 1985).

Identification of Fossils. As far as I am aware, no characters have been given for the identification of this species as a fossil.

FAMILY BUFONIDAE GRAY, 1825

The large family Bufonidae is cosmopolitan except for the Australian, Madagascan, and Oceanic regions (Frost, 1985) and is represented by 31 genera and about 370 species. Only the genus *Bufo*, currently represented by about 211 species (Duellman, 1993), is known from the North American fossil record. An osteological definition of the Bufonidae follows.

In the vertebral column, there are five to eight holochordal, procoelous presacral vertebrae with the imbricate condition. *Pseudobufo* lacks ossified intervertebral bodies. Variation in the number of presacral vertebrae results from fusion of presacrals I and II, such as in *Atelopus, Leptophryne, Oreophrynella,* and *Pelophryne.* Cervical cotyles of presacral I are juxtaposed. Ribs are not present. The sacrum has dilated diapophyses and a bicondylar articulation with the urostyle, except in taxa with a forward shift of the sacral articulation. In this situation the original sacral vertebra is incorporated into the urostyle and there is a monocondylar articulation or a fusion of the urostyle and functional sacral vertebra, such as in *Didynamipus*, *Laurentophryne*, *Mertensophryne*, *Nectophryne*, *Pelophryne*, *Wolterstorffina*, and a few *Rhamphophryne*. Transverse processes are not present on the urostyle. The pectoral girdle is either arciferal or pseudofirmisternal, which is caused by fusion of the epicoracoids such as occurs in *Atelopus* and *Osornophryne*. The omosternum is absent in most taxa but is present in *Nectophrynoides*, *Werneria*, and some *Bufo*. A bony sternum occurs, and the scapula is not overlain anteriorly by the clavicle. Palatines are present except in *Nectophryne* and *Pelophryne*, the parahyoid is absent, and the cricoid ring is complete. Both the premaxillae and the maxillae are toothless. The astragalus and calcaneum are fused only proximally and distally, two tarsalia are present, and the phalangeal formula is normal in most taxa.

There are many fossil records of *Bufo* sp. in North America, especially from the Pleistocene. These records are not detailed here because the large North American content of the genus renders these identifications rather meaningless. References to temporal occurrences and localities of *Bufo* sp. can be found in Sanchiz (1998, p. 75).

GENUS *BUFO* LAURENTI, 1768

TOADS

The content of the genus *Bufo* has been given in the introduction to the family above. The genus is cosmopolitan, except for Arctic regions, New Guinea, and Australia and adjacent islands (Frost, 1985).

**BUFO ALIENUS* TIHEN, 1962

(FIG. 43)

Holotype. An almost complete right ilium (National Museum of Natural History: USNM 22233).

Etymology. The specific name is from the Latin *alienus*, "alien" (alluding to the tentative reference of the species to the European *Bufo calamita* group).

Locality and Horizon. Quarry E, near Long Island, Phillips County, Kansas: Late Miocene (Hemphillian NALMA). Ogallala Formation.

Modified Diagnosis. A *Bufo* of moderate size, tentatively referred to the *Bufo calamita* group on the basis of the triangular dorsal prominence, the posterior slope of the prominence being decidedly steeper than the anterior, and the prominence projecting somewhat medially from the axis of the shaft.

Description. This description is modified from Tihen (1962b). The length from the dorsal tip of the dorsal acetabular expansion to the anterior end of the shaft is 17.7 mm, but it is estimated that 1 or 2 mm of the anterior end is missing. The shaft is rather strongly curved and is said

FIGURE 43. Holotype right ilium of *Bufo alienus* from the Late Miocene (Hemphillian) of Kansas. Scale bar = 5 mm.

to have a low dorsal crest along the central two-thirds of this length; but this is not shown in Tihen's illustration of the type specimen (see Fig. 43, this volume). The posterior part of the shaft is slightly compressed. The prominence is small, and its height is slightly less than 25% of the length of its base, but the base itself appears to be of limited longitudinal extent. The posterior slope of the prominence is sharply sigmoidal, and the anterior slope is more even.

General Remarks. Tihen's (1962b) tentative referral of this ilium to the *Bufo calamita* group is disputed here. Actually, the only ilial character that appears to be diagnostic of the *B. calamita* group as a whole (*Bufo calamita, Bufo raddei,* and *Bufo viridis*) is a thickened, sometimes swollen, sometimes rather bladelike structure on the ventral portion of the posterior part of the shaft referred to as the *calamita* blade (see Holman, 1998). The *calamita* blade is variable within the species in which it is most highly developed, *B. calamita,* and when the blade occurs it seems to be sexually dimorphic. I studied ilia of 39 *B. calamita* specimens from western and central Europe and found that the "blade" occurred in 27 of these 39 specimens (69.2%) (8 were not sexed). Of 15 identified females, only 2 of the largest 3 had recognizable blades, and these were poorly developed. Of the 16 males, only 1 lacked a blade.

I suggest that *Bufo alienus* is not in the *Bufo calamita* group, on the basis of the following characters: (1) *calamita* blade lacking (it's present in most males of *B. calamita* and in *Bufo raddei* and *Bufo viridis*); (2) ilial shaft moderately curved (it's slightly curved in *B. calamita* and some east-central European *B. raddei* and *B. viridis*); (3) ilial prominence deflected posteriorly (it's deflected anteriorly or perpendicular to the long axis of the shaft in *B. calamita, B. raddei,* and *B. viridis*). More material is needed to determine the relationship of this taxon.

Bufo alvarius Girard, 1859
Colorado River Toad

Fossil Localities. **Pliocene (Blancan NALMA):** Benson locality Cochise County, Arizona—Tihen (1962b), Gehlbach (1965) (recorded as *Bufo* cf. *Bufo alvarius* by both authors). **Pleistocene (Irvingtonian NALMA):** El Golfo locality, Sonora, Mexico—Lindsay (1984), Sanchiz (1998). **Pleistocene (Rancholabrean NALMA):** Rancho la Brisca, Sonora, Mexico—Van Devender et al. (1985), Sanchiz (1998).

This is the largest toad in the western United States: it ranges from 100 to 190 mm in SVL (Stebbins, 1985). The Colorado River Toad has long, kidney-shaped parotoid glands. The smooth skin of the legs of this species is highlighted by several large warts. The coloration above can be olive, gray, or brown. In the United States, this species ranges from mesquite–creosote bush lowlands to oak–sycamore–walnut-dominated canyons in the mountains. In Mexico, this species may occupy tropical thorn forests (Stebbins, 1985). *Bufo alvarius* occurs from extreme southeastern California, southern Arizona, and extreme southwestern New Mexico, south through Sonora and extreme southwestern Chihuahua, Mexico, to northern Sinaloa, Mexico (Frost, 1985).

Identification of Fossils. Van Devender et al. (1985, p. 29) commented on the identification of *Bufo alvarius* frontoparietals, a scapula, and ilia from the Pleistocene (Rancholabrean) locality at Rancho la Brisca in Sonora, Mexico: "The fossils are from toads of 110–160 mm snout–vent length (SVL). Today only *B. alvarius* and *B. marinus* (Marine Toad) [= Cane Toad] reach this and larger sizes. *Bufo woodhousei bexarensis* (Friesenhahn Cave Toad), an extinct late Pleistocene form in Texas, reached body lengths of 160 mm but had a narrow frontoparietal with tubular dorsal ornamentation (Mecham 1959; Tihen 1962a) quite different than the la Brisca fossil. *Bufo alvarius* and *B. marinus* are in the *B. valliceps* (Gulf Coast Toad) species group of Tihen (1962b). Both species have wide frontoparietals shaped similar to the fossils, but *B. marinus* has heavier, more rugose dorsal ornamentation with prominent ridges oriented obliquely to the anterior–posterior axis of the bone (fig. 5A). The anterior edge of the frontoparietal in our reference specimens is more squared off than in the fossils. The ilia referred to *B. alvarius* have a broader dorsal prominence without the pronounced knob of *B. marinus* (fig. 5B). The scapula of *B. alvarius* differs from that of *B. marinus* in that the articular surface is slightly smaller, and that the acromial portion is relatively narrow." Van Devender et al. assigned dentaries, a nasal, vertebrae, humeri, and a radio-ulna to *Bufo* cf. *Bufo alvarius* but warned that *Bufo marinus* could not be eliminated from consideration as a part of this material.

General Remarks. This is, surprisingly, the only Rancholabrean record for *Bufo alvarius*.

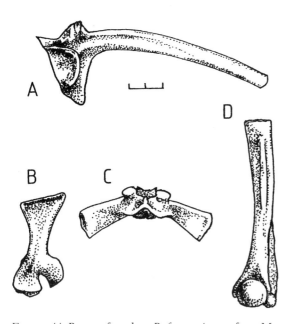

FIGURE 44. Bones of modern *Bufo americanus* from McLean County, Illinois. (A) Right ilium in lateral view. (B) Right scapula in medial view. (C) Sacrum in dorsal view. (D) Right humerus in ventral view. Scale bar = 4 mm and applies to all figures.

Bufo americanus Holbrook, 1836
American Toad
(Fig. 44)

Fossil Localities. **Pleistocene (Irvingtonian NALMA):** Albert Ahrens Local Fauna, Nuckolls County, Nebraska — Ford (1992), Holman (1995b), Sanchiz (1998). Cumberland Cave, Allegany County, Maryland — Holman (1977c, 1980), Holman and Winkler (1987), Sanchiz (1998). Hamilton Cave, Pendleton County, West Virginia — Holman and Grady (1989), Holman (1995b), Sanchiz (1998). Hanover Quarry No. 1 fissure, Adams County, Pennsylvania — L. P. Fay (written communication, May 12, 1993),[1] Holman (1995b), Sanchiz (1998). Trout Cave, Pendleton County, West Virginia — Holman (1982b, 1995b), Sanchiz (1998). **Pleistocene (Rancholabrean NALMA):** Baker Bluff Cave, Sullivan County, Tennessee — Van Dam (1978), Fay (1988), Holman (1995b), Sanchiz (1998). Bell Cave, Colbert County, Alabama — Holman et al.

(1990), Holman (1995b), Sanchiz (1998). Bootlegger Sink, York County, Pennsylvania—Guilday et al. (1966), Holman (1995b), Sanchiz (1998). Cheek Bend Cave, Maury County, Tennessee—Klippel and Parmalee (1982), Holman (1995b), Sanchiz (1998) (recorded as *Bufo* cf. *Bufo americanus* by all authors). Clark's Cave, Bath County, Virginia—Holman (1986b, 1995b), Fay (1988), Sanchiz (1998). Frankstown Cave, Blair County, Pennsylvania—Fay (1988), Holman (1995b), Sanchiz (1998). Guy Wilson Cave, Sullivan County, Tennessee—L. P. Fay (written communication, May 12, 1993),[2] Holman (1995b), Sanchiz (1998). Kelso Cave, Halton County, Ontario, Canada—Churcher and Dods (1979), Holman (1995b, 2001b), Sanchiz (1998) (recorded as *Bufo* cf. *B. americanus* by all authors). Kingston Saltpeter Cave, Bartow County, Georgia—Fay (1988), Holman (1995b), Sanchiz (1998). Ladds Quarry, Bartow County, Georgia—Holman (1967b, 1985a, b, 1995b), Wilson (1975), Sanchiz (1998). Meskill Road site, St. Clair County, Michigan—Holman et al. (1986), Holman (1995b, 2001b), Sanchiz (1998). Moscow Fissure, Iowa County, Wisconsin—Foley (1984), Holman (1995b), Sanchiz (1998). Natural Chimneys site, Augusta County, Virginia—Fay (1988), Holman (1995b), Sanchiz (1998). New Paris 4 site, Bedford County, Pennsylvania—Fay (1988), Holman (1995b), Sanchiz (1998). New Trout Cave, Pendleton County, West Virginia—Holman and Grady (1987), Holman and Winkler (1987), Holman (1995b), Sanchiz (1998). Peccary Cave, Newton County, Arkansas—Davis (1973), Holman (1995b), Sanchiz (1998). Robinson Cave, Overton County, Tennessee—Guilday et al. (1969), Holman (1995b), Sanchiz (1998). Sheriden Pit Cave, Wyandot County, Ohio—Holman (1995b, 1997, 2001b), Sanchiz (1998). Strait Canyon Fissure, Highland County, Virginia—Fay (1984), Holman (1995b), Sanchiz (1998).

The American Toad (*Bufo americanus*) is the common toad of the northeastern United States. This animal somewhat resembles *Bufo fowleri*, with which it sometimes hybridizes, but can usually be separated from that species in having only one or two large warts in each of the large spots on the back, having the chest and abdomen spotted with dark markings, having large warts on the tibial area, and having the parotoid gland separated from the ridge posterior to the eye or connected to it only by a short extension of the ridge. *Bufo americanus* is an animal of ubiquitous habitats, ranging from urban gardens and vacant lots to wild and remote areas. The main requirements seem to be shallow, still water in which to breed, a moist hiding place in which to retire, and abundant insects and other invertebrates to eat. This common toad occurs from the Maritime provinces to southeastern Manitoba in Canada; south to Mississippi and adjacent Louisiana and east to Kansas, with isolated colonies in southeastern North Dakota, northeastern North Carolina, and Newfoundland, Canada (Conant and Collins, 1998).

Identification of Fossils. In North America, most fossil *Bufo* species have been identified on the basis of individual elements. In eastern North America, Pleistocene specimens of the three common large species, *Bufo americanus*, *Bufo fowleri*, and *Bufo terrestris*, have been separated from one another on the basis of frontoparietals, squamosals, nasals, and ilia, although not always without difficulty (Wilson, 1975). In the frontoparietal, *B. americanus* and *B. fowleri* have a cranial crest that is less rounded and knoblike than in *B. terrestris*. In the squamosal, the anterior dorsal portion of the bone is proportionally smaller in *B. americanus* and *B. fowleri* than in *B. terrestris*. In the nasal, the dorsal ridges are less clearly defined in *B. americanus* and *B. fowleri* than in *B. terrestris*.

Turning to the ilium, we find that the ilia of *Bufo americanus* and *Bufo fowleri* have a higher, more narrowly based ilial prominence than in *Bufo terrestris*. Separating the ilia of *B. americanus* and *B. fowleri* can be somewhat more difficult. Although the ilial prominences of both species are relatively high, the base of the ilial prominence in *B. americanus* is wider than that of equal-sized ilia in *B. fowleri*. Determining this character depends upon having an adequate sample of modern skeletons of both *B. americanus* and *B. fowleri* to compare with one another, as some intraspecific variation occurs.

General Remarks. As far as I have been able to determine, the only reliable records of *Bufo americanus* are from Pleistocene localities. The questionable record of the cf. *Bufo americanus* group (Sanchiz, 1998) from the Miocene (Hemphillian NALMA) of Long Island, Kansas, probably represents an unidentified extinct taxon.

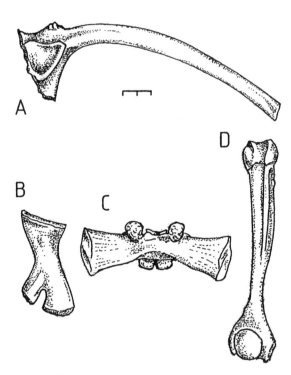

FIGURE 45. Bones of modern *Bufo boreas* from Lake Quinault, Washington. (A) Right ilium in lateral view. (B) Left scapula in medial view. (C) Sacrum in dorsal view. (D) Right humerus in ventral view. Scale bar = 4 mm and applies to all figures.

BUFO BOREAS BAIRD AND GIRARD, 1852
WESTERN TOAD
(FIG. 45)

Fossil Localities. **Pleistocene (Rancholabrean NALMA):** Costeau Pit, Orange County, California—Hudson and Brattstrom (1977), Holman (1995b), Sanchiz (1998). Haystack Cave, Gunnison County, Colorado—Mead and Bell (1994), Sanchiz (1998) (recorded as *Bufo boreas* or *Bufo woodhousii* by both). Newport Beach Mesa, Orange County, California—Hudson and Brattstrom (1977), Holman (1995b), Sanchiz (1998). Potter Creek Cave, Shasta County, California—Brattstrom (1958b), Tihen (1962b), Holman (1995b), Sanchiz (1998). Rancho La Brea, Los Angeles County, California—Camp (1917), Brattstrom (1953a), Stock (1956), Tihen (1962b), Gehlbach (1965), Holman (1995b), Sanchiz (1998). Smith Creek Cave, White Pine County, Nevada—Mead et al. (1982), Holman (1995b), Sanchiz (1998).

This toad may usually be distinguished within its range in the western United States on the basis of having a white or cream-colored, narrow stripe down the middle of the back and of lacking cranial crests (Stebbins, 1985). The coloration is gray or greenish above, with warts within dark blotches. The parotoid glands are oval and well separated from each another. This species occupies a wide variety of habitats ranging from the border of desert streams and springs, through grasslands and woodlands, to mountain meadows. It tends to be active at night in warm habitats and diurnal in the mountains. The Western Toad occurs in western North America from southern Alaska through western Canada and the western United States to southern Colorado, Utah, Nevada, and northern Baja California, Mexico (Frost, 1985).

Identification of Fossils. Pleistocene remains of *Bufo boreas* should be identifiable, especially if two or more key elements are available. The frontoparietals lack cranial crests (see Tihen, 1962a, fig. 8, p. 164). The distal one-third of the humeral shaft appears to be remarkably narrow in some individuals (see Fig. 45, this volume), which may indicate that this is a sexually dimorphic character. The ilium (see Fig. 45) is moderately curved, and the anterior margin of the ventral acetabular expansion makes an almost hemispherical curve with the posterior part of the shaft. The apices of both the dorsal and the ventral acetabular expansions are pointed rather than curved. The ventral acetabular expansion extends beyond the anterior margin of the acetabular cup. The acetabular cup is deep, with strong borders. The dorsal prominence is low, and its dorsal border is roughened or has two or three small tubercles.

General Remarks. The lack of fossil records of this now common and widespread species is rather difficult to explain, as is its restriction to the Rancholabrean.

Bufo campi Brattstrom, 1955

(Fig. 46)

Holotype. A left tibiofibula (Los Angeles County Museum of Natural History: LACM 276/5120). The specimen was formerly at the California Institute of Technology.

Etymology. The specific name recognizes North American vertebrate paleontologist Charles L. Camp.

Locality and Horizon. Patterson Field locality, Arroyo de los Burros, Chihuahua, Mexico: Late Miocene (early Hemphillian NALMA). Yepomera Formation.

Diagnosis. The species was diagnosed as a new species based on the tibiofibula having a thin and wide longitudinal ridge along the central portion of the postaxial surface (Brattstrom, 1955b; Tihen, 1962b).

Discussion. Tihen (1962b), Sanchiz (1998), and Roček and Rage (2000b) have continued to questionably retain *Bufo campi* as a species. This was done on the basis that, even though the species is known from an element of little diagnostic value, its morphology (i.e., the wide longitudinal ridge) differs from that of all other living or fossil taxa in North America in which this element is known (Sanchiz, 1998). I propose that the ridge is pathological.

In preparing hundreds of anuran skeletons and examining numerous individual elements of single species from archaeological sites (e.g., Holman, 1992a), I have often seen a ridged condition similar to that exhibited

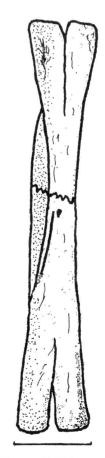

FIGURE 46. Holotype left tibiofibula of *Bufo campi* from the Late Miocene (Hemphillian) of Chihuahua, Mexico. Scale bar = 7 mm.

by the tibiotarsus of *Bufo campi*. This condition is a manifestation of a typical anuran healing process after breakage of the bone. Probably, owing to the fragility of anuran bones, especially the tibiofibula, a longitudinal (greatly swollen to rather thin) ridge develops as a brace on the external surface of the bone in the area of the fracture. Sometimes the actual fracture can be seen on the bone if the healing process has been recent. Other times only the brace itself is evident. In other words, once formed, the ridge remains a part of the structure of the bone itself.

The figure of the holotype of *Bufo campi* (Brattstrom, 1955b, p. 2; redrawn in Sanchiz, 1998, p. 77) indicates a diagonal break through both the bone and the brace, which might indicate to some that the ridge is natural rather than pathological. On the other hand, a photograph of a cast of the type specimen (Tihen, 1962b, fig. 25, p. 16) indicates that the two portions of the bone had been cemented together in the laboratory. It may have been along this zone of weakness that the original fracture occurred. In any case, the fact that such ridges occur occasionally in tibiae of modern anurans leads me to suggest that *B. campi* be considered a nomen vanum *sensu* Sanchiz (1998). In other words, the name was validly proposed under zoological nomenclatorial rules, but it cannot be defined on the basis of the available specimens or diagnosis.

BUFO COGNATUS SAY, 1823
GREAT PLAINS TOAD
(FIG. 47)

Fossil Localities. **Miocene (Clarendonian NALMA):** WaKeeney Local Fauna, Trego County, Kansas—Wilson (1968), Holman (1975); Sanchiz (1998) (recorded as *Bufo* cf. *Bufo cognatus*). **Miocene (Hemphillian NALMA):** Driftwood Creek, Hitchcock County, Nebraska—Tihen (1962b), Sanchiz (1998) (recorded as *Bufo* cf. *B. cognatus* by both authors). Quarry E, near Long Island, Phillips County, Kansas—Tihen (1962b), Gehlbach (1965), Sanchiz (1998) (recorded as *Bufo* cf. *B. cognatus* by all authors). Santee site, Knox County, Nebraska—Parmley (1992), Sanchiz (1998) (recorded as *Bufo cognatus* or *Bufo speciosus* by both authors). **Pliocene (Blancan NALMA):** Beck Ranch Local Fauna, Scurry County, Texas—Rogers (1976), Sanchiz (1998). Big Springs Quarry, Antelope County, Nebraska—Rogers (1984), Sanchiz (1998). Hornets Nest Quarry, Knox County, Nebraska—Rogers (1984), Sanchiz (1998). Rexroad Local Fauna, Meade County, Kansas—Tihen (1962b), Gehlbach (1965), Sanchiz (1998) (recorded as *Bufo* cf. *B. cognatus*

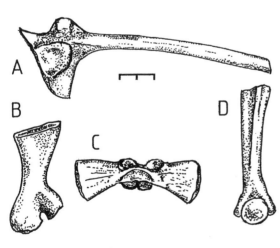

FIGURE 47. Bones of modern *Bufo cognatus*. (A) Right ilium in lateral view. (B) Right scapula in medial view. (C) Sacrum in dorsal view. (D) Left humerus (upper portion broken away, distal portion eroded) in ventral view. Scale bar = 4 mm and applies to all figures.

by all authors). White Rock Fauna, Republic County, Kansas—Eshelman (1975), Sanchiz (1998) (recorded as *Bufo* aff. *Bufo cognatus* by both authors). **Pleistocene (Irvingtonian NALMA):** Albert Ahrens Local Fauna, Nuckolls County, Nebraska—Ford (1992), Holman (1995b), Sanchiz (1998). Hansen Bluff Chronofauna, Alamosa County, Colorado—Rogers et al. (1985), Rogers (1987), Holman (1995b), Sanchiz (1998). **Pleistocene (Rancholabrean NALMA):** Butler Spring Local Fauna, Meade County, Kansas—Tihen (1962b), Gehlbach (1965), Holman (1986a, 1995b), Sanchiz (1998). Cragin Quarry Local Fauna, Meade County, Kansas—Tihen (1962b), Holman (1995b), Sanchiz (1998). Friesenhahn Cave, Bexar County, Texas—Mecham (1959), Tihen (1962b), Gehlbach (1965), Holman (1969c, 1995b), Sanchiz (1998) (recorded as *Bufo* cf. *B. cognatus* by all authors). Howard Ranch Local Fauna, Hardeman County, Texas—Holman (1964, 1969c, 1995b), Sanchiz (1998) (recorded as *Bufo cognatus* or *Bufo speciosus* by both authors). Jinglebob Local Fauna, Meade County, Kansas—Tihen (1954, 1962b), Holman (1995b), Sanchiz (1998). Jones Fauna, Meade County, Kansas—Tihen (1962b), Gehlbach (1965), Holman (1995b), Sanchiz (1998). Lubbock Lake site, Lubbock County, Texas—Johnson (1987), Holman (1995b), Sanchiz (1998). Rancho la Brisca, Sonora, Mexico—Van Devender et al. (1985), Sanchiz (1998) (recorded as *Bufo* cf. *Bufo cognatus* by both authors). Sandahl Local Fauna, McPherson County, Kansas—Holman (1971, 1995b), Sanchiz (1998). Slaton Local Fauna, Lubbock County, Texas—Holman (1969a, c, 1995b), Sanchiz (1998) (recorded as *Bufo cognatus* or *Bufo speciosus* by both authors).

The Great Plains Toad, *Bufo cognatus*, is a moderately large species with very large blotches that contain many warts. These blotches have distinct light borders. The coloration ranges from brown to gray and from green to yellowish. The cranial crests are widely separated posteriorly, but they converge anteriorly to form a boss on the snout. This is a grassland species that burrows well and is normally nocturnal in its habits. At present, this species occurs in the Great Plains in southern Manitoba and Saskatchewan, Canada; south through Texas to San Luis Potosi, Mexico; west to southeastern California; and then south to Sinaloa, Mexico (Frost, 1985).

Identification of Fossils. Tihen (1962b) gave skeletal characteristic that are helpful in the identification of *Bufo cognatus* fossils. The frontoparietals are distinctive: the supraorbital crests converge anteriorly, obliquely traversing the dorsal surface of these dermal roofing bones of the skull. A nasal boss is a prominent feature on completely articulated skulls but is less noticeable on individual nasal bones, where it can be recognized mainly as a thickening of the bone near its medial border. The ilium is weakly curved. The dorsal prominence is large and strongly developed. The prominence varies in height from about 40% the length of its base in small specimens to about 50% in large ones. The anterior and posterior slopes of this prominence are often subequal, but the anterior slope is occasionally distinctly steeper. The ilia of *Bufo cognatus*, *Bufo compactilis*, and *Bufo speciosus*, however, are difficult to distinguish from one another (see Tihen, 1962b, pp. 26–31).

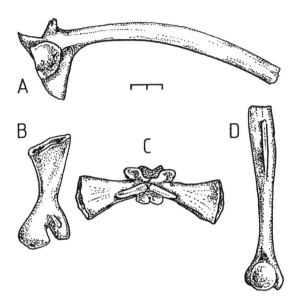

FIGURE 48. Bones of modern *Bufo fowleri* from Jefferson County, Alabama. (A) Right ilium in lateral view. (B) Right scapula in medial view. (C) Sacrum in dorsal view. (D) Right humerus in ventral view. Scale bar = 4 mm and applies to all figures.

BUFO FOWLERI HINCKLEY, 1882

FOWLER'S TOAD

(FIG. 48)

Fossil Localities. **Pleistocene (Irvingtonian NALMA):** Conard Fissure, Newton County, Arkansas—Tihen (1962b), Holman (1995b), Sanchiz (1998). Cumberland Cave, Allegany County, Maryland—Holman (1977c, 1995b), Sanchiz (1998). Hamilton Cave, Pendleton County, West Virginia—Holman and Grady (1989), Holman (1995b), Sanchiz (1998). **Pleistocene (Rancholabrean NALMA):** Baker Bluff Cave, Sullivan County, Tennessee—Van Dam (1978), Fay (1988), Holman (1995b), Sanchiz (1998). Bootlegger Sink, York County, Pennsylvania—Guilday et al. (1966), Holman (1995b), Sanchiz (1998). Bradenton site, Manatee County, Florida—Tihen (1962b), Holman (1995b), Sanchiz (1998). Cheek Bend Cave, Maury County, Tennessee—Klippel and Parmalee (1982), Holman (1995b), Sanchiz (1998) (recorded as *Bufo* cf. *Bufo fowleri* by all authors). Clark's Cave, Bath County, Virginia—Fay (1988), Holman (1995b), Sanchiz (1998). Devil's Den chamber 3, Levy County, Florida—Holman (1978b, 1995b), Sanchiz (1998). Guy Wilson Cave, Sullivan County, Tennessee—L. P. Fay (written communication, May 12, 1993),[3] Holman (1995b), Sanchiz (1998). Haile (Rancholabrean complex), Alachua County, Florida—Tihen (1962b), Holman (1995b), Sanchiz (1998). Ichetucknee River site, Columbia County, Florida—Tihen (1962b), Holman (1995b), Sanchiz (1998). Kanapaha I site, Alachua County, Florida—Tihen (1962b), Holman (1995b). Kendrick IA site, Marion County, Florida—Tihen (1962b), Holman (1995b), Sanchiz (1998). Kingston Saltpeter Cave, Bartow County, Georgia—Fay (1988), Holman (1995b), Sanchiz (1998). Natural Chimneys site, Augusta County, Virginia—Fay (1988), Holman (1995b), Sanchiz (1998). New Paris 4 site, Bedford County, Pennsylvania—Fay (1988), Holman (1995b), Sanchiz (1998). Reddick I site, Marion County, Florida—Tihen (1962b), Wilson (1975), Holman (1995b), Sanchiz (1998). Saltville site, Saltville Valley, Virginia—Holman and McDonald (1986), Holman (1995b), Sanchiz (1998). Strait Canyon Fissure, Highland County, Virginia—Fay (1984), Holman (1995b), Sanchiz (1998). Zoo Cave, Taney County, Missouri—Holman (1974, 1995b), Hood and Hawksley (1975), Sanchiz (1998).

Bufo fowleri was long recognized as a subspecies of *Bufo woodhousii* until it was finally realized that the two former subspecies, *Bufo woodhousii woodhousii* and *Bufo woodhousii fowleri*, are distinct species, given their skeletal and external morphological characters. This is reflected in the Pleistocene literature, where these were the only North American anuran subspecies that could consistently be identified as fossils (see Holman, 1995b).

Living *Bufo fowleri* are sometimes difficult to distinguish from *Bufo americanus*, with which they occasionally hybridize. Nevertheless, the two species can usually be separated on the basis of the number of warts in its largest dark spots: *B. fowleri* has three or more warts in each, whereas *Bufo americanus* has only one or two. The call, however, easily distinguishes the two.

Bufo fowleri tends to occupy a wide variety of habitats in the eastern coastal plain of the United States, but in inland situations it tends to occur in sandy areas near ponds, lakes, and rivers. The modern range of *B. fowleri* is from central New England to the Gulf Coast of the United States, and west to Michigan, northwestern Arkansas, and eastern Louisiana; it is absent from the southern part of the Atlantic coastal plain and from most of Florida (Conant and Collins, 1998).

Identification of Fossils. Ilial characters that separate *Bufo fowleri* from *Bufo americanus* were given in the section on *B. americanus*. In the southeastern United States where *B. americanus* is replaced by *Bufo terrestris*, *B. fowleri* can be separated from *B. terrestris* on the basis of having a much higher and better developed ilial prominence.

BUFO HEMIOPHRYS COPE, 1886
CANADIAN TOAD

Fossil Localities. **Pleistocene (Irvingtonian NALMA):** Courland Canal Fauna, Jewell County, Kansas—Rogers (1982), Holman (1995b), Sanchiz (1998). Cudahy Fauna, Meade County, Kansas—Tihen (1962b), Gehlbach (1965), Holman (1995b), Sanchiz (1998) (recorded as *Bufo* cf. *Bufo hemiophrys* by all authors). Hall Ash Fauna, Jewell County, Kansas—Rogers (1982), Holman (1995b), Sanchiz (1998). Vera Local Fauna, Knox County, Texas—Parmley (1988a), Holman (1995b), Sanchiz (1998) (recorded as *Bufo* cf. *B. hemiophrys* by all authors). **Pleistocene (Rancholabrean NALMA):** Medicine Hat Fauna 4, Medicine Hat, Alberta, Canada—Harington (1978), Sanchiz (1998).

The Canadian Toad, *Bufo hemiophrys*, is a rather large species with a boss between the eyes that is so large that it obscures the cranial crests and extends from the snout back to the posterior edge of the eyelids. The coloration of the upper body is usually brownish or greenish but can sometimes be reddish. The margins of the parotoid glands are indistinct and cause the glands to blend in with the skin itself. This toad is much more aquatic than most North American *Bufo* species and is often found at the edge of ponds or small lakes, often swimming out into the water when disturbed. This species occurs from eastern South Dakota and western Minnesota to western Ontario, eastern Alberta, and the extreme southern Northwest Territories, Canada; isolated populations occur in southeastern Wyoming (Frost, 1985).

Identification of Fossils. Tihen (1962b) provided information that is helpful in the identification of fossils of this species. The frontoparietals are distinctive: they bear high, thin crests, with a unique transverse crest that extends medially from and at right angles to the supraorbital crest at about the level of its junction with the postorbital crest. In the ilium, the dorsal prominence varies in height from 30% of the length of its base in small individuals to nearly 45% of the length of its base in some larger individuals. The anterior slope of the dorsal prominence is distinctly steeper than the posterior slope, and this steepness is more pronounced in large individuals than in small ones.

General Remarks. The occurrence of this northern species in the Irvingtonian sites in Kansas and Texas, far south of its modern range, is noteworthy.

Bufo hibbardi Taylor, 1936
(Fig. 49)

Holotype. A sacral vertebra, complete except for the terminal portions of the diapophyses (Museum of Natural History, University of Kansas: KU 1437).

Etymology. The specific name recognizes North American vertebrate paleontologist Claude W. Hibbard.

Locality and Horizon. Edson beds, Sherman County, Kansas: Late Miocene (Hemphillian NALMA). Ogallala Formation.

Other Material. Referred elements from the type locality include frontoparietals, presacral vertebrae II and IV, sacra, a scapula, a humerus, a radio-ulna, ilia, a femur, and a tibiofibula referred to KU 6395–6417 (Tihen, 1962b). Records of *Bufo hibbardi* are extensive compared with those of other fossil species of *Bufo* and include material from the following Miocene sites. Egelhoff Local Fauna, Keya Paha County, Nebraska (medial Barstovian NALMA)—one left ilium (University of Michigan Museum of Paleontology, Vertebrate Collection: UMMPV 59955) and two right ilia (UMMPV 59956), referred to *Bufo* cf. *Bufo hibbardi* by Chantell (1971); and two left and one right ilia (Michigan State University, Museum Vertebrate Paleontology Collection: MSUVP 1100), referred to *Bufo hibbardi* by Holman (1987). Hottell Ranch rhino quarries, Banner County, Nebraska (medial Barstovian NALMA)—two right ilia (University of Nebraska State Museum: UNSM 93591), referred to *Bufo* cf. *B. hibbardi* by Voorhies et al. (1987). Norden Bridge Local Fauna, Brown County, Nebraska (medial Barstovian NALMA)—two ilia and fragments of limb bones (UNSM 61018), referred to *Bufo* cf. *B. hibbardi* by Estes and Tihen (1964); and one right ilium (MSUVP 980), referred to *Bufo hibbardi* by Holman (1982a). Niobrara River site (University of Chicago: UC locality V-3218, probably one of the "Valentine Quarries"), Cherry County, Nebraska (late Barstovian NALMA)—13 ilia, 1 sacrum, and limb fragments (UC 65852), referred

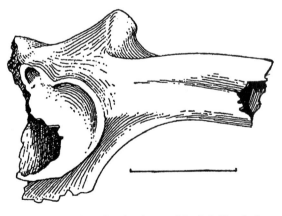

Figure 49. Referred right ilium of *Bufo hibbardi* from the Late Miocene (Hemphillian) of Kansas. Scale bar = 5 mm.

to *Bufo* cf. *B. hibbardi* by Estes and Tihen (1964). WaKeeney Local Fauna, Trego County, Kansas (Clarendonian NALMA)—two left and three right ilia (MSUVP 759), referred to *Bufo hibbardi* by Holman (1975). Lemoyne Quarry, Keith County, Nebraska (Hemphillian NALMA)—one right ilium (MSUVP 1234) and one right ilium (UNSM 96400), assigned to *Bufo* cf. *B. hibbardi* by Parmley (1992).

Diagnosis. This diagnosis is slightly modified from those of Sanchiz (1998) and Tihen (1962b). A large species of the *Bufo americanus* species group with heavy, porous to granular supraorbital and postorbital crests; almost all the posterior portion of the frontoparietal, including the otic plate, is covered by the material of the crests; dorsal ilial prominence high and about 45% of the length of its base, with the anterior slope of the prominence much steeper than the posterior slope; sacral centrum relatively long and narrow.

Description. This description is modified from Tihen (1962b). All of this material is from the type locality. The two frontoparietals both have heavy crests that are apparently typical of this species. The entire dorsal surface of these crests is roughened, and almost the entire surface of the bone at the level of and posterior to the postorbital crests is covered by material of the crests. The supraorbital and postorbital crests meet at a slightly obtuse angle. There is no occipital crest as such, but the thick supraorbital and postorbital crests themselves occupy the area adjacent to and including that normally occupied by the occipital crest. The anterior part of the supraorbital crests is thinner than the posterior part, but both the anterior and the posterior parts in *Bufo hibbardi* are still heavier than those parts in *Bufo woodhousii* or *Bufo americanus.*

Turning to the sacral vertebrae, we find that the width of the sacral vertebrae at the condyles varies from 116 to 127% of the length of the sacral centrum. The total variation in width of the anterior end of the centrum, expressed as a proportion of the length of the centrum, ranges from 88 to 91%; the height of the anterior end of the centrum varies from 61 to 68% of its length. In one specimen, the crests in the neural arch meet at a definite but obtuse angle, although they are somewhat broadly thickened medially, and their posterior border forms a broadly curved angle. In other specimens they meet in a broad, slightly angular arch. There is a low median longitudinal crest anterior to the transverse crest. There is no longitudinal crest or only a faint one posterior to the transverse ones.

In the ilium, the dorsal shaft is moderately curved. In all specimens that have a complete dorsal prominence, the height of the prominence varies only from 43 to 48% of the length of its base. All have a characteristic shape that is somewhat similar to that of *Bufo americanus.* The posterior slope of the prominence is very even and not particularly steep. The anterior slope of the prominence is steep dorsally, with a sharp inflection about halfway between the peak of the prominence and the dorsal edge of the shaft, becoming much less steep at this point and forming a tiny web between the ventral half of the prominence and the shaft.

General Remarks. Tihen (1962b) synonymized *Bufo arenarius* Taylor, 1936 with *Bufo hibbardi* and incorporated all of the elements originally referred to *B. arenarius* into *B. hibbardi*.

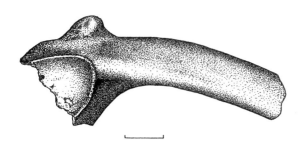

FIGURE 50. Holotype right ilium of *Bufo holmani* from the Late Miocene (Hemphillian) of Nebraska. Scale bar = 5 mm.

Bufo holmani PARMLEY, 1992
(FIG. 50)

Holotype. A right ilium (University of Nebraska State Museum: UNSM 56902).

Etymology. The specific name recognizes North American vertebrate paleontologist J. Alan Holman.

Locality and Horizon. Devils Nest Airstrip site, Knox County, Nebraska: Late Miocene (late Hemphillian NALMA). Ash Hollow Formation.

Other Material. A right ilium UNSM 56943 from the Late Miocene (late Hemphillian NALMA) Santee Local Fauna of Knox County, Nebraska, was designated as a paratype by Parmley (1992).

Diagnosis. The diagnosis is directly from Parmley (1992, pp. 275–276): "A distinctive, large-sized bufonid that differs from all other known living and fossil North American bufonid species in the following combination of ilial characters: (1) ilium large and extremely robust; (2) dorsal prominence–protuberance complex [= dorsal prominence in this volume] long, moderately high and smooth; (3) acetabulum large and moderately shallow; and (4) ilial shaft compressed along upper two-thirds, but wider and ridge-like along the lower one-third."

Description. The description is slightly modified from Parmley (1992). In lateral view, the holotype has the acetabulum large and moderately shallow, with well-extruded margins. The major part of the dorsal acetabular expansion is missing. The distal end of the ventral acetabular expansion is missing, but the portion ventral to the acetabular cup is wide. A well-marked foramen occurs at the edge of the ventral acetabular expansion immediately anterior to the acetabular cup (this foramen is not obvious in the rendering of the holotype in Parmley (1992); see Fig. 50, this volume). The dorsal prominence is smooth, moderately high, and long and rounded, with the anterior angle steep and the posterior angle long and gentle in its descent. The height of the prominence is 44% of the length of its base. A knoblike, teardrop-shaped protuberance extends laterally from the prominence. The ilial shaft is compressed along the upper two-thirds but wider along the lower one third. In dorsal view, the ilial shaft is compressed laterally. In this view the knob-like protuberance of the dorsal prominence is laterally deflected. Measurements of the holotype are as follows: estimated length of straight line, 72.2 mm; height of acetabular fossa, 10.2 mm; width of acetabular fossa, 7.5 mm; height of dorsal prominence, 7.0 mm; length of dorsal prominence, 3.1 mm; height of ilial shaft, 6.3 mm; width of ilial shaft, 3.7 mm.

Turning to the paratype, we find that the ilium is smaller and more worn than the holotype, but it is similar in having a wide, robust ilial shaft, a moderately high and long dorsal prominence, and a large, shallow acetabular cup. It differs from the holotype in having a less angular dorsal prominence that is 36% of the length of the base. Measurements of the paratype are as follows: estimated length of straight line, 68.1 mm; height of acetabular fossa, 10.5 mm; width of acetabular fossa, 7.2 mm; height of dorsal prominence, 6.9 mm; length of dorsal prominence, 2.5 mm; height of ilial shaft, 5.9 mm; width of ilial shaft, 3.6 mm.

General Remarks. Parmley (1992) pointed out that *Bufo holmani* does not bear any clear relationship to any of Tihen's species groups of the genus *Bufo*. Actually, it is difficult to do this comparison in many extinct species of *Bufo*, as the most diagnostic characters that Tihen used were the cranial differences that occur in various species groups, and good cranial elements of *Bufo* are relatively rare in the North American fossil record.

Nevertheless, Parmley (1992) pointed out the following differences between *Bufo holmani* and other large, extinct and living species of North American *Bufo*. *Bufo holmani* differs from the large Miocene (Barstovian NALMA) species *Bufo kuhrei* Holman in having a higher dorsal prominence, a smaller acetabular cup, and a wider ventral acetabular expansion. It differs from large individuals of *Bufo marinus* in having a higher, less rugose dorsal prominence–protuberance complex; and from large *Bufo woodhousii* (including the large extinct subspecies *Bufo woodhousii bexarensis* Mecham) in having a lower, less-angled dorsal prominence and a larger acetabular cup. *Bufo holmani* differs from the very large, living southwestern species *Bufo alvarius* in having a higher and longer dorsal prominence with a steeper posterior angle and in having a less angular ilial shaft.

BUFO CF. *BUFO KELLOGGI* TAYLOR, 1938 "1936"
LITTLE MEXICAN TOAD

Fossil Locality. **Pleistocene (Rancholabrean NALMA):** Rancho la Brisca locality, Sonora Mexico—Van Devender et al. (1985), Sanchiz (1998).

The Little Mexican Toads, which are a little over an inch long (about 25–35 mm SVL), are the smallest toad of the *Bufo punctatus* group of Ferguson and Lowe (1969), which also includes *Bufo punctatus*, *Bufo retiformis*, and *Bufo debilis*. At present, *Bufo kelloggi* occurs in Mexico in coastal subtropical lowlands from Nayarit to as far north as central Sonora in the area of Hermosillo, and with an outlying population near Santa Ana (Hulse, 1977).

Identification of Fossils. Van Devender et al. (1985) commented on the identification of humeri from the Rancho la Brisca locality above as modified here. The fossils are from specimens of 30–35 mm SVL. The humeri of *Bufo retiformis* and *Bufo punctatus* are not as well ossified as those as *Bufo kelloggi* of the same size. The size of *Leptodactylus melanonotus* humeri is similar to that of *B. kelloggi* humeri but have better developed medial crests, more flattened shafts between the middle and radial condyles, and equally developed the lateral crests and epicondyle.

General Remarks. The Rancho la Brisca locality is northwest of the present known range of the species.

**BUFO KUHREI* HOLMAN, 1973
(FIG. 51)

Holotype. A left ilium (Michigan State University, Museum Vertebrate Paleontology Collection: MSUVP 719).

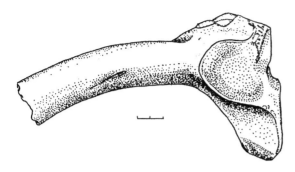

FIGURE 51. Holotype left ilium of *Bufo kuhrei* from the Middle Miocene (medial Barstovian) of Nebraska. Scale bar = 2 mm.

Etymology. The specific name recognizes Loring Kuhre of Ainsworth, Nebraska, host of many Michigan State University collecting trips.

Locality and Horizon. Norden Bridge Local Fauna, Brown County, Nebraska: Middle Miocene (medial Barstovian NALMA). Valentine Formation.

Diagnosis. This diagnosis is modified from Holman (1973b) A large *Bufo* that can be readily distinguished from other fossil and Recent species by the following combinations of ilial characters: (1) ilial prominence long and low, having its height only 17% the length of its base; (2) two tubercles occurring on the summit of the ilial prominence, a smaller anterior one and a larger posterior one, neither produced laterally; (3) dorsal prominence with small web anterior to the most anterior tubercle; (4) acetabular fossa extending almost to the anterior edge of the ventral acetabular expansion; (5) acetabular cup moderately excavated, with a pronounced lip occurring on approximately its ventral half; (6) a large, elongate, roughened tubercular bridge produced from the posterodorsal margin of the acetabular cup and extending posterodorsally; and (7) in medial view, a large tubercle occurring on the medial part of the acetabular region in the posterodorsal part of the bone.

Description. The description is modified from Holman (1973b). In lateral view, the acetabular cup is large and extends just slightly posterior to the anterior edge of the ventral acetabular expansion. This cup is moderately well excavated and has a produced lip that occurs on approximately its ventral half. The dorsal acetabular expansion is mainly missing. The ventral acetabular expansion is truncated anteroventrally. The dorsal ilial prominence is long and low, and its height is only 17% of the length of its base. Two marked tubercles occur on the summit of the ilial prominence: a smaller, anterior one and a larger, posterior one (the drawing of *Bufo kuhrei* in Sanchiz, 1998, fig. 116, p. 79, is somewhat incorrect, as it depicts a broken-off tubercular area). These tubercles are dorsally rather than laterally produced, as is the single tubercle of *Bufo holmani*. The part of the dorsal prominence anterior to the most anterior tubercle forms a small web. The anterior and posterior slopes of the ilial prominence are gentle and about equal in length. The ilial shaft is without a crest or blade. The ilial shaft is somewhat rounded laterally but is more flattened medially. In medial view, there is a large tubercle that occurs in the medial part of the acetabular region in the posterodorsal part of the bone. Measurements are as follows: length of ilial prominence, 5.8 mm; height of ilial prominence, 1.0 mm; distance from end of ventral acetabular expansion through summit of ilial prominence, 11.3 mm; height of acetabular cup, 5.2 mm; height of ilial shaft, 3.7 mm.

General Remarks. It is difficult to suggest the relationships of *Bufo kuhrei* to other *Bufo* groups. The low ilial prominence of *B. kuhrei* is suggestive of the *Bufo valliceps* group of Tihen (1962b). Evolutionary diversity of taxa swells and shrinks from time to time in response to climatic change and other volatile conditions, and obviously some taxa that evolve during the swell phase can become quickly extinct; see the account of the genus *Terrapene* in Florida during the Pleistocene and comments on this by Holman (1995b). For this reason it is quite possible that both *B. kuhrei* and *Bufo holmani* were short-lived, dead-end forms. Obviously, if specialized species come and go, as I strongly believe the fossil record indicates, it will continue to be difficult to assign fossil taxa to specific modern species groups, especially if these novelties of evolution are based on individual elements. One can, of course, be ultra-conservative and assign these forms to *Bufo* sp., but this tends to bury such fossils in the literature and obscure the recognition of the dynamic nature of the evolutionary process.

BUFO MARINUS (LINNAEUS, 1758)
CANE TOAD
(FIG. 52)

Fossil Locality. **Miocene (Clarendonian NALMA):** WaKeeney Local Fauna, Trego County, Kansas—Wilson (1968), Holman (1975).

The Cane Toad, *Bufo marinus*, is an immense toad that reaches a length of 238 mm in South America but probably not more than 178 mm in the United States (Conant and Collins, 1998). It can be immediately recognized by its huge, deeply pitted parotoid glands, which extend far down the sides of the body. These glands are highly toxic to many mammals that are unfortunate enough to bite these toads. The Cane Toad, which barely reaches extreme southern Texas in the United States at present, extends southward through the Amazon Basin in South America. It has been widely introduced into the tropics of the Old World and can be an unwanted pest. In the tropics, at least, the Cane Toad is an animal that occupies a wide variety of habitats.

Identification of Fossils. Wilson (1968) and Holman (1975) identified this semitropical and tropical animal from the Miocene (Clarendonian NALMA) of Kansas. Wilson (1968) identified a frontoparietal, a temporal plate, the distal portion of a right ilium, a left ilium, and two nasal fragments from the WaKeeney Local Fauna as belonging to *Bufo marinus*. The most diagnostic element in this identification was the frontoparietal, discussed as follows by Wilson (1968, p. 95): "Ventral aspects of the frontoparietal are morphologically identical to the Recent species of *B. marinus* that were examined. The slope of the ventral supraorbital shelf, the presence of a groove for the occipital artery at the margin of the inward sloping frontoparietal laminae, terminating at the lateral margin of the occipital crest, and the extreme width of the orbital shelf are characteristic of the fossil element and of *B. marinus*. The fossil frontoparietal and those of *B. marinus* appear identical in size and shape. Separation of *B. marinus* and *B. valliceps* Wiegmann is based on the frontoparietal, particularly by the absence

FIGURE 52. Left ilium of modern *Bufo marinus*. Scale bar = 15 mm.

of a definite supraorbital crest and instead a dorsal outward slope in this region in the former. In *B. valliceps* the ventral surface of the supraorbital areas has no pronounced occipital groove as noted in *B. marinus*. The fossil frontoparietal has a greater width than comparable-sized individuals of *B. alvarius* Girard [another giant toad from southwestern United States], and is more massive and with less slope to the supraoccipital crest. Dermal encrustation is only slightly developed in the medial region of the fossil element, whereas in specimens of *B. alvarius* the ornamentation is well developed toward the center of the skull. However, ventrally the fossil specimen and the frontoparietal of *B. alvarius* are similar except the occipital groove in *B. marinus* trends more dorsad and laterad from the laminae than in *B. alvarius*." Remarking on the other specimens from the site, Wilson (1968, p. 95) stated that "although the number of specimens do not permit determination of a size range they do indicate a body length of about 110 mm."

Holman (1975) assigned five left and six right ilia, three distal humeri, and one puboischium from the WaKeeney Local Fauna to *Bufo marinus*. Commenting on the frontoparietal identified by Wilson (1968), Holman (1975) stated that although this frontoparietal was smaller than in Recent adult *B. marinus*, it did compare remarkably well with this species, even in minute detail. Relative to both the ilia of Wilson (1968) and Holman (1975), Holman stated that the ilia (Michigan State University, Museum Vertebrate Paleontology Collection: MSUVP 757; and University of Michigan Museum of Paleontology, Vertebrate Collection: UMMPV55441) were identical to *B. marinus* except that the fossils represent smaller animals. Comparisons of the measurements of fossil *B. marinus* and those of modern species are as follows: height from ventral border of acetabular cup through tip of dorsal prominence of ilium in fossil *B. marinus* (Holman, 1975), 7.0–10.1 mm (mean, 8.7 mm; $n = 9$); and in Recent *B. marinus*, 10.1–13. 2 mm (mean, 10.98 mm; $n = 12$); width of the frontoparietal in fossil *B. marinus* (Wilson, 1968), 6.9 mm; and in Recent *B. marinus*, 8.4–10.0 mm (mean, 9.32 mm; $n = 13$).

General Remarks. The presence of a tortoise (*Hesperotestudo orthopygia*) of giant proportions in the Clarendonian WaKeeney Local Fauna of Kansas indicates a climate with mild winters and temperatures seldom if ever reaching the freezing point (Hibbard, 1960). If this is true, then the Cane Toad would have had no trouble adjusting to the climate of the time, as we know that extreme southern Texas, where *Bufo marinus* currently occurs, is not spared freezing temperatures from time to time. It is somewhat odd that this species has not been reported from other Miocene localities in the High Plains area. Nevertheless, I will take the devil's advocate stance here until the Kansas *B. marinus* identification can be proven incorrect.

BUFO MAZATLANENSIS TAYLOR, 1940 "1939"
SINALOA TOAD

Fossil Locality. **Pleistocene (Rancholabrean NALMA):** Rancho la Brisca locality, Sonora, Mexico—Van Devender et al. (1985), Sanchiz (1998).

The Sinaloa Toad is rather large, reaching a length of about 80 or

90 mm. It is a member of the *Bufo valliceps* species group. This species is restricted to the tropical lowlands of northwestern Mexico, from southern Sinaloa northward to central Sonora (Van Devender et al., 1985). This is the first fossil record of *Bufo mazatlanensis* as a fossil. The specific identification is based on a right basioccipital–frontoparietal, two left humeri, and two left and two right ilia (Van Devender et al., 1985).

Identification of Fossils. Comments on the identification of the fossil material from the Rancho la Brisca locality are modified from Van Devender et al. (1985). The fused basioccipital–frontoparietal portion of the skull represents a toad with an SVL of about 90 mm. The frontoparietal is broad and moderately rugose and does not have a well-developed ridge for cranial crests. *Bufo alvarius* also has a broad frontoparietal without a crest but has a strongly papillose or echinate (spiny) dorsal surface. The frontoparietal of *Bufo cognatus* is narrow and has a well-developed cranial crest. The condyloid fossa lateral to the occipital condyle is smaller in *Bufo mazatlanensis* than in *B. alvarius*. The humeri identified as *B. mazatlanensis* are from a female with an SVL of about 65 mm and a male with an SVL of about 82 mm. The radial condyle is well rounded, and the ulnar (medial) condyle is well developed. The lateral crest and lateral epicondyle are present but not strongly produced.

The four ilia that are identified as *Bufo mazatlanensis* are all from toads that would have had an SVL of about 75–88 mm. All these ilia have low, broad dorsal prominences that are fairly flat on top and are either rugose or non-rugose. The ilia have a broad, deep ventral acetabular expansion. The dorsal prominences of *Bufo alvarius*, *Bufo cognatus*, *Bufo woodhousii*, and *Bufo microscaphus* are higher. The ventral acetabular expansion of *B. alvarius* is deep, but it is not broad as in *B. mazatlanensis*.

General Remarks. Other bones (ethmoids, a basioccipital, an atlas, a sacral vertebra, a urostyle, a humerus, radio-ulnae, and tibiofibulae) that were identified as *Bufo* sp. from the Rancho la Brisca locality could all be from *Bufo mazatlanensis* (Van Devender et al., 1985).

BUFO PLIOCOMPACTILIS WILSON, 1968
(FIG. 53)

Holotype. A right frontoparietal (University of Michigan Museum of Paleontology, Vertebrate Collection: UMMPV 55415).

Etymology. The specific name refers to the supposed Pliocene (actually Miocene, Clarendonian) relationships of the fossil species to the modern *Bufo compactilis*.

Locality and Horizon. WaKeeney Local Fauna, Trego County, Kansas: Miocene (Clarendonian NALMA). Ogallala Formation.

Other Material. Wilson (1968) designated a nearly complete left frontoparietal

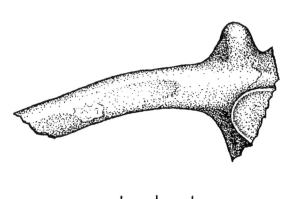

FIGURE 53. Left ilium of *Bufo pliocompactilis* from the Late Miocene (Hemphillian) of Nebraska. Scale bar = 2 mm.

(UMMPV 55431) as a paratype and 15 left and 9 right ilia (under UMMPV 55432, 55433, 55434) as referred material, all of this material being from the type locality. Holman (1975) collected an additional 40 left and 47 right ilia (all under Michigan State University, Museum Vertebrate Paleontology Collection: MSUVP 765) from the type locality and assigned these to B. pliocompactilis. Holman (1973a) also assigned two right and one left ilia (South Dakota School of Mines: SDSM 68149) from the Mission Local Fauna, Mellette County, South Dakota (Miocene, Clarendonian NALMA) to *Bufo pliocompactilis*. Parmley (1992) extended the temporal range of this species forward in the Miocene to the middle Hemphillian NALMA on the basis of two left ilia (University of Nebraska State Museum: UNSM 96410) from the Lemoyne Quarry, Keith County, Nebraska. Parmley and Peck (2002) extended both the temporal and the geographic range of *Bufo compactilis* on the basis of a late Hemphillian record of two left ilia (University of Arizona Laboratory of Paleontology: UALP 23405, 23417) from the White Cone Local Fauna (Bidahochi Formation) of Navajo County, Arizona.

Emended Diagnosis. This diagnosis is from Wilson (1968) and from my observations on the ilia of this species. Frontoparietals and ilia represent a small *Bufo* with an SVL of about 25–40 mm. Frontoparietals with supra- and postorbital ridges absent and with dermal sculpturing most pronounced toward the lateral border. Orbital shelf with its greatest overlap at the posterior corner of the orbit and covering the anterior opening to the occipital canal. Ventrally, frontoparietal includes an oval depression surrounded by a low ridge where a portion of the prootic is attached to these roofing bones. Perpendicular laminae of both frontoparietals define the shape of the supra- and postorbital area and form an angle of about 120 degrees. Ilia small and compact with no evidence of juvenile pitting in the area in and around acetabular cup. Ilial prominence very high, with its central portion an elongated tubercle that is straight up or slightly reflected anteriorly and with webbed portions forming its anterior and posterior slopes. Anterior margin of ventral acetabular expansion extends well beyond anterior margin of acetabular cup. Ilial shaft slightly ventrally curved to almost straight.

General Remarks. This is one of the most distinctive extinct North American fossil *Bufo* species. The ilial morphology most closely resembles that of the living southeastern species *Bufo quercicus* and the extinct Miocene (Hemphillian) *Bufo tiheni* from Florida. The complete ossification in all the recovered elements of *Bufo pliocompactilis* shows that the fossil assemblages represented adult rather than juvenile populations. The abundance of *B. pliocompactilis* in the Miocene (Clarendonian NALMA) WaKeeney Local Fauna of Kansas (a minimum of at least 56 individuals is represented) is noteworthy. Of interest is the fact that adult *Bufo quercicus* currently occur in dense populations in the southeast.

These populations can be so dense, in fact, that in Florida groups of cattle egrets sometimes assemble to feed on them in grassy areas, the same way that these birds assemble to feed on swarms of grasshoppers and other insects (personal observations). *Bufo pliocompactilis* ranges in the Miocene from Clarendonian to mid-Hemphillian NALMAs and is the most easily identified anuran species in the fossil localities in which it occurs.

Sanchiz (1998) included *Bufo pliocompactilis* in his nomina dubia section, suggesting that it was similar to and probably synonymous with *Bufo compactilis*, a form that is currently restricted in its range to the central highlands of Mexico. I reject this designation here, as *B. pliocompactilis* should be recognized as a valid species. Parmley and Peck (2002) supported the retention of *B. pliocompactilis* as a distinct taxon on the basis that it has a higher, more robust dorsal prominence–protuberance complex, longer subacetabular expansions, and slightly larger acetabular fossae than Recent *B. compactilis*.

Bufo praevius Tihen, 1951

(Fig. 54)

Holotype. The distal 13 mm of a right ilium (Museum of Comparative Zoology, Harvard University: MCZ 1991).

Etymology. The specific name is from the Latin *praevius*, "going before, preceding."

Locality and Horizon. Thomas Farm Local Fauna, Gilchrist County, Florida: Miocene (Hemingfordian NALMA). Torreya Formation.

Other Material. Tihen (1951) assigned the following material from the type Thomas Farm Local Fauna to *Bufo praevius*: eight ilia (MCZ 1992), two tibiofibulae (MCZ 2000), five humeri (MCZ 2002), three urostyles (MCZ 1995), one femur (MCZ 2005), and one atlas and three other presacral vertebrae (MCZ 1996). Other elements listed by Tihen (1951, p. 230) may belong to this species. Tihen (1962b) reported that numerous other elements of that species had been recovered, most of which were in the collections of the University of Florida and the Museum of Comparative Zoology. Auffenberg (1956) briefly discussed *B. praevius* and reported a few additional elements. Holman (1967a) discussed more material of *B. praevius* and reported on an additional 391 ilia (175 left and 216 right) and hundreds of additional postcranial bones, mainly from the Florida Geological Survey (these bones now reside in the Florida Museum of Natural History), but some were from the Museum of Comparative Zoology.

Diagnosis. This diagnosis is quoted from Sanchiz (1998, p. 80) after Tihen (1962a, b): "Moderately-sized *Bufo* of the *B. valliceps* group. Skull with high, granular supraorbital crests, oriented almost vertically and forming a sharp angle with the postorbital crests; frontoparietals fused with prootics, ilium with tuber superior [dorsal prominence] with a height about 20% of its base length."

Description. Tihen (1951) gave a thorough description of the most diagnostic elements of *Bufo praevius* known at that time, but his comments based on the additional material (Tihen, 1962b) are more relevant to the relationships of that species. Modifications of those comments follow. In texture, the cranial crests are distinctly granular as in most of the *Bufo valliceps* group and few of the *Bufo americanus*

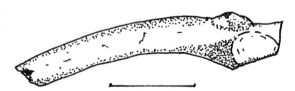

FIGURE 54. Holotype left ilium of *Bufo praevius* from the Early Miocene (Hemingfordian) Thomas Farm locality of Florida. Scale bar = 5 mm.

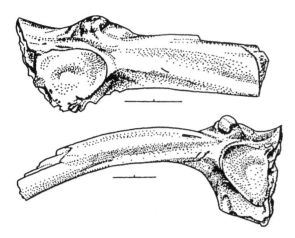

FIGURE 55. Two ilia (in lateral view) from the Early Miocene (Hemingfordian) Thomas Farm locality, assigned to *Bufo praevius* by Holman (1967a). These ilia have rounded tubercles on the dorsal prominences, unlike those in the type series figured by Tihen (1951), and may represent another species. Scale bars = 2 mm.

group. The dorsal surface of the frontoparietals between these crests is heavily ornamented as is typical of the Caribbean section of the *B. valliceps* group and some members of other sections of this group. Several specimens are sufficiently complete to demonstrate conclusively that the frontoparietals and prootics were fused as in the *B. valliceps* group. The crest of the urostyle is low, in contrast to the most usual condition in the *B. americanus* group. In the numerous ilia that became available by 1962 it was possible to ascertain that the height of the dorsal prominence is consistently about 20% the length of its base. The anterior and posterior slopes of this prominence are subequal, and their angles each average about 28 degrees; the peak angle varies between 115 and 130 degrees, and the dorsal acetabular angle varies between 135 and 145 degrees. These measurements and angles are typical of the Caribbean section of the *B. valliceps* group.

General Remarks. It could be argued that the abundant *Bufo* material at the Thomas Farm Local Fauna represents more than a single species because of the variation in the ilial prominence and a more massive ilium in some individuals (Fig. 55).

BUFO PUNCTATUS BAIRD AND GIRARD, 1852
RED-SPOTTED TOAD
(FIG. 56)

Fossil Localities. **Pleistocene (Rancholabrean NALMA):** Deadman Cave, Pima County, Arizona—Mead et al. (1984), Holman (1995b), Sanchiz (1998). Dry Cave Fauna, Eddy County, New Mexico—Holman (1970b, 1995b), Applegarth (1980), Harris (1987), Sanchiz (1998). Gypsum Cave, Clark County, Nevada—Brattstrom (1958a), Tihen (1962b), Holman (1995b), Sanchiz (1998). Redtail Peaks site, San Bernardino County, California—Van Devender and Meade (1978), Holman (1995b), Sanchiz (1998). Tunnel Ridge site, San Bernardino County, California—Van Devender and Meade (1978), Holman (1995b), Sanchiz (1998). Wolcott Peak, Pima County, Arizona—Van Devender and Mead (1978), Holman (1995b), Sanchiz (1998).

Bufo punctatus is a small toad with a pointed snout and a flattened head and body that bears orange to reddish warts (Stebbins, 1985). Cranial crests are weak or absent in this species. This toad lives in habitats that range from desert scrub to Mexican pine–oak woodlands (Van Devender and Lowe, 1977). It is adapted to xerophytic habitats. Its present-day distribution is southeastern California, Utah, and western Kansas, south to southern Baja California, Sinaloa, Guanajuato, San Luis Potosi, and Tamaulipas, Mexico (Frost, 1985).

Identification of Fossils. Tihen (1962b) description of definitive osteological aspects of *Bufo punctatus* is slightly modified as follows. The skull is very low and broad (reflecting the general flatness of the head and body), its height barely, if at all, exceeding 40% of its length. The cranial crests are obsolete, but there is extensive sculpturing on the roofing bones and on the maxillae, with coossification of the dermis occurring. A limited flange is present on the pterygoid, and a slightly more extensive one occurs on the squamosal. The shaft of the squamosal is short and fails to reach the quadratojugal. The quadratojugal itself is also short and in limited contact with the maxilla, even though the jaw articulation is set far forward. The centra of the vertebrae are slightly depressed. The ilial prominence is rounded and moderately low, and the limb bones are fairly long and slender. The scapula, however, is short and relatively broad.

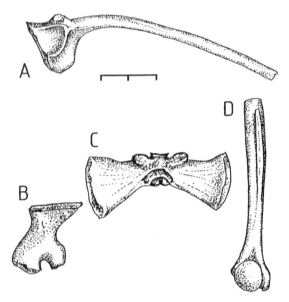

FIGURE 56. Bones of modern *Bufo punctatus* from Uvalde County, Texas. (A) Right ilium in lateral view. (B) Right scapula in medial view. (C) Sacral vertebra in dorsal view. (D) Right humerus in ventral view. Scale bar = 4 mm and applies to all figures.

General Remarks. Using the above criteria, we can easily identify this toad as a Pleistocene fossil within its modern range.

BUFO QUERCICUS HOLBROOK, 1840

OAK TOAD

(FIG. 57)

Fossil Locality. **Pleistocene (Rancholabrean NALMA):** Reddick I site, Marion County, Florida — Tihen (1962b), Holman (1995b), Sanchiz (1998).

Bufo quercicus is a very small toad that usually has an SVL of about 25 mm. This anuran has a conspicuous light stripe down the middle of the back, usually with several dark spots on either side. The name Oak Toad is somewhat of a misnomer, as the animal is most abundant in rather poorly drained pine woodlands. Its present range is in southeastern Louisiana and southeastern Virginia along the coastal plain to peninsular Florida (Frost, 1985).

Identification of Fossils. Tihen (1962b) provided a discussion of the osteology of this diminutive anuran, which is modified as follows. *Bufo quercicus* is osteologically distinct. The skull is not at all depressed. The cranial crests are distinct, but they are low. The sculpturing of the cranium is neither prominent nor extensive. The otoparietal and the prootic are fused, but in the orbit the perpendicular laminae of the frontoparie-

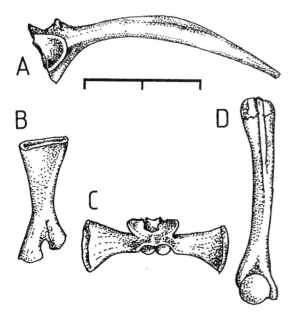

FIGURE 57. Bones of modern *Bufo quercicus* from Orange County, Florida. (A) Right ilium in lateral view. (B) Right scapula in medial view. (C) Sacrum in dorsal view. (D) Right humerus in ventral view. Scale bar = 4 mm and applies to all figures.

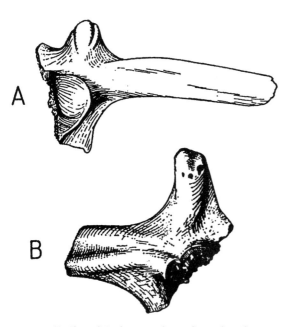

FIGURE 58. Ilia of *Bufo rexroadensis* from the Pliocene (Blancan) of Kansas. (A) Right ilium referred to an immature individual. (B) Left ilium referred to a large adult. Scale not available.

tal are often distinct from the prootic, giving an impression that these elements are not fused. The supraorbital and orbital crests meet at a rounded angle, rather than the broad curve of some species of *Bufo*. The scapula is of normal proportions. The centrum is depressed dorsoventrally. The ilium, however, has a high prominence: the height of the prominence is 27–35% the length of its base (35.4% in Fig. 57). I have noticed that a moderately well defined dorsal crest occurs in the medial part of the ilial shaft.

General Remarks. It is probable that this species is known from only a single site in Florida because of the small size of its bones. Its closest fossil relatives appear to be *Bufo pliocompactilis* of the Miocene (Clarendonian and Hemphillian NALMAs) of Kansas and Nebraska, respectively, and *Bufo tiheni* from the Miocene (Hemphillian) of Florida.

**BUFO REXROADENSIS* TIHEN, 1962

Fig. 58

Holotype. A nearly complete right frontoparietal (University of Michigan Museum of Paleontology, Vertebrate Collection: UMMPV 40139).

Etymology. The specific name reflects the Rexroad Formation of Meade County, Kansas.

Locality and Horizon. Fox Canyon locality, Meade County, Kansas: Pliocene (Blancan NALMA). Rexroad Formation.

Other Material. The museum location and the origin of other elements referred to *Bufo rexroadensis* were given by Tihen (1962b, p. 38): "The majority of the known specimens are from the Fox Canyon locality of the Rexroad formation, Meade County, Kansas. The sacra and ilia from this locality tentatively referred to *B. rexroadensis* include UMMPV Nos. 40140–40146; similarly referred specimens from Rexroad Locality No. 3, also of Meade County, are UMMVP Nos. 40147–40149. In addition to these, unidentifiable elements undoubtedly including representatives of this form are catalogued under UMMVP No. 27673 and Nos. 40156–40160." *Bufo rexroadensis* has also been reported from the Pliocene (Blancan) of the Beck Ranch Local Fauna, Scurry County, Texas (Rogers, 1976; two ilia—Midwestern University: MU 258) and the Pliocene (Blancan) of the Big Springs Quarry, Antelope County, Nebraska (Rogers, 1984; eight ilia—University of Nebraska: UNSM 52054).

Diagnosis. This diagnosis is slightly modified from Tihen (1962b). A moderately large *Bufo* of the *Bufo americanus* species group with heavy supraorbital and postorbital crests meeting each other at an obtuse angle; crests greatly thickened at this point of juncture, but not covering the entire portion of the bone. The ilium has a very high prominence that has subequal anterior and posterior slopes or has the anterior slope slightly steeper. The centrum of the sacral vertebra is short, and the crests of the sacral neural arch usually meet in a broad, subangular arc.

Description. The holotype specimen lacks only the extreme anterior tip and possibly the extreme lateral tip of the otic plate. The length of the bone is 12.4 mm, and the greatest width, including the otic portion, is 8.0 mm. The crests are extensive but slightly less so than in *Bufo hibbardi*. An articular surface may be recognized on the ventral face of the otoparietal, and the perpendicular lamina is mainly intact, with no broken edge, thus indicating that this element was independent from the prootic. Some of the sacra referred to *Bufo rexroadensis* are shorter and wider than those observed in any living forms by Tihen (1962b). The crests on the neural arch of the sacral vertebrae meet medially in a broad but often slightly angular arc.

General Remarks. Tihen (1962b) stated that *Bufo rexroadensis* may have been derived from *Bufo hibbardi* and may have given rise to *Bufo woodhousii*.

Bufo (?) *speciosus* Girard, 1854
Texas Toad
(Fig. 59)

The Texas Toad, *Bufo speciosus* (which occurs in southwestern Kansas and southeastern New Mexico and almost all of Texas down into northern Mexico) is closely related and osteologically similar to the Great Plains Toad, *Bufo cognatus*, with which it sometimes hybridizes. In fact, the Miocene and Pleistocene records that deal with *B. speciosus* refer to this occurrence as *Bufo cognatus* or *Bufo speciosus*. The localities for these records are as follows. **Miocene (Hemphillian NALMA):** Santee site, Knox County, Nebraska—Parmley (1992), Sanchiz (1998). **Pleistocene (Rancholabrean NALMA):** Howard Ranch Local Fauna, Hardeman County, Texas—Holman (1964, 1969c, 1995b), Sanchiz (1998). Slaton Local Fauna, Lubbock County, Texas— Holman (1969a, 1995b), Sanchiz (1998).

FIGURE 59. Left ilium of modern *Bufo speciosus* (Georgia College Collection of Herpetological Skeletons, specimen No. 3011). Scale not available.

Tihen (1962b) tentatively referred ilia to *Bufo speciosus* from three Late Pliocene (Blancan NALMA) localities in Meade County, Kansas (Fox Canyon locality, Rexroad 3 locality, and Wendell Fox Pasture locality).

**BUFO SPONGIFRONS* TIHEN, 1962
(FIG. 60)

Holotype. A virtually complete left frontoparietal (National Museum of Natural History: USNM 22234).

Etymology. The specific name is from the Latin *spongia*, "sponge," and *frons*, "forehead" (in reference to the spongy appearance of the dorsal surface of the type frontoparietal).

Locality and Horizon. Quarry E, near Long Island, Phillips County, Kansas: Miocene (Hemphillian NALMA). Ogallala Formation.

Other Material. Tihen (1962b) referred a right frontoparietal of about the same size as the holotype and an incomplete third specimen (both from the type locality) to *Bufo spongifrons*. USNM numbers were not given for these specimens. Three ilia (all registered under USNM 22236) from the type locality were referred to this taxon. Rogers (1984) recorded *Bufo* cf. *Bufo spongifrons* from the Pliocene (Blancan NALMA) Big Springs Local Fauna, Antelope County, Nebraska, based on eight ilia (University of Nebraska State Museum: UNSM 52057).

Diagnosis. This diagnosis is from Sanchíz (1998, p. 82) after Tihen (1962b): "A moderately sized *Bufo* of the *B. americanus* group. Cranial crests high but not especially heavy; the supraorbital, postorbital, and to a lesser extent the frontoparietal surfaces have a spongy texture; sacral centrum presumably short and broad; tuber superior [dorsal ilial prominence] high, between 45 and 50% of the length of its base, with posterior slope steeper than the anterior one."

Description. According to Tihen (1962b), the three ilia from the type locality referred to *Bufo spongifrons* differ from those referred to *Bufo cognatus* in having a steeper slope to the prominence and, correlatively, having a more acute dorsal acetabular angle. The height of the prominences in the three ilia is 45–50% of the length of the base. The anterior angle is about 45 degrees in all of them; the posterior angle ranges between 55 and 65 degrees; and the peak angle varies between 70 and 80 degrees. The dorsal acetabular expansion angle varies between 95 and 105 degrees.

General Remarks. Tihen (1962b) commented that the affinities of *Bufo spongifrons* are certainly with the western section of the *Bufo americanus* group but that the fossil is not closely related to any particular species within this group.

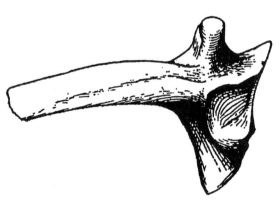

FIGURE 60. Left ilium referred to *Bufo spongifrons* from the Late Miocene (Hemphillian) of Kansas. Scale not available.

BUFO SUSPECTUS TIHEN, 1962

(FIG. 61)

Holotype. A partial right ilium (University of Michigan Museum of Paleontology, Vertebrate Collection: UMMPV 40155.

Etymology. The specific name is from the Latin *suspicio*, "imperfect conception" (probably because the new taxon was based on a single partial ilium).

Other Material. No other material is known.

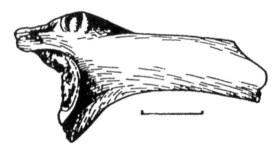

FIGURE 61. Holotype right ilium of *Bufo suspectus* from the Pliocene (Blancan) of Kansas. Scale bar = 2 mm.

Locality and Horizon. Fox Canyon locality, Meade County, Kansas: Pliocene (Blancan NALMA). Rexroad Formation.

Emended Diagnosis. Most similar to *Bufo valentinensis* Estes and Tihen, 1964, from the North American Miocene but differing as follows: (1) the ilial prominence is lower and longer (height of prominence about 17–18% of the length of its base, whereas it's about 25–30 % of the length of the base in *B. valentinensis*); (2) anterior edge of ilial prominence slopes less steeply into the ilial shaft, and (3) anterior edge of acetabular cup extends well behind the anterior edge of the ilial prominence (anterior edge of acetabular cup extends from slightly beyond to slightly behind the anterior edge of the ilial prominence in *B. valentinensis*).

General Remarks. Sanchiz (1998) relegated *Bufo suspectus* to the synonymy of *Bufo valentinensis*. I follow Estes and Tihen (1964) in recognizing this poorly known Pliocene species as distinct from the Miocene *B. valentinensis*, because even though the ilial prominence is relatively low as in *B. valentinensis*, it appears that it is readily separable from that of the Miocene form (see diagnosis above). It is quite possible that *B. suspectus* was derived from *B. valentinensis*.

BUFO TERRESTRIS (BONNATERRE, 1789)

SOUTHERN TOAD

(FIG. 62)

Fossil Localities. **Pleistocene (Irvingtonian NALMA):** Leisey Shell Pit Fauna, Hillsborough County, Florida—Hulbert and Morgan (1989), Holman (1995b), Meylan (1995), Sanchiz (1998). **Pleistocene (Rancholabrean NALMA):** Arredondo site, Alachua County, Florida—Tihen (1962b), Lynch (1965), Holman (1995b), Sanchiz (1998). Bradenton site, Manatee County, Florida—Tihen (1962b), Holman (1995b), Sanchiz (1998). Cheek Bend Cave, Maury County, Tennessee—Klippel and Parmalee (1982), Holman (1995b), Sanchiz (1998) (recorded as *Bufo* cf. *Bufo terrestris* by all authors). Devil's Den chamber 3, Levy County, Florida—Holman (1978b, 1995b), Sanchiz (1998). Haile (Rancholabrean complex) site, Alachua County, Florida—Tihen (1962b), Holman (1995b), Sanchiz (1998). Hornsby Spring site, Alachua County, Florida—Tihen (1962b), Holman (1995b), Sanchiz (1998). Ichetucknee River site, Columbia County, Florida—Tihen (1962b), Holman (1995b), Sanchiz

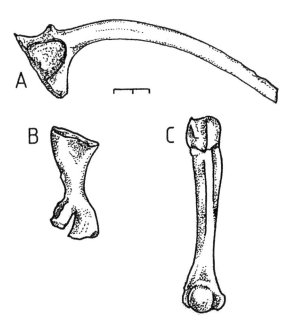

FIGURE 62. Bones of modern *Bufo terrestris* from Alachua County, Florida. (A) Right ilium in lateral view. (B) Left scapula in medial view. (C) Right humerus in ventral view. Scale bar = 4 mm and applies to all figures.

(1998). Kanapaha I site, Alachua County, Florida—Tihen (1962b), Holman (1995b). Ladds Quarry site, Bartow County, Georgia—Holman (1967a, 1985a, b, 1995b), Wilson (1975), Sanchiz (1998). Orange Lake site, Marion County, Florida—Holman (1959b, 1995b), Tihen (1962b), Sanchiz (1998). Reddick I site, Marion County, Florida—Tihen (1962b), Gut and Ray (1963), Holman (1995b), Sanchiz (1998). Williston IIIA site, Levy County, Florida—Holman (1959a, 1995b, 1996c), Sanchiz (1998). Winter Beach site, Saint Lucie County, Florida—Tihen (1962b), Holman (1995b), Sanchiz (1998).

The Southern Toad, *Bufo terrestris*, is a medium-sized toad with large, knoblike structures produced from the posterior parts of the interorbital crests. The anterior parts of the interorbital crests are thinner and tend to converge medially. The most common coloration is some shade of brown, but some populations are reddish or can be almost black. A light mid-dorsal stripe may be present but often is faint. This toad tends to favor sandy areas, where it can be quite common.

At present, this toad occurs from extreme southern Virginia and southeastern Louisiana along the coastal plain to peninsular Florida (Frost, 1985).

Identification of Fossils. If frontoparietal bones are present, the large knobs or evidence of large knobs should distinguish them from bones of other species of *Bufo*. The dorsal prominence of the ilium is lower than that of the northern form, *Bufo americanus*.

BUFO TIHENI AUFFENBERG, 1957
(FIG. 63)

Holotype. A sacral vertebra (Florida Museum of Natural History: UF 5203).

Etymology. The specific name recognizes North American anuran paleontologist Joseph A. Tihen.

Locality and Horizon. Haile locality VIA fauna, Alachua County, Florida: Miocene (Hemphillian NALMA).

Other Material. Referred specimens, all from the type locality, represent an additional sacral vertebra (UF 5095) and five ilia (assigned to UF 6363 and 6479).

Emended Diagnosis. A small *Bufo* with a sacral vertebra having a relatively long, dorsoventrally depressed centrum with a shallow ventral groove between the condyles; a horizontally oval glenoid cavity; and the condyles closely juxtaposed. Ilium with a high dorsal prominence, with the height 33–42% of the length of its base and with pronounced webbing anterior and posterior to the elongated tubercular portion. Ilium differing from both *Bufo pliocompactilis* and the living *Bufo quercicus* in having an anteriorly–posteriorly oriented knob capping the elongated tubercular portion of the ilium, with this knob reflected anteriorly. Differing from *Bufo quercicus* in lacking a small dorsal crest in the middle part of the ilial shaft.

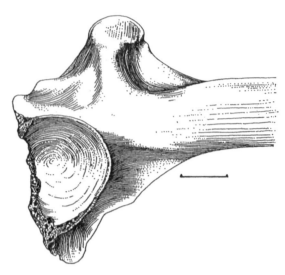

FIGURE 63. Right ilium referred to *Bufo tiheni* from the Late Miocene (Hemphillian) of Florida. Scale bar = 1 mm.

General Remarks. Bufo tiheni appears to be closely related to *Bufo pliocompactilis* of the Miocene (Clarendonian and Hemphillian NALMAs) of Kansas and Nebraska in its small size and high dorsal prominence that is characterized by a central, elongated tubercular portion set off anteriorly and posteriorly by pronounced webbing; and to modern *Bufo quercicus* on the basis of its small size and high dorsal prominence, as well as the very depressed centrum of the sacral vertebra.

BUFO VALENTINENSIS ESTES AND TIHEN, 1964
(FIG. 64)

Holotype. A right frontoparietal (University of Nebraska State Museum: UNSM 6109).

Etymology. The specific name refers to the Valentine Formation, which yielded the fossil species.

Locality and Horizon. Norden Bridge Quarry, Brown County, Nebraska: Miocene (medial Barstovian NALMA). Valentine Formation.

Other Material. This is an important and abundant species that has been found in both the United States and Canada. Originally referred material from the type locality (Estes and Tihen, 1964) includes two ilia and one fragmentary frontoparietal (UNSM 61020) and another ilium (UMMPV 42192, in part). Another ilium (UNSM 61021) from the Miocene (probably late Barstovian NALMA) from Notre Dame University (locality ND V-337), Cherry County, Nebraska, was assigned to *Bufo valentinensis* by Estes and Tihen (1964).

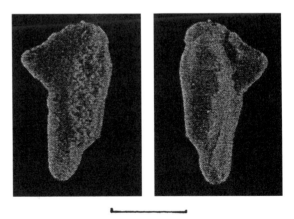

FIGURE 64. Holotype right frontoparietal in dorsal (left) and ventral (right) views of *Bufo valentinensis* from the Middle Miocene (medial Barstovian) of Nebraska. Scale bar = 4 mm.

Sites (type locality included) that have yielded additional material of this species since Estes and Tihen (1964) are as follows (an extensive list of literature references to these sites is available in Sanchiz, 1998). Egelhoff Local Fauna, Keya Paha County, Nebraska (Miocene, medial Barstovian NALMA). Hottell Ranch rhino quarries, Banner County, Nebraska (Miocene, medial Barstovian). Kleinfelder Farm locality, near Rockglen, Saskatchewan, Canada (Miocene, medial Barstovian). Railway Quarry A, Cherry County, Nebraska (Miocene, late Barstovian). Lemoyne Quarry site, Keith County, Nebraska (Miocene, Hemphillian NALMA). WaKeeney Local Fauna, Trego County, Kansas (Miocene, Clarendonian NALMA).

Emended Diagnosis. Frontoparietal somewhat thickened laterally, without prominent crests but with finely granular ornamentation over the entire surface. Similar to *Bufo suspectus* of the Pliocene (Blancan NALMA) of Kansas in the relatively low dorsal prominence of the ilium, but differing in having (1) the ilial prominence somewhat higher and shorter (height of ilial prominence about 25–30% of the length of its base in *Bufo valentinensis* vs. 17–18% in *B. suspectus*; (2) anterior edge of ilial prominence sloping much more steeply into the ilial shaft in *B. valentinensis* than in *B. suspectus*; and (3) anterior edge of acetabular cup extending anteriorly from slightly beyond to slightly behind the anterior edge of the ilial prominence in *B. valentinensis* and extending well beyond the anterior edge of the ilial prominence in *B. suspectus*.

General Remarks. It seems probable that *Bufo valentinensis* is the direct ancestor of *Bufo suspectus*.

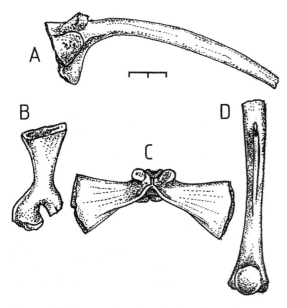

FIGURE 65. Bones of modern *Bufo valliceps* from San Luis Potosi, Mexico. (A) Right ilium in lateral view. (B) Right scapula in medial view. (C) Sacrum in dorsal view. (D) Right humerus in ventral view. Scale bar = 4 mm and applies to all figures.

BUFO VALLICEPS WIEGMANN, 1833
GULF COAST TOAD
(FIG. 65)

Fossil Localities. **Pliocene (Blancan NALMA):** Beck Ranch Local Fauna, Scurry County, Texas—Rogers (1976) (recorded as *Bufo* cf. *Bufo valliceps*). **Pleistocene (Rancholabrean NALMA):** Fowlkes Cave, Culberson County, Texas— Parmley (1988b), Holman (1995b), Sanchiz (1998).

The Gulf Coast Toad, *Bufo valliceps*, is moderately large, with a distinct, dark lateral stripe bordered by a light stripe above the dark one. The cranial crests are long and prominent, with a marked depression in the head between them. This morphology gives the face a characteristic elfin look. This is a "running toad" that can make mouselike dashes. The distribution of *B. valliceps* is from Louisiana, southern Arkansas, and eastern and southern Texas south through eastern Mexico to northern Costa Rica on the Atlantic versant; and from the Isthmus of Tehuantepec to south-central Guatemala on the Pacific slope; a specimen is also known from the Atlantic versant in Costa Rica (Frost, 1985).

Identification of Fossils. In his identification of two *Bufo valliceps* from the extralimital Fowlkes Cave late Pleistocene locality in the Trans-Pecos region of Texas, Parmley (1988b, p. 357) gave the following criteria for his identification: "The ilia of *Bufo valliceps* and the fossils may be distinguished from those of *B. woodhousii*, *B. cognatus*, *B speciosus*, *B. debilis*, and *B. punctatus* on the basis of (1) a low dorsal prominence that is hooked anteriorly and deflected laterally (dorsal prominence much higher in *B. cognatus* and *B. speciosus*, and more symmetrical and positioned more parallel to the long axis of the ilial shaft in *B. woodhousii* of similar size), and (2) dorsal acetabular expansion relatively reduced in size (but larger than in *B. punctatus* and *B. debilis*)."

General Remarks. Bufo valliceps ranges no closer to the Fowlkes Cave locality today than about 150 km to the south-southeast, in Brewster County, Texas (Parmley, 1988b).

BUFO WOODHOUSII GIRARD, 1854
WOODHOUSE'S TOAD
(FIG. 66)

Fossil Localities. **Miocene (Hemphillian NALMA):** White Cone Local Fauna, Navajo County, Arizona—Parmley and Peck (2002). **Pliocene (Blancan NALMA):** Beck Ranch Local Fauna, Scurry County, Texas—Rogers (1976), Sanchiz (1998). Benson Fauna, Cochise County, Arizona—Brattstrom (1955a), Tihen (1962b), Gehlbach (1965). White Rock Fauna, Republic County, Kansas—Eshelman (1975), Sanchiz (1998). **Late Pliocene or early Pleistocene (Blancan NALMA):** Borchers Fauna, Meade County, Kansas—Tihen (1962b), Gehlbach (1965), Sanchiz (1998). **Pleistocene (Irvingtonian NALMA):** Gilliland Fauna, Knox County, Texas—Parmley (1988a), Holman (1995b), Sanchiz (1998). Hansen Bluff Chronofauna, Alamosa County, Colorado—Rogers et al. (1985), Rogers (1987), Holman (1995b), Sanchiz (1998). Java Local Fauna, Walworth County, South Dakota—Holman (1977a). **Pleistocene (Rancholabrean NALMA):** Clear Creek Fauna, Denton County, Texas—Holman (1963b, 1995b), Sanchiz (1998) (recorded as *Bufo* cf. *Bufo woodhousii* by both authors). Dark Canyon Cave,

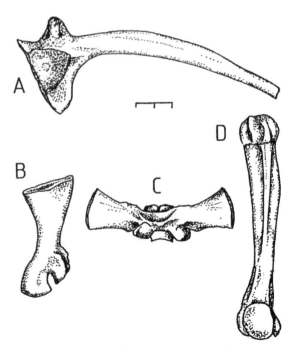

FIGURE 66. Bones of modern *Bufo woodhousii* from Trego County, Kansas. (A) Right ilium in lateral view. (B) Right scapula in medial view. (C) Sacral vertebra in dorsal view (anterior–posterior position is the reverse of that of other figures in this series). (D) Right humerus in ventral view. Scale bar = 4 mm and applies to all figures.

Eddy County, New Mexico—Applegarth (1980), Sanchiz (1998). Deadman Cave, Pima County, Arizona—Mead et al. (1984), Holman (1995b), Sanchiz (1998) (recorded as *Bufo* cf. *B. woodhousii* by all authors). Dry Cave Fauna, Eddy County, New Mexico—Holman (1970b, 1995b), Harris (1987), Sanchiz (1998). Duck Creek Local Fauna, Ellis County, Kansas—Holman (1984, 1995b), Sanchiz (1998). Easley Ranch Local Fauna, Foard County, Texas—Holman (1962b, 1995b), Sanchiz (1998). Fowlkes Cave, Culberson County, Texas—Parmley (1988b), Holman (1995b), Sanchiz (1998). Friesenhahn Cave, Bexar County, Texas—Mecham (1959), Holman (1995b), Sanchiz (1998) (recorded as *Bufo woodhousii bexarensis* by all authors [see general remarks section below]). Jinglebob Local Fauna, Meade County, Kansas—Tihen (1962b), Holman (1995b), Sanchiz (1998). Jones Fauna, Meade County, Kansas—Tihen (1962b), Gehlbach (1965), Holman (1995b), Sanchiz (1998). Lubbock Lake site, Lubbock County, Texas—Johnson (1987), Holman (1995b), Sanchiz (1998). Sandahl Local Fauna, McPherson County, Kansas—Holman (1971, 1995b), Sanchiz (1998). Schulze Cave, Edwards County, Texas—Holman (1969b, c), Parmley (1986) (recorded as *Bufo* cf. *Bufo woodhousii* by both authors); Sanchiz (1998). Slaton Local Fauna, Lubbock County, Texas—Holman (1969a, c, 1995b) (recorded as *Bufo* cf. *B. woodhousii*); Sanchiz (1998). Smith Creek Cave, White Pine County, Nevada—Mead et al. (1982), Holman (1995b), Sanchiz (1998) (recorded as *Bufo* cf. *B. woodhousii* by all authors). Williams Local Fauna, Rice County, Kansas—Holman (1984), Sanchiz (1998).

Woodhouse's Toad, *Bufo woodhousii*, a rather large species, has few definitive marks that distinguish it from other North American toads. Consistent characters are a light mid-dorsal stripe, prominent cranial crests, and elongate parotoid glands (Conant and Collins, 1998). This toad is often quite abundant in some localities within its range. It occurs from the midland prairies to the arid southwest and occupies many habitats, as long as they are fairly close to a source of water. At present, this animal ranges from the Dakotas and Montana to southern Texas and northern Mexico and westward through the Rocky Mountains, with isolated colonies in the far west.

Identification of Fossils. Tihen (1962b) gave characters that are helpful in identifying fossils of this species if the proper elements are recovered. The cranial crests are fairly robust, are striated or cancellous in nature, and are especially thick where the supraorbital and postorbital crests meet. The occipital groove is open in young specimens, but the anterior part of this groove has its anterior part roofed by the cranial crests in adults. The vertebral centra are short and broad. The width of the sacral centrum at the condyles is usually at least 130–150% of its length, and the width at the anterior end is 95–115% of its length. The height of this centrum is about 70% of its length. The neural arch crests usually form a broad arc.

Turning to the ilium, I add here that the dorsal ilial prominence of *Bufo woodhousii* is almost always higher than in *Bufo americanus*, *Bufo fowleri*, and *Bufo terrestris*; thus *B. woodhousii* can be rather easily distinguished from these species if a series of modern skeletons of these forms is available for comparison. Tihen (1962b) also presented the following ilial characteristics of *B. woodhousii*. The height of the ilial prominence is about 45% of the length of its base, slightly less in young specimens. The anterior angle varies from 45 degrees in young individuals up to almost 60 degrees in large individuals. The posterior angle is about 45 degrees in all specimens, and the dorsal acetabular angle is 115–130 degrees.

General Remarks. The identification of *Bufo woodhousii* from the White Cone Local Fauna in Arizona extends the range of this species back to the Late Miocene (Hemphillian NALMA). Whether the two ilia assigned to this species represent *B. woodhousii* or one of the extinct species of *Bufo* reported from the Miocene and Pliocene remains to be seen. A fossil subspecies, *Bufo woodhousii bexarensis* Mecham (1959), that differs from modern *B. woodhousii* primarily in estimated adult size (about 100–160 mm SVL) was described from the late Pleistocene (Ranchola-brean NALMA) Friesenhahn Cave deposit in Bexar County, Texas (Mecham, 1959). This large subspecies was also reported from the Clovis period (11.1 ka BP) from the Lubbock Lake site, Lubbock County, Texas, in association with a large, extinct subspecies of the modern box turtle species *Terrapene carolina* (Johnson, 1987).

FAMILY HYLIDAE GRAY, 1825 (1815)

At present, the huge family Hylidae occurs in North and South America, the West Indies, and the Australo-Papuan region; one species group of *Hyla* occurs in temperate Eurasia, extreme northern Africa, and the Japanese Archipelago (Frost, 1985). Forty-three genera and about 719 living species were recognized in 1992 (Duellman, 1993), and more have been described since that date.

Osteological characters used to define the family are as follows. Eight procoelous, holochordal presacral vertebrae with the non-imbricate condition (imbricate in phyllomedusines) are present. Presacral vertebrae I and II are not fused, and the cervical cotyles of presacral I are widely separated. Ribs are not present. The sacrum has dilated diapophyses (rounded in *Acris* and some Neotropical hylines) and a bicondylar articulation with the urostyle. The urostyle lacks transverse processes. The pectoral girdle has the arciferal condition and a cartilaginous omosternum (absent in *Allophryne*) and sternum; the clavicle does not overlap the scapula anteriorly. Palatines are usually present. A parahyoid is absent, and the cricoid ring is complete. Teeth are present on both the premaxilla and the maxilla (*Allophryne* is toothless). The calcaneum and astragalus are only proximally and distally fused. Two tarsalia are present. The phalangeal formula is added to by the presence of a short, cartilaginous intercalary element between the penultimate and terminal phalanges. In *Cyclorana* this element is variable and may be cartilaginous, ossified, or absent.

Identification of fossil hylids to the specific level is extremely time consuming and difficult, especially since most hylid fossils consist of tiny, broken individual elements. Most tentative identifications of pre-Pleistocene modern species of hylid genera are so questionable (e.g., Holman, 1987) that I shall not list these individual records here. Moreover, because of the present widespread occurrence and large number of species of *Hyla* in North America today (see Stebbins, 1985; Conant and Collins, 1998), Pleistocene records of "*Hyla* sp." are not especially useful, as they all lie within the present range of the genus. Thus, the reader is referred to Sanchiz (1998), where North American Pleistocene occurrences, and references to the occurrences, of "*Hyla* sp.," "*Acris* sp.," and "*Pseudacris* sp." may be found.

North American Miocene records of *Hyla* sp., sometimes tentatively referred to modern species (e.g., Chantell, 1964, 1965; Holman, 1975, 1976), include the following. Thomas Farm Local Fauna, Gilchrist County, Florida (Miocene, Hemingfordian NALMA)—Holman (1967a). Hottell Ranch rhino quarries, Banner County, Nebraska (Miocene, medial Barstovian NALMA)—Voorhies et al. (1987). Norden Bridge Local Fauna, Brown County, Nebraska (Miocene, medial Barstovian)—Chantell (1964). Bijou Hills Local Fauna, Charles Mix County, South Dakota (Miocene, late Barstovian)—Holman (1978a). Glenn Olson Quarry, Charles Mix County, South Dakota (Miocene, late Barstovian)—Green and Holman (1977). WaKeeney Local Fauna, Trego County, Kansas (Miocene, Clarendonian NALMA)—Holman (1975). Lemoyne Quarry site, Keith County, Nebraska (Miocene, medial Hemphillian NALMA)—Parmley (1992). White Cone Local Fauna, Navajo County, Arizona—Parmley and Peck (2002).

North American Miocene and Pliocene records of *Acris* sp., sometimes referred to modern species or species groups (e.g., Tihen, 1954, 1960a; Chantell, 1964, 1965, 1966; Holman, 1976), include the following. Quarry A, Martin Canyon Local Fauna, Logan County, Colorado (Miocene, Hemingfordian NALMA)—Chantell (1965). Egelhoff Local Fauna, Keya Paha County, Nebraska (Miocene, medial Barstovian NALMA)—Holman (1976, 1987). Norden Bridge Local Fauna, Brown County, Nebraska (Miocene, medial Barstovian)—Chantell (1964, 1966), Holman (1976). WaKeeney Local Fauna, Trego County, Kansas (Miocene, Clarendonian NALMA)—Holman (1975). Rexroad Local Fauna, Meade County, Kansas (Pliocene, Blancan NALMA)—Chantell (1966). White Rock Fauna, Republic County, Kansas (Pliocene, Blancan)—Eshelman (1975).

North American Miocene and Pliocene records of *Pseudacris* sp., sometimes referred to modern species or species groups (e.g., Chantell, 1964, 1965, 1966, 1970; Holman, 1975, 1976, 1978a), include the following. Egelhoff Local Fauna, Keya Paha County, Nebraska (Miocene, medial Barstovian NALMA)—Holman (1976, 1987). Norden Bridge Local Fauna, Brown County, Nebraska (Miocene, medial Barstovian)—Chantell (1964), Holman (1976). Bijou Hills Local Fauna, Charles Mix County, South Dakota (Miocene, late Barstovian)—Holman (1978a). Glad Tidings Prospect, Knox County, Nebraska (Miocene, late Barstovian)—Holman (1996a). WaKeeney Local Fauna, Trego County, Kansas (Miocene, Clarendonian NALMA)—Holman (1975). Hagerman locality (county not given), Idaho (Pliocene, Blancan NALMA)—Chantell (1970). Sanders Local Fauna, Meade County, Kansas (Pliocene, Blancan)—Chantell (1966).

GENUS *ACRIS* DUMÉRIL AND BIBRON, 1841

CRICKET FROGS

Living species of *Acris* occur in the United States east of the Rocky Mountains and the extreme southeastern part of southern Ontario (Point Pelee), Canada; they also occur in northern Coahuila, Mexico (Frost, 1985). Only two living species of *Acris* occur, *Acris crepitans* and *Acris gryllus*, and both of these are known in the North American fossil record. Chantell (1968a) provided a detailed study of the osteology of *Acris* in which he demonstrated that the genus was well differentiated from other hylid genera in North America but that the two *Acris* species were similar to each another. Some characters from Chantell (1968a) that separate *Acris* from the North American genera *Pseudacris* and *Hyla* are slightly modified as follows.

The ethmoid lacks an anterior elongation. The premaxilla has the distal portion of the dorsal process directed posteriorly and lacks a lingual shelf expansion. The coracoid lacks a lateral ridge. The scapula has a well-developed anterior projection. The humerus has a long ventral crest; a medial epicondylar flange is present, but an epicondyle is absent. The sacrum has the condyles well separated, and the sacral diapophyses are essentially cylindrical. The urostylar nerve foramina are large and round, and the urostylar cotyles are separated, have short necks, and are directed posteroventrally. The ilium has a distinctive positional relationship between the dorsal prominence and the acetabular border. The ventral acetabular expansion is narrow and nearly perpendicular to the shaft. The dorsal acetabular expansion is concave. A low dorsal crest on the ilial shaft is present.

ACRIS BARBOURI HOLMAN, 1967

(FIG. 67)

Holotype. A right ilium (Florida Museum of Natural History: UF 10208).

Etymology. The specific name recognizes North American zoologist Thomas Barbour.

Locality and Horizon. Thomas Farm Local Fauna, Gilchrist County, Florida: Miocene (Hemingfordian NALMA). Torreya Formation.

Other Material. A right ilium (Florida Geological Survey: FGS V6088; currently at the Florida Natural History Museum) was designated as a paratype by Holman (1967a).

Emended Diagnosis. A Miocene *Acris* that is distinguished from the living species *Acris crepitans* and *Acris gryllus* in having (1) a low dorsal crest (ilial shaft ridge of some authors) gently arising from the anterior border of the dorsal prominence and having its dorsal margin straight (low crest arising well anterior to dorsal prominence and slightly curved dorsally in A. *crepitans* and A. *gryllus*); (2) dorsal tubercle (dorsal protuberance of some authors) rounded

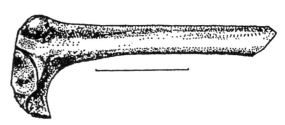

FIGURE 67. Holotype right ilium of *Acris barbouri* from the Early Miocene (Hemingfordian) of Florida. Scale bar = 1 mm.

and relatively smooth (dorsal tubercle oval and sometimes roughened in *A. crepitans* and *A. gryllus*); (3) ilial shaft almost straight (ilial shaft somewhat curved in *A. crepitans* and *A. gryllus*).

Description. This description covers only those features of the *Acris barbouri* ilia that were not featured in the diagnosis. In the holotype, the dorsal acetabular expansion is broken just posterior to the dorsal tubercle. The dorsal tubercle is anterior in position, having its posterior border about even with the anterior border of the acetabular fossa. The acetabular fossa has its posterior portion broken. The tip of the ventral acetabular expansion is broken, but the portion of this expansion anterior to the anterior margin of the acetabular fossa is narrow. The acetabular fossa itself is moderately excavated and has its lower border well produced from the acetabular region. The ilial shaft is compressed and has a moderately excavated area on its lateral surface just anterior to the dorsal protuberance.

Turning to the paratype, we find that this element is from a somewhat larger individual than the holotype and shows some slight difference that may be attributed to individual or ontogenetic variation. The dorsal tubercle is somewhat rougher than in the holotype, and the angle between the ilial shaft and the anterior border of the ventral acetabular expansion is slightly greater than in the holotype. The ilial shaft is more complete than in the holotype and indicates a slightly greater development of the low dorsal crest (ilial shaft ridge) than in the holotype.

General Remarks. Acris barbouri is much more distinct from the living species *Acris crepitans* and *Acris gryllus* than the latter two forms are distinct from one another.

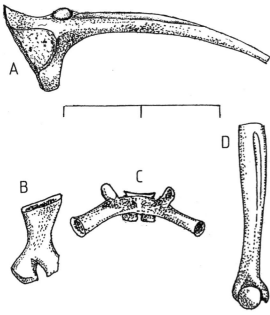

FIGURE 68. Bones of modern *Acris crepitans* from Douglas County, Kansas. (A) Right ilium in lateral view. (B) Right scapula in medial view. (C) Sacral vertebra in dorsal view. (D) Right humerus in ventral view. Scale bar = 4 mm and applies to all figures.

ACRIS CREPITANS BAIRD, 1854
NORTHERN CRICKET FROG
(FIG. 68)

Fossil Localities. **Pliocene (Blancan NALMA):** Beck Ranch Local Fauna, Scurry County, Texas—Rogers (1976), Sanchiz (1998). **Pleistocene (Irvingtonian NALMA):** Vera Local Fauna, Knox County, Texas—Holman (1995b), Parmley (1988a) (recorded as *Acris* cf. *Acris crepitans* by both authors); Sanchiz (1998). **Pleistocene (Rancholabrean NALMA):** Butler Spring Local Fauna, Meade County, Kansas—Holman (1986a, 1995b), Sanchiz (1998). Clear Creek Fauna, Denton County, Texas—Holman (1963b, 1995b), Sanchiz (1998) (recorded as *Acris* cf. *A. crepitans* by both authors). Easley Ranch Local Fauna, Foard County, Texas—Holman (1962b, 1995b) (recorded as Acris cf. Acris crepitans); Sanchiz (1998). Howard Ranch (sometimes called Groesbeck Creek) Local Fauna, Hardeman County, Texas—Holman (1964, 1995b), Sanchiz (1998). Kanapolis Local Fauna, Ellsworth County, Kansas—Holman (1972c, 1995b), Sanchiz (1998). Lubbock Lake site (Clovis-period level), Lubbock County, Texas—Johnson (1987), Holman (1995b), Sanchiz (1998). Slaton Local Fauna, Lubbock County, Texas—Holman (1969a, c, 1995b), Sanchiz (1998). Williams Local Fauna, Rice County, Kansas—Holman (1984, 1995b), Sanchiz (1998).

Both living species of Cricket Frogs, *Acris crepitans* (Northern Cricket Frog) and *Acris gryllus* (Southern Cricket Frog), are very small, warty anurans with tiny discs on the end of the toes, and both are non-climbing species. Both are quite varied in coloration, except that they have a dark triangular or V-shaped marking between the eyes and a long dark stripe or stripes on the thighs. If you have specimens in hand, the best way to tell the two species apart is by their toes: *A. crepitans* has its toes nearly fully webbed, whereas *A. gryllus* has the webbing reduced. Both escape from predators by erratic zigzag hops. Both live in the vicinity of permanent shallow water, where they are often found sitting in the open sunlight on muddy, sandy, or sometimes gravelly banks. The Northern Cricket Frog occurs in the United States east of the Rocky Mountains (except in the deep southeastern states); in extreme southern Ontario (Point Pelee), Canada; and in northern Coahuila, Mexico. The Southern Cricket Frog inhabits the southeastern United States (see Conant and Collins, 1998, maps, p. 529).

Identification of Fossils. Most of the identifications of one or the other of these species in the fossil record have been based on zoogeography, as the two species are osteologically similar to each other.

General Remarks. Based on both the fossil record of these species and their morphological similarity to one another, I would suggest that they did not evolve until latest Pliocene times.

ACRIS GRYLLUS (LeCONTE, 1825)
SOUTHERN CRICKET FROG
(Fig. 69)

Fossil Locality. **Pleistocene (Rancholabrean NALMA):** Arredondo site, Alachua County, Florida—Lynch (1965) (recorded as *Acris* cf. *Acris gryllus*); Holman (1995b), Sanchiz (1998).

General Remarks. Holman (1995b) referred to this specimen as *Acris gryllus* on zoogeographic grounds.

GENUS *HYLA* LAURENTI, 1768
TREEFROGS
Living species of *Hyla* occur in central and southern Europe; eastern Asia; northwest-

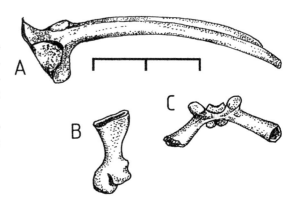

FIGURE 69. Bones of modern *Acris gryllus* from Alachua County, Florida. (A) Right ilium in lateral view. (B) Right scapula in medial view. (C) Sacral vertebra in dorsal view. Scale bar = 4 mm and applies to all figures.

ern Africa; North, Central, and South America; and the Greater Antilles in the West Indies (Frost, 1985).

Two hundred and eighty-one species of *Hyla* were recognized in 1992, in contrast to the 258 species that were known in 1985 (Duellman, 1993). There is so much osteological variation at all levels in this group that identification of species of the genus on the basis of individual fossil elements has proven daunting, to say the least.

HYLA ARENICOLOR COPE, 1886
CANYON TREEFROG
(FIG. 70)

Fossil Localities. **Pleistocene (Rancholabrean NALMA):** Howell's Ridge Cave, Grant County, New Mexico—Van Devender and Worthington (1977), Holman (1995b), Sanchiz (1998) (recorded as *Hyla* cf. *Hyla arenicolor* by all authors). Rancho la Brisca locality, Sonora, Mexico—Van Devender et al. (1985), Sanchiz (1998).

Hyla arenicolor is a Treefrog of moderate size with a somewhat granular skin and a rather drab, brownish yellow color; it usually lacks a dark stripe through the eye. It is mainly a terrestrial species but will sometimes climb in bushes and trees. It is associated with arroyos and canyons in arid regions and often hides in rock outcrops and ledges near water. At present, this species occurs in mountain and plateau areas of the United States (southern Utah, western Colorado, and southward through eastern Arizona, western New Mexico, and western Texas) and ranges southward in Mexico to Michoacan, Guerrero, and Oaxaca (Van Devender et al., 1985).

Identification of Fossils. Van Devender et al. (1985, pp. 32–33) gave characters they used to identify ilia from the Pleistocene (Rancholabrean NALMA) Rancho la Brisca locality, Sonora, Mexico, as belonging to *Hyla* arenicolor: "The ilia are referred to *H. arenicolor* because: (1) the dorsal and ventral acetabular expansions (posterior and anterior pelvic spines of Gaup 1896) are subequal (the dorsal acetabular expansion is relatively small in *H. regilla*); (2) larger size for maturity of the bone than in *H. eximia*; (3) relatively larger acetabulum for size of ilial head than in *H. cadaverina*; (4) a relatively low, broad dorsal prominence without a well-developed 'knob' compared to *H. regilla* and *H. eximia*. *Pternohyla fodiens* is a larger treefrog which has the dorsal prominence oriented more dorsally than in *Hyla*."

General Remarks. The lack of fossil records of this widespread western species is odd.

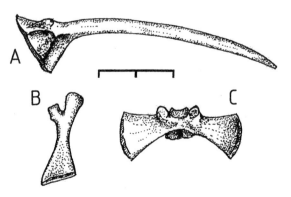

FIGURE 70. Bones of modern *Hyla arenicolor* from Arizona. (A) Right ilium in lateral view. (B) Right scapula in medial view (top and bottom positions are the reverse of those of other bones in this series). (C) Sacrum in dorsal view. Scale bar = 4 mm and applies to all figures.

[*]*Hyla baderi* Lynch, 1965
(Fig. 71)

Holotype. The distal 5.4 mm of a right ilium (Florida Natural History Museum: UF 9113).

Etymology. The specific name recognizes North American vertebrate paleontologist Robert S. Bader.

Locality and Horizon. Arredondo Pitt II site, Alachua County, Florida: Pleistocene (Rancholabrean NALMA).

Diagnosis. This diagnosis is from Lynch (1965, p. 75): "A species of *Hyla* whose fossil ilium is separable from those of all other species examined in possessing an ilial prominence very close to the fossa [acetabular fossa]; dorsal protuberance [or tubercle] situated laterally, rounded; ilial prominence posterior to anterior border of fossa [acetabular fossa] as in many species of *Hyla*."

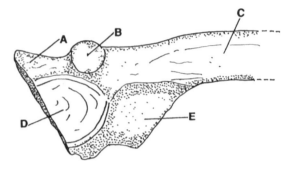

FIGURE 71. Holotype right ilium of *Hyla baderi* from the late Pleistocene (Rancholabrean) of Florida: A, dorsal acetabular expansion; B, knoblike dorsal protuberance produced laterally from dorsal prominence; C, ilial shaft; D, acetabular fossa; E, ventral acetabular expansion. No scale was provided, but the specimen is 5.4 mm long.

Description. This description is modified from Lynch (1965). The acetabular fossa is suboval and relatively shallow. The ilial tubercle is lateral, rounded, and very close to the border of the acetabular fossa. An ilial shaft ridge is lacking. The subacetabular fossa (depressed area on the ventral acetabular expansion) is deep. The ventral acetabular expansion has its ventral tip broken off. The acetabular fossa shows little erosion.

General Remarks. Lynch (1965) compared the ilium of *Hyla baderi* with a series of ilia of *Hyla cinerea* of various age groups to determine the extent of ontogenetic variation in the hylid ilium. He determined that although the dorsal prominence and its tubercle become more rugose with age, the relationship of these structures to the acetabular fossa remains essentially the same except in very young specimens. I have not seen such a round dorsal ilial tubercle (protuberance) in any modern species of North American *Hyla*. Sanchiz (1998, p. 127) included the distinctive *H. baderi* in his nomina dubia section.

Hyla chrysoscelis Cope, 1880 or *Hyla versicolor* LeConte, 1825
Cope's Gray Treefrog, or Gray Treefrog
(Fig. 72)

Hyla chrysoscelis and *Hyla versicolor* are cryptic, sibling species that can only be distinguished by call (sometimes with difficulty), genetic karyotypes, or cell volume. The range of each species within the composite range of both (which extends from southern Canada and the Dakotas south to Louisiana and Florida) is poorly understood. Thus, fossil elements of the *Hyla chrysoscelis–Hyla versicolor* complex are best identified as "*Hyla chrysoscelis* or *H. versicolor*," as no one has been able to find osteological differences between the two, although the complex itself appears to be osteologically distinct.

Fossil Localities. **Pleistocene (Irvingtonian NALMA):** Albert Ahrens

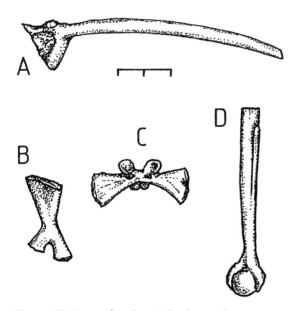

FIGURE 72. Bones of modern *Hyla chrysoscelis* or its cryptic sibling species, *Hyla versicolor*, from Brown County, Indiana. (A) Right ilium in lateral view. (B) Left scapula in medial view. (C) Sacral vertebra in dorsal view. (D) Right humerus in ventral view. Scale bar = 4 mm and applies to all figures.

Local Fauna, Nuckolls County, Nebraska—Ford (1992), Holman (1995b), Sanchiz (1998). Hamilton Cave, Pendleton County, West Virginia—Holman and Grady (1989), Holman (1995b), Sanchiz (1998). **Pleistocene (Rancholabrean NALMA):** Easley Ranch Local Fauna, Foard County, Texas—Holman (1962b, 1995b), Sanchiz (1998). Frankstown Cave, Blair County, Pennsylvania—Fay (1988), Holman (1995b), Sanchiz (1998). Howard Ranch (sometimes called Groesbeck Creek) Local Fauna, Hardeman County, Texas—Holman (1964, 1995b); Lynch (1966) (recorded as *Hyla holmani* [see Holman, 1995b, p. 204]); Sanchiz (1998) (recorded as *Hyla* cf. *Hyla cinerea* or *Hyla versicolor*). Kanapolis Local Fauna, Ellsworth County, Kansas—Holman (1972c, 1995b), Sanchiz (1998). Ladds Quarry site, Bartow County, Georgia—Holman (1985a, b, 1995b), Sanchiz (1998). Schulze Cave Fauna, Edwards County, Texas—Parmley (1986), Holman (1995b), Sanchiz (1998).

The sibling Gray Treefrog species (*Hyla chrysoscelis* and *Hyla versicolor*) are moderately large species (as North American Treefrogs go), with a coloration that is normally gray or green (both species can change their color from gray to green); the hidden surface of their hind legs is orangish, mottled with dark pigment. The skin is rather warty, but not as warty as that of toads. These frogs normally stay in trees and bushes and may even frequent drainage gutters in buildings. During the breeding season they may be seen near the edge of aquatic habitats.

Identification of Fossils. A definitive study of the osteology of *Hyla chrysoscelis* and *Hyla versicolor* is badly needed. If someone finds a difference between the two species, even on the basis of one skeletal element, it will be a herpetological breakthrough.

HYLA CINEREA (SCHNEIDER, 1799)
GREEN TREEFROG
(FIG. 73)

Fossil Localities. **Pleistocene (Irvingtonian NALMA):** Inglis IA site, Citrus County, Florida—Meylan (1982), Holman (1995b), Sanchiz (1998). **Pleistocene (Rancholabrean NALMA):** Arredondo site, Alachua County, Florida—Lynch (1965), Holman (1995b), Sanchiz (1998). Reddick I site, Marion County, Florida—Wilson (1975), Holman (1995b), Sanchiz (1998).

As its vernacular name implies, *Hyla cinerea* is usually bright green. A conspicuous light or yellow stripe is usually present along the entire side of the body (including the head) but may be missing in some individuals. The tiny golden spots that occur on the backs of many individuals make this frog a jewel of the wet lowlands and swampy areas of the southeastern United States. Beyond the southeastern United States this species has been introduced into Puerto Rico (Frost, 1985).

Identification of Fossils. In the identification of a *Hyla cinerea* ilium from the Pleistocene (Rancholabrean NALMA) Reddick I site in Florida, Wilson (1975, p. 15) made this comment: "The large size and oval shape of the dorsal protuberance [tubercle], its relatively great distance from the acetabular fossa, and the relatively acute angle with which the anterior edge of the ventral acetabular expansion meets the ilial shaft are characteristic of *Hyla cinerea*."

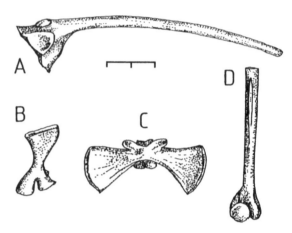

FIGURE 73. Bones of modern *Hyla cinerea* from Levy County, Florida. (A) Right ilium in lateral view. (B) Right scapula in medial view. (C) Sacrum in dorsal view. (D) Right humerus in ventral view. Scale bar = 4 mm and applies to all figures.

General Remarks. In Wilson's Pleistocene specimen of *Hyla cinerea* from Florida, the dorsal protuberance is oval and the protuberance is farther from the acetabular fossa than in *Hyla baderi*.

Hyla femoralis Bosc, 1800
Pine Woods Treefrog
(Fig. 74)

Fossil Localities. **Pleistocene (Rancholabrean NALMA):** Haile (Rancholabrean complex) site, Alachua County, Florida—Lynch (1964), Holman (1995b), Sanchiz (1998). Williston IIIA site, Levy County, Florida—Holman (1959a, 1995b, 1996c), Sanchiz (1998).

Hyla femoralis is a small, brownish Treefrog that is nondescript, except for the row of light or orangish spots on the thigh. This arboreal species is most commonly found in flat pine land and near cypress swamps. This southeastern species occurs on the Atlantic coastal plain from Virginia to Louisiana (Frost, 1985).

Identification of Fossils. In the identification of a *Hyla femoralis* ilium from the Pleistocene (Rancholabrean NALMA) Williston IIIA site in Florida, Holman (1959a) pointed out that in *Hyla squirella* the dorsal tubercle (protuberance) is very

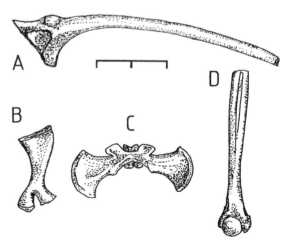

FIGURE 74. Bones of modern *Hyla femoralis* from Alachua County, Florida. (A) Right ilium in lateral view. (B) Right scapula in medial view. (C) Sacrum in dorsal view. (D) Right humerus in ventral view. Scale bar = 4 mm and applies to all figures.

small and directed dorsally, whereas in *H. femoralis* it is larger and laterally directed. In *Hyla versicolor* [also *Hyla chrysoscelis*], *Hyla cinerea*, and *Hyla avivoca* the tubercle is much closer to the border of the acetabular fossa and much more oval than in *H. femoralis*. *Hyla gratiosa* is a much larger species than *H. femoralis*, and the tubercle is more oval and dorsally directed. Interestingly, *Pseudacris crucifer* has a rounded tubercle as in many *H. femoralis*, but the tubercle is anterior to the acetabular fossa in *P. crucifer* whereas it lies somewhat posterior to this fossa in *H. femoralis*.

General Remarks. The presence of *Hyla femoralis* in a Pleistocene fauna probably indicates the present of pine flatwoods or cypress swamps in the ancient habitat represented by the frog.

Hyla gratiosa LeConte, 1857 "1856"

Barking Treefrog

(Fig. 75)

Fossil Localities. **Pleistocene (Rancholabrean NALMA):** Arredondo site, Alachua County, Florida—Lynch (1965), Holman (1995b), Sanchiz (1998). Bell Cave unit I, Colbert County, Alabama—Holman et al. (1990), Holman (1995b), Sanchiz (1998). Haile (Rancholabrean complex) site, Alachua County, Florida—Lynch (1964), Holman (1995b), Sanchiz (1998).

The Barking Treefrog, *Hyla gratiosa*, is one of the larger North American Treefrogs and can usually be distinguished on the basis of its distinct spots, which usually remain visible even during color changes. This species is arboreal but is known to burrow from time to time, especially during dry spells. The modern range of this species is in the United States from North Carolina to eastern Louisiana, with isolated records from southern Kentucky, Tennessee, and northern Alabama (Frost, 1985).

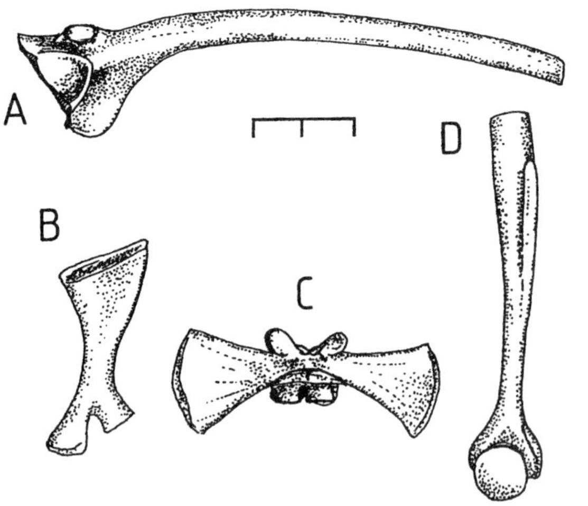

Identification of Fossils. Holman et al. (1990, p. 524), in their identification of *Hyla gratiosa* from the Pleistocene of Bell Cave unit I, northern Alabama, characterized the ilium as follows: "*Hyla gratiosa* has an ilium that is distinguishable from other *Hyla* [North American] in being quite large; with an extensive ventral acetabular expansion; rounded dorsal protuberance [tubercle] that occurs posterior to the edge of the acetabular cup." In addition, this species often has a long, deep groove on the lower part of the ilial shaft anterior to the anterior border of the acetabular fossa.

General Remarks. The lack of fossil records of this large species is surprising.

Figure 75. Bones of modern *Hyla gratiosa* from Levy County, Florida. (A) Right ilium in lateral view. (B) Right scapula in medial view. (C) Sacrum in dorsal view. (D) Right humerus in ventral view. Scale bar = 4 mm and applies to all figures.

Hyla miocenica Holman, 1966
(Fig. 76)

Holotype. A left ilium (Shuler Museum of Paleontology, Southern Methodist University: SMPSMU 61871).

Etymology. The specific name refers to the age (Miocene) of the species.

FIGURE 76. Holotype left ilium of *Hyla miocenica* from the Middle Miocene (early Barstovian) of Texas. Scale bar = 2 mm.

Locality and Horizon. Trinity River Local Fauna, San Jacinto County, Texas: Miocene (early Barstovian NALMA). Fleming Formation.

Other Material. A sacral vertebra (SMPSMU 63677) from the Trinity River Local Fauna represented a *Hyla* about the same size as the individual represented by the holotype and was tentatively assigned to *Hyla miocenica* by Holman (1977b).

Emended Diagnosis. This emendation of the diagnosis of Holman (1966b) is modified from Holman (1996b). A moderately large species of *Hyla* whose ilium is most similar to those of the sibling species *Hyla chrysoscelis* and *Hyla versicolor*, but differing in that the dorsal tubercle (protuberance) is more dorsally produced from the shaft than in either of these species.

Description. The description of the type ilium is slightly modified from Holman (1966b). The ilium is complete except that part of the dorsal surface of the dorsal tubercle is broken, and the tips of the dorsal and ventral acetabular expansions are broken. The dorsal prominence that supports the tubercle is poorly developed. The dorsal tubercle is oval in outline in lateral view, is well produced laterally, and is longer than it is high. The tubercle is moderately distant from the border of the acetabular fossa. The ventral acetabular expansion is extensive, and its anterior border makes a wide angle with the shaft. The shaft is moderately curved and lacks either a crest or a ridge. A long, deep groove is not present on the lower lateral part of the shaft anterior to the acetabular fossa.

General Remarks. Sanchiz (1998) was evidently unaware of Holman's (1996b) paper, where it was pointed out that *Hyla miocenica* was most closely related to *Hyla chrysoscelis* and *Hyla versicolor*, as he based his slightly modified diagnosis of the species on Holman (1966b).

Hyla miofloridana Holman, 1967
(Fig. 77)

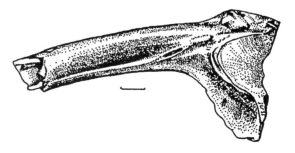

Holotype. A right ilium (Florida Museum of Natural History: UF 10209).

Etymology. The specific name refers to the age (Miocene) of the type specimen and the state (Florida) where it was collected.

Locality and Horizon. Thomas Farm Local Fauna, Gilchrist County, Florida: Miocene (Hemingfordian NALMA). Torreya Formation.

FIGURE 77. Holotype left ilium of *Hyla miofloridana* from the Early Miocene (Hemingfordian) of Florida. Scale bar = 1 mm.

Emended Diagnosis. A moderately large *Hyla* similar to *Hyla gratiosa*, but differing in having an ilium with dorsal tubercle less laterally produced, less distinct from the ilial prominence, and less rounded; acetabular fossa less excavated and its border weaker; ventral acetabular expansion more excavated anterodorsally.

Description. The description is slightly modified from Holman (1967a). The dorsal acetabular expansion has its tip broken off, and the portion remaining has its surface worn rather smooth. The dorsal tubercle is roughly rounded in shape, rather weakly produced laterally from the ilial shaft, and not highly distinct from the prominence. The anterior edge of the dorsal prominence lies slightly posterior to the anterior edge of the acetabular fossa. The acetabular fossa is only moderately excavated, and its border is rather weak. The ventral acetabular expansion is well developed and wide. The anterior edge of the ventral acetabular expansion makes an angle of much greater than 90 degrees with the ilial shaft. The ilial shaft is compressed and lacks a ridge or crest on its dorsal border. A long, deep grove extends anteriorly on the ventral part of the ilial shaft a few millimeters anterior to the acetabular fossa; this deep groove also occurs in ilia of many modern *H. gratiosa* and appears to be rather rare in other *Hyla* species in North America.

General Remarks. Sanchiz (1998) considers this a questionable species (nomen dubium).

HYLA SQUIRELLA BOSC, 1800
SQUIRREL TREEFROG
(FIG. 78)

Fossil Locality. **Pleistocene (Rancholabrean NALMA):** Haile (Rancholabrean complex) site, Alachua County, Florida—Lynch (1964), Holman (1995b), Sanchiz (1998).

This attractive little Treefrog is said to be like a chameleon in its many variations of color and patterns (Conant and Collins, 1998). Often, a dark spot or bar occurs between the eyes, and almost all the ones I have seen in Florida have a light stripe along the side of the body.

This species is common in some parts of the southeast and can occur in almost any moist habitat where there is breeding water available. It is an excellent climber. The modern range of *Hyla squirella* is the coastal plain and Mississippi Valley of southeastern North America; it was introduced onto Grand Bahama and other Bahama islands (Frost, 1985).

General Remarks. Auffenberg (1956) described *Hyla goini* as an extinct species from the Miocene (Hemingfordian) Thomas Farm Local Fauna of Florida on the

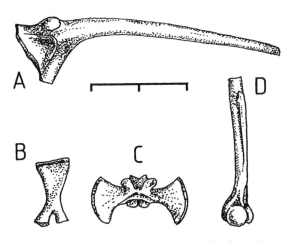

FIGURE 78. Bones of modern *Hyla squirella* from Harris County, Texas. (A) Right ilium in lateral view. (B) Right scapula in medial view. (C) Sacrum in dorsal view. (D) Right humerus in ventral view. Scale bar = 4 mm and applies to all figures.

basis of four ilia. Holman (1967a) was able to study an additional 14 left and 20 right ilia from the type locality.

In this study (see Holman, 1967a, pp. 134–135), it was found that the ilia of *Hyla goini* most closely resembled those of *Hyla squirella* and, in fact, that no single character consistently separated the two species. Sanchiz (1998) included *H. goini* in his synonyms section and suggested that the fossil might be considered synonymous with, or a direct ancestor to, *H. squirella*. I would favor the *direct ancestor* term, but it is apparent that more fossil material is needed to establish the exact relationships of *H. goini*.

Hyla swanstoni Holman, 1968
(Fig. 79)

Holotype. A left ilium (Saskatchewan Museum of Natural History: SMNH 1435).

Etymology. The specific name recognizes North American paleontologist A. E. Swanston.

Locality and Horizon. Calf Creek Local Fauna, near East End, Saskatchewan, Canada: Eocene (Chadronian NALMA). Cypress Hills Formation.

Other Material. A left ilium (SMNH 1436) from the same locality was designated as a paratype by Holman (1968). Three partial tibiofibulae (all under SMNH 1437) were designated as referred material (Holman, 1968).

Emended Diagnosis. Most similar to *Hyla miofloridana* of the Early Miocene (Hemingfordian NALMA) of Florida in having a rounded tubercle, laterally produced from the dorsal prominence; ventral acetabular expansion very wide, its anterior margin extending well beyond the anterior margin of the acetabular fossa; and a groove on the lower part of the ilium anterior to the acetabular fossa. Differs from *H. miofloridana* in having iliac tubercle more rounded, dorsal ilial prominence less extensive, acetabular fossa smaller with its anterior hemispherical rather than oval, and iliac groove much less well developed.

Description. The description of the *Hyla swanstoni* elements is slightly modified from Holman (1968). In the holotype, the anterior border of the dorsal prominence ends anterior to the level of the anterior border of the acetabulum. The tubercle of the prominence is generally rounded but is slightly longer than it is wide. The tubercle is somewhat eroded,

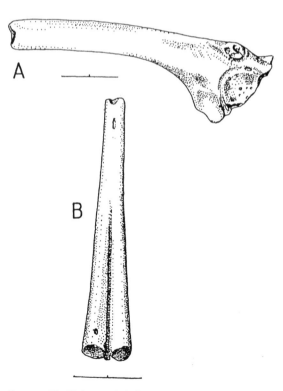

FIGURE 79. Holotype left ilium (A) and referred tibiofibula (B) of *Hyla swanstoni* of the Late Eocene (Chadronian) of Saskatchewan, Canada. Scale bars = 2 mm.

but it projects more laterally than dorsally. The distance between the ventral border of the prominence and the border of the acetabulum is only about one-fourth the length of the prominence.

The dorsal acetabular expansion has its tip broken. The acetabular area is somewhat worn, but it is apparent that the acetabular fossa is weakly excavated. The acetabular border is worn. The ventral acetabular expansion is relatively wide, and its anterior border makes an angle of much greater than 90 degrees with the long axis of the shaft. The tip of the ventral acetabular expansion is broken. No dorsal ridge or crest occurs on the dorsal border of the ilial shaft. A shallow groove lacking the distinct ventral border of *Hyla miofloridana* lies on the lower lateral face of the ilial shaft just anterior to the acetabulum. Measurements are as follows: greatest height of ilial shaft, 1.3 mm; height of acetabular fossa, 1.7 mm; length of dorsal tubercle, 1.0 mm.

Turning to the other material of *Hyla swanstoni*, we find that the paratype ilium is more eroded than the holotype, but the position of what remains of the dorsal prominence and its tubercle appears to be generally the same as in the holotype. The lateral iliac groove of the paratype is somewhat more distinct than in the holotype. Measurements are as follows: greatest height of ilial shaft, 1.4 mm; height of acetabular, fossa 1.6 mm. The referred tibiofibulae represent frogs of about the same size as those represented by the ilia and have the same gracile and elongate proportions as modern hylid frogs.

General Remarks. Hyla swanstoni is the oldest hylid known in North America, as it occurs in the Late Eocene (Chadronian NALMA) (Prothero and Emry, 1996). Sanchiz (1998) mistakenly referred to this occurrence as being Lower Oligocene.

GENUS #*PROACRIS* HOLMAN, 1961

Genotype. Proacris mintoni Holman, 1961.

Etymology. The generic name comprises the Latin *pro*, "before, in front of," and *Acris*, a modern hylid frog. The specific name recognizes North American herpetologist Sherman A. Minton.

Diagnosis. The diagnosis is the same as for the genotype and only known species, *Proacris mintoni* Holman, 1961.

#*PROACRIS MINTONI* HOLMAN, 1961

(FIG. 80)

Holotype. A left ilium (Florida Geological Survey: FGS V5950). This type specimen is now housed at the Florida Museum of Natural History (UF), Gainesville.

Locality and Horizon. Thomas Farm Local Fauna, Gilchrist County, Florida: Miocene (Hemingfordian NALMA). Torreya Formation.

Emended Diagnosis. A hylid species differing from other North American species in having (1) a very large acetabular fossa, (2) a ventral acetabular expansion that is almost completely ventral to the dorsal acetabular expansion, (3) a very large ball-like dorsal tubercle supported by a relatively small dorsal ilial prominence, and (4) an ilial shaft very high along posterior two-thirds of its extent and then becoming abruptly lower anteriorly.

Description. This description deals with ilial characters other than the ones presented in the above emended diagnosis. The dorsal acetabular expansion is pointed. An ilial shaft ridge occurs just anterior to the dorsal tubercle (not posterior to it as stated in Holman, 1967a). The dorsal margin of the ilial shaft is straight, and the ventral margin is slightly curved dorsally. The ventral acetabular expansion is narrow and ends in a point that is beveled on both sides.

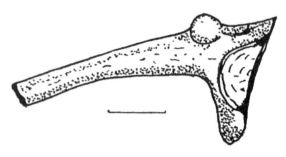

FIGURE 80. Holotype left ilium of *Proacris mintoni* from the Early Miocene (Hemingfordian) of Florida. Scale bar = 1 mm.

General Remarks. Proacris mintoni has an ilial morphology that is quite different from that of any hylid genus I have seen in North or South America. The generic name is unfortunate, as it is apparent now that *Proacris* is not ancestral to *Acris*. Hopefully, additional fossil material will help elucidate the relationships of *Proacris* among the Hylidae.

GENUS *PSEUDACRIS* FITZINGER, 1843

CHORUS FROGS

Living species of *Pseudacris* occur east of the Rocky Mountains from Hudson Bay to the Gulf of Mexico (Frost, 1985). As of 1992, 11 species were recognized (Duellman, 1993).

Chantell (1968b) studied the osteology of *Pseudacris*, but since that study, the taxa *Hyla crucifer, Hyla regilla,* and *Limnaoedus* have been transferred to the genus *Pseudacris*. Species that Chantell studied included *Pseudacris brachyphona, Pseudacris brimleyi, Pseudacris clarkii, Pseudacris nigrita, Pseudacris ornata, Pseudacris streckeri,* and *Pseudacris triseriata.*

PSEUDACRIS CLARKII (BAIRD, 1854)

SPOTTED CHORUS FROG

Fossil Localities. **Pleistocene (Irvingtonian NALMA):** Albert Ahrens Local Fauna, Nuckolls County, Nebraska—Ford (1992), Holman (1995b), Sanchiz (1998) (recorded as *Pseudacris* cf. *Pseudacris clarkii* by all authors). **Pleistocene (Rancholabrean NALMA):** Easley Ranch Local Fauna, Foard County, Texas—Holman (1962b, 1969c, 1995b), Lynch (1964), Sanchiz (1998). Slaton Local Fauna, Lubbock County, Texas— Holman (1969a, 1995b), Sanchiz (1998).

The Spotted Chorus Frog, *Pseudacris clarkii*, is an attractive species that can be distinguished by its bright green blotches, which are outlined in black. The surrounding body is usually grayish, and the belly is white. This anuran inhabits grassland regions where moist habitats occur. At present, this species occurs in the central United States from Kansas to the Gulf of Mexico and the Rio Grande Valley, both in Texas and in Tamaulipas, Mexico (Frost, 1985).

Identification of Fossils. Chantell (1968b) reported that the scapula of *Pseudacris clarkii* is much shorter and stouter than that of the other species of *Pseudacris* he studied. The radio-ulna is shorter and stouter in *P. clarkii* and *Pseudacris streckeri* than in any of the others; and the sacral condyles are more closely juxtaposed in *P. clarkii* than in the others. The ilium of *P. clarkii* has the entire anterior margin of the ventral acetabular expansion anteriorly convex, whereas it varies from straight to concave to concave–convex in the other species studied by Chantell.

PSEUDACRIS CRUCIFER (WIED-NEUWIED, 1838)
SPRING PEEPER
(FIG. 81)

Fossil Localities. **Pleistocene (Irvingtonian NALMA):** Cumberland Cave, Allegany County, Maryland—Lynch (1966), Holman (1977c, 1995b), Sanchiz (1998). Hamilton Cave, Pendleton County, West Virginia—Holman and Grady (1989), Holman (1995b), Sanchiz (1998). **Pleistocene (Rancholabrean NALMA):** Cheek Bend Cave, Maury County, Tennessee—Klippel and Parmalee (1982), Holman (1995b), Sanchiz (1998). Clark's Cave, Bath County, Virginia—Fay (1988), Holman (1995b), Sanchiz (1998). Frankstown Cave, York County, Pennsylvania—Fay (1988), Holman (1995b), Sanchiz (1998). Kingston Saltpeter Cave, Bartow County, Georgia—Fay (1988), Holman (1995b), Sanchiz (1998). Ladds Quarry site, Bartow County, Georgia—Holman (1967b, 1985a, b, 1995b), Sanchiz (1998). New Paris 4 site, Bedford County, Pennsylvania—Lynch (1966), Fay (1988), Holman (1995b), Sanchiz (1998). New Trout Cave, Pendleton County, West Virginia—Holman and Grady (1987), Holman (1995b), Sanchiz (1998).

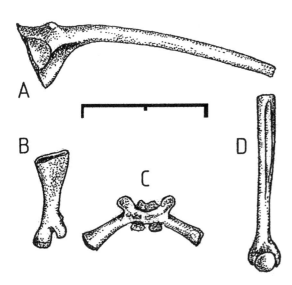

FIGURE 81. Bones of modern *Pseudacris crucifer* from Alachua County, Florida. (A) Right ilium in lateral view. (B) Right scapula in medial view. (C) Sacrum in dorsal view. (D) Right humerus in ventral view. Scale bar = 4 mm and applies to all figures.

The Spring Peeper, *Pseudacris crucifer*, was in the genus *Hyla* for many years. This little frog, one of the earliest harbingers of spring, is usually easily distinguished by the dark cross on its back. Its other dorsal coloration is generally light brownish gray, and there are no stripes or mottled blotches present. This species is a woodland frog that is often active on the forest floor when moist conditions are present. This species occurs in eastern North America except in extremely cold areas.

Identification of Fossils. The ilium of *Pseudacris crucifer* is generally similar to that of small species of *Hyla*, except that the dorsal tubercle of the ilium is more anteriorly placed relative to the anterior border of the acetabular fossa, as in *Acris* and *Pseudacris* (see Lynch, 1966, table 1, p. 266).

General Remarks. The presence of *Pseudacris crucifer* in the fossil localities above indicates the presence of closed-canopy woodlands.

PSEUDACRIS NIGRITA (LeCONTE, 1825)
SOUTHERN CHORUS FROG
(FIG. 82)

Fossil Locality. **Pleistocene (Rancho-labrean NALMA):** Reddick I site, Marion County, Florida—Wilson (1975), Holman (1995b), Sanchiz (1998).

Pseudacris nigrita, the southern counterpart of the Northern Chorus Frog, *Pseudacris triseriata,* can be separated from the latter on this basis: *P. nigrita* has either three rows of broad (usually broken) stripes or three rows of closely spaced spots on the back; *P. triseriata* has either three broad stripes or three thin (often broken) stripes on the back (Conant and Collins, 1998). Flat wetlands are the preferred habitat of *P. nigrita.*

Identification of Fossils. Wilson (1975, p. 16) identified *Pseudacris nigrita* from the Reddick I Pleistocene (Rancholabrean NALMA) site, northern Florida, on the criterion that the "ilium of *Pseudacris nigrita* exhibits a wider ventral acetabular expansion than is found in either *P. ornata* (Holbrook) or *P. triseriata* (Wied) and can be separated from these two species on that basis."

General Remarks. Wilson (1975) is the only one who has identified this species as a fossil.

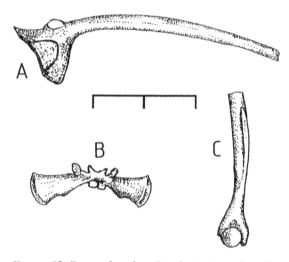

FIGURE 82. Bones of modern *Pseudacris nigrita* from Alachua County, Florida. (A) Right ilium in lateral view. (B) Sacrum in dorsal view. (C) Right humerus in ventral view. Scale bar = 4 mm and applies to all figures.

PSEUDACRIS NORDENSIS CHANTELL, 1964
(FIG. 83)

Holotype. The distal 6 mm of a left ilium (University of Nebraska State Museum: UNSM 61008).

Etymology. The specific name refers to the Norden Bridge Quarry, Brown County, Nebraska, where the type material was collected.

Locality and Horizon. Norden Bridge Quarry, Brown County, Nebraska: Miocene (medial Barstovian NALMA). Valentine Formation.

Other Material. Holman (1987) reported a complete right ilium (see Fig. 83, this volume) (Michigan State University, Museum Vertebrate Paleontology Collection: MSUVP 1105) from the Egelhoff site, Keya Paha County, Nebraska: (Miocene, medial Barstovian NALMA, Valentine Formation). This is the only other known material of *Pseudacris nordensis.*

Emended Diagnosis. A *Pseudacris* with the dorsal acetabular expansion very small and narrow; the dorsal tubercle very robust and elevated far from the acetabular fossa; the ventral acetabular expansion very narrow dorsally and making a wide angle with the long axis of the shaft; ilial size equaled only by that of large *Pseudacris streckeri*; well-developed dorsal ilial shaft ridge present.

Description. The description is based on the complete specimen in

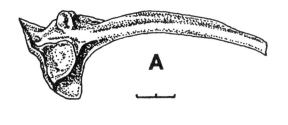

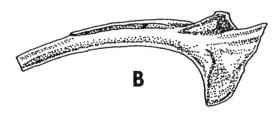

FIGURE 83. Right ilium in lateral (A) and medial (B) views, from the Middle Miocene (medial Barstovian), referred by Holman (1987) to *Pseudacris nordensis*. Scale bar = 2 mm and applies to both figures.

Holman (1987), since the type is quite incomplete (see Chantell, 1964, fig. 3a, p. 217). The ilium (MSUVP 1105) is complete and strongly built. Both the dorsal prominence and the laterally produced dorsal tubercle are quite robust. The dorsal tubercle is located about one-half of its length behind the anterior border of the acetabular fossa. It is dorsoventrally ovoid, and its middle part is roughened in lateral view. The dorsal prominence slopes gently into the tubercle anteriorly and abruptly into the very small dorsal acetabular expansion posteriorly. The acetabular fossa is well excavated, and its border is sharply defined and roughly hemispherical. The ilial shaft is strong and has a well-defined ilial shaft ridge that can be seen both laterally and medially. The ventral acetabular expansion is rather small, but its anterior edge extends moderately well beyond the anterior edge of the acetabular fossa. The surface of the ventral acetabular fossa is well excavated in lateral view.

Measurements of MSUVP 1105 compared with those of a large *Pseudacris streckeri* (the largest modern species of the genus) from Bath County, Illinois, and those of the type specimen are as follows: greatest length of ilium, 13.5 mm (in *P. streckeri*, 13.5 mm; unavailable in type specimen); length through posterior tip of dorsal acetabular expansion to anterior end of dorsal tubercle, 2.3 mm (in *P. streckeri*, 2.0 mm; in type specimen, 2.4 mm); height of ilial shaft just anterior to dorsal tubercle, 1.0 mm (in *P. streckeri*, 0.8 mm; in type specimen, 1.1 mm); height of dorsal tubercle through ventral border of acetabular fossa, 3.0 mm (in *P. streckeri*, 2.9 mm; in type specimen, 2.9 mm). These measurements indicate that rather than being larger than any living *Pseudacris* species as reported by Chantell (1964), MSUVP 1105 is actually about the same size as a modern living *P. streckeri*.

General Remarks. With more material, *Pseudacris nordensis* will likely be recognized as a new genus. Chantell (1964) suggested that *P. nordensis* exhibits characters intermediate between those of *Hyla* and those of *Pseudacris.* Sanchiz (1998, p. 129) relegated *P. nordensis* to his nomina dubia section.

PSEUDACRIS ORNATA (HOLBROOK, 1836)
ORNATE CHORUS FROG
(FIG. 84)

Fossil Locality. **Pleistocene (Rancholabrean NALMA):** Arredondo site, Alachua County, Florida—Lynch (1965), Holman (1995b), Sanchiz (1998). Isle of Hope site, Chatham County, Georgia—Hulbert and Pratt (1998).

The Ornate Chorus Frog, *Pseudacris ornata*, is, as its scientific name suggests, the most ornately colored of all the Chorus Frogs. At its best, it has a greenish back, with light pink spots and circles, as well as large black spots bordered by white rings. Some variations are darker and less colorful. This frog is found in the vicinity of wetlands harboring pine flatwoods, as well as cypress ponds and flooded meadows. This species inhabits the southeastern coastal plain of the United States from southeastern North Carolina to extreme eastern Louisiana.

Identification of Fossils. Lynch (1965) identified a *Pseudacris ornata* ilium from the Pleistocene (Rancholabrean NALMA) Arredondo site of Alachua County, Florida, on the basis that the acetabular fossa of *P. ornata* is a little larger than that of the closely related species *Pseudacris streckeri* and the angle of the ventral acetabular expansion is a little smaller.

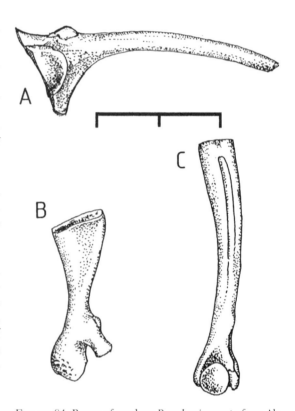

FIGURE 84. Bones of modern *Pseudacris ornata* from Alachua County, Florida. (A) Right ilium in lateral view. (B) Right scapula in medial view. (C) Right humerus in ventral view. Scale bar = 4 mm and applies to all figures.

PSEUDACRIS REGILLA (BAIRD AND GIRARD, 1852)

PACIFIC TREEFROG

(FIG. 85)

Fossil Locality. **Pleistocene (Rancholabrean NALMA):** Newport Beach Mesa site, Orange County, California—Hudson and Brattstrom (1977), Holman (1995b), Sanchiz (1998).

The Pacific Treefrog, *Pseudacris regilla*, is a moderately small, slightly plump species with a marked black or brown eye stripe but no other really significant stripes or blotches on the body. Moderately developed toe pads are present on the feet. This species occurs in many habitats where moisture and breeding sites are available, from sea-level flats to high mountain ranges. At present, the species occurs in western North America from southern British Columbia, Canada, and western Montana to southern Baja California, Mexico (Frost, 1985).

Identification of Fossils. Chantell (1970, pp. 655–656) presented a succinct summary of characters of the ilium of this species: "A well-developed dorsal protuberance [tubercle] with its anterior edge slightly anteriad to anterior edge of acetabulum; dorsal protuberance situated dorsolaterad, relatively near dorsal rim of acetabulum; acetabulum large with subtriangular outline; supraacetabular expansion [dorsal acetabular expansion] short, straight, directed dorsoposteriad; subacetabular expansion [ventral acetabular expansion] wide, forms relatively sharp angle with ilial shaft."

About the separation of the *Pseudacris regilla* from other modern

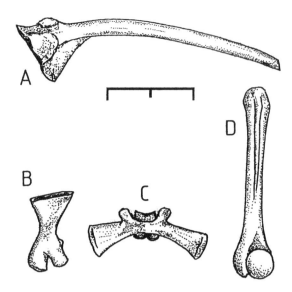

FIGURE 85. Bones of modern *Pseudacris regilla* from Riverside County, California. (A) Right ilium in lateral view. (B) Right scapula in medial view. (C) Sacrum in dorsal view. (D) Left humerus in ventral view. Scale bar = 4 mm and applies to all figures.

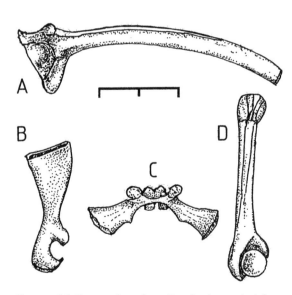

FIGURE 86. Bones of modern *Pseudacris streckeri* from Mason County, Illinois. (A) Right ilium in lateral view. (B) Right scapula in medial view. (C) Sacrum in dorsal view. (D) Left humerus in ventral view. Scale bar = 4 mm and applies to all figures.

hylids in the area, Chantell (1970, p. 656) stated, "The anterior edge of the dorsal protuberance [tubercle] is usually situated posterior to the anterior edge of the acetabulum in *H.* [*Hyla*] *wrightorum*, usually anterior in *H. californiae* and *H. arenicolor*, and usually even with the edge in *H.* [*Pseudacris*] *regilla*. In *H. californiae* and *H. arenicolor* the supra-acetabular [dorsal acetabular] expansion is longer, its dorsal surface more concave and the subacetabular [ventral acetabular] expansion is narrower as compared to *H.* [*Pseudacris*] *regilla*."

General Remarks. The value of *Pseudacris regilla* as an indicator of paleohabitats is diminished because of the present ubiquity of the species.

PSEUDACRIS STRECKERI A. A. WRIGHT AND A. H. WRIGHT, 1933
STRECKER'S CHORUS FROG
(FIG. 86)

Fossil Localities. **Pleistocene (Rancholabrean NALMA):** Howard Ranch (sometimes called Groesbeck Creek) Local Fauna, Hardeman County, Texas — Lynch (1966), Holman (1995b), Sanchiz (1998). Miller's Cave Fauna, Llano County, Texas — Holman (1966a, 1969c), Sanchiz (1998).

Strecker's Chorus Frog, *Pseudacris streckeri*, is rather toadlike for a Chorus Frog, with a chubby body and stout forelimbs. It is the largest living frog in the genus. Although its color is variable, it has a dark, wide stripe that runs from its rostrum to its shoulder. This frog is a hardy species that occupies a wide variety of habitats. It has a discontinuous range in the south-central United States and occurs in west-central and southern Illinois, southeastern Missouri, northeastern Arkansas, and from Oklahoma and west-central Arkansas south to the Gulf of Mexico (Conant and Collins, 1998).

Identification of Fossils. Holman (1966a, pp. 373–374) discussed his identification of two ilia of *Pseudacris streckeri* from the Pleistocene (Rancholabrean NALMA) Miller's Cave Fauna, Texas, as follows: "The ilium of *P. streckeri* is quite characteristic. It is larger and has its dorsal protuberance [tubercle] in a more posterior position on the shaft than in *Acris crepitans* and *A. gryllus*. It may be separated from *Hyla cinerea, H. squirella,* and *H. versicolor* by its rounder, more anteriorly placed protuberance. *Pseudacris streckeri* is larger and has a more anteriorly placed dorsal protuberance than *P. nigrita, P. ornata,* and *P. triseriata.* It differs from *P. clarkii* (the species sympatric with it in the area today) in being larger and having a straight border to its ventral acetabular expansion."

General Remarks. Some *Pseudacris streckeri* have a small dorsal ilial blade (see Fig. 86), which is a useful character in the identification of the species as a fossil.

PSEUDACRIS TRISERIATA (WIED-NEUWIED, 1838)
WESTERN CHORUS FROG
(FIG. 87)

Fossil Localities. **Pleistocene (Irvingtonian NALMA):** Cudahy Fauna, Meade County, Kansas—Chantell (1966), Holman (1995b), Sanchiz (1998) (recorded as *Pseudacris* cf. *Pseudacris triseriata* by all authors). Cumberland Cave, Allegany County, Maryland—Lynch (1966), Holman (1977c, 1995b), Sanchiz (1998). Hansen Bluff Chronofauna, Alamosa County, Colorado—Rogers et al. (1985), Rogers (1987), Holman (1995b), Sanchiz (1998). **Pleistocene (Rancholabrean NALMA):** Cragin Quarry Local Fauna, Meade County, Kansas—Tihen (1960b), Gehlbach (1965), Holman (1995b), Sanchiz (1998) (recorded as *Pseudacris* cf. *P. triseriata* by all authors). Dark Canyon Fauna, Eddy County, New Mexico—Applegarth (1980), Sanchiz (1998). Doby Springs Local Fauna, Beaver County, Oklahoma—Chantell (1966), Holman (1995b), Sanchiz (1998). Dry Cave Fauna unit II, Eddy County, New Mexico—Holman (1970b, 1995b), Sanchiz (1998). Jinglebob Local Fauna, Meade County, Kansas—Chantell (1966), Holman (1995b), Sanchiz (1998) (recorded as *Pseudacris* cf. *P. triseriata* by all authors). Jones Fauna, Meade County, Kansas (Rancholabrean NALMA)—Chantell (1966), Holman (1995b), Sanchiz (1998) (recorded as *Pseudacris* cf. *P. triseriata* by all authors). Sandahl Local Fauna, McPherson County, Kansas—Holman (1971, 1995b), Sanchiz (1998). Sheriden Pit Cave, Wyandot County, Ohio—Holman (1997).

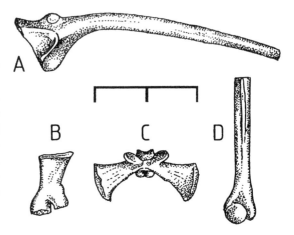

It really makes no sense to call *Pseudacris triseriata* the Western Chorus Frog, as it occurs almost everywhere in the eastern United States except peninsular Florida and does not occur west of the Rocky

FIGURE 87. Bones of modern *Pseudacris triseriata* from Ingham County, Michigan. (A) Right ilium in lateral view. (B) Right scapula in medial view. (C) Sacrum in dorsal view. (D) Right humerus in ventral view. Scale bar = 4 mm and applies to all figures.

Mountains. This small Chorus Frog has three dark stripes down the back; the stripes are broad and uninterrupted in some geographic areas and narrow and interrupted in others. These frogs occur in many habitats but need available shallow water in which to breed. The present range of this species is in North America: east of the Rocky Mountains, from Great Bear Lake and southeastern Quebec, Canada, southward to northern Florida and Texas (Frost, 1985).

Identification of Fossils. Holman (1971) separated *Pseudacris triseriata* from *Pseudacris clarkii*, a form with which it is sympatric in part of its western extent, on the basis that in *P. triseriata* the free margin of the ventral acetabular expansion is alternatively concave and convex, whereas in *P. clarkii* this margin is convex only.

General Remarks. The fact that almost all fossil records of *Pseudacris triseriata* are from the Pleistocene of the central plains of the United States is noteworthy, considering the extensive range this species has in the United States and Canada (see Conant and Collins, 1998, map, p. 543).

GENUS *PTERNOHYLA* BOULENGER, 1882

BURROWING TREEFROGS

This genus has the skin of the head co-ossified with the dermal bones of the head, forming a so-called casque. A prominent horizontal ridge is present between the nostrils and the eyes, and there is a prominent horizontal fold of skin just in back of the head. Only one species, *Pternohyla fodiens*, occurs in the United States, just getting into extreme southern Arizona. Only one other species occurs in this odd hylid genus that currently ranges from extreme southern Arizona southward to Michoacan and Aguascalientes, Mexico.

PTERNOHYLA FODIENS BOULENGER, 1882

LOWLAND BURROWING TREEFROG

Fossil Locality. **Pleistocene (Rancholabrean NALMA):** Rancho la Brisca locality, Sonora, Mexico—Van Devender et al. (1985), Sanchiz (1998).

Pternohyla fodiens is now called the Lowland Burrowing Treefrog. This animal lives in open grassy country and in tropical scrub forests and occurs from sea level to about 5000 feet (1500 m) above sea level (Stebbins, 1985). This odd anuran remains in its burrow during the daytime and becomes active at night. At present it occurs from extreme southern Arizona southward to Sonora and Michoacan in western Mexico (Van Devender et al., 1985).

Identification of Fossils. Characters on the basioccipital, dentary, and humerus used by Van Devender et al. (1985) to identify *Pternohyla fodiens* from the Pleistocene (Rancholabrean NALMA) of Sonora, Mexico are as follows. The basioccipital was referred to *P. fodiens* on the basis that the occipital condyle is a rounded knob rather than a flattened, elongate structure; the epiotic eminence is a prominent knob; the prootic–transverse process is expanded; and the basioccipital is not fused to the frontoparietal as it is in adult *Eleutherodactylus augusti* and *Bufo maz-*

atlanensis. The fossil dentary is referred to *Pternohyla fodiens* because it is larger than in *Leptodactylus melanonotus* and *Bufo kelloggi* but smaller than in female *Eleutherodactylus augusti, Bufo alvarius, Bufo cognatus,* and *Bufo mazatlanensis*; the medial flange is well developed on the ventral border of Meckel's canal just anterior to the articular surface as in *Hyla* (but not in *Bufo punctatus* and *Bufo retiformis*); and the dentary is thicker and curves more laterally than in *Hyla arenicolor.*

Turning to the fossil humerus, we find that this element has the broad distal end with the narrow shaft of the hylids but represents a larger individual (about 60 mm SVL) and is stouter than adult *Hyla arenicolor* (45 mm SVL). The humerus of *Smilisca baudinii* (a large tropical Treefrog) is not well ossified at 65 mm SVL and has a broader distal end.

General Remarks. As far as I can determine, the above locality has yielded the only fossil record of the genus.

SUPERFAMILY RANOIDEA FITZINGER, 1826

The superfamily Ranoidea contains the North American families Ranidae and Microhylidae.

FAMILY RANIDAE GRAY, 1825

The exceedingly large family Ranidae is cosmopolitan except for southern South America and most of Australia (Frost, 1985). Forty-six genera and about 625 species were recognized in 1992 (Duellman, 1993). However, only one genus, *Rana,* occurs in North America. Osteological characters used to define the family Ranidae follow.

Eight procoelous, holochordal presacral vertebrae, usually with the non-imbricate condition (imbricate in astylosternines), are present. In most ranid species the eighth presacral vertebra is biconcave, and the sacrum is biconvex. The first and the second presacral are not fused in most ranids except *Hemisus.* The atlantal cotyles of the first presacral are well separated. The sacrum has cylindrical diapophyses and a bicondylar articulation with the urostyle. The urostyle itself lacks transverse processes. The pectoral girdle is firmisternal except for a few species of *Rana* that have the arciferal condition. The omosternum is usually ossified, and the sternum is ossified in most groups; it is usually cartilaginous in petropedetines. Postzonal sternal elements are present except for *Hemisus.* These elements are cartilaginous in arthroleptines and astylosternines and bony in the other ranids. The scapula is not overlain by the clavicle in any ranids. The parahyoid is absent, and the cricoid ring is complete.

Both the premaxilla and the maxilla are tooth-bearing except for the hemisines, the mantellines, and some arthroleptines. The calcaneum and astragalus are fused only proximally and distally. Two tarsalia occur except in the astylosternines and some petropedetines, which have three. The phalangeal formula is normal, except in the mantellines, in which it is increased by the addition of short, cartilaginous intercalary elements between the penultimate and terminal phalanges.

Identification of fossil ranids below the generic level is often daunting, as osteological characters are exceedingly similar among many ranid species. Moreover, except for a few instances, most North American ranid fossils consist of isolated skeletal elements that are often broken and poorly preserved. Species in the *Rana pipiens* group are especially difficult to identify on the basis of individual fossil elements, and in fact, almost all identifications of pre-Pleistocene fossils in this group have been listed as "*Rana pipiens* group" or "*Rana pipiens* complex." Sanchiz (1998) has referred to this type of reference as *Rana* (*pipiens*) sp. Below I shall list North American localities where fossils of *Rana pipiens* complex have been reported. Literature references to these localities may be found in Sanchiz (1998).

In some cases, identification of species of the *Rana pipiens* group, mainly based on zoogeographic grounds, have been made in North American Pleistocene localities. I shall list these here with references. I shall not list fossil records referred to only as *Rana* sp., because of the widespread occurrence and large number of species of *Rana* that occur in North America today (see Stebbins, 1985; Conant and Collins, 1998). These records are not especially useful, as they all lie within the present range of the genus. The reader is again referred to Sanchiz (1998), where North American records of "*Rana* sp." localities and references to the localities can be found.

RANA PIPIENS COMPLEX LOCALITIES

Miocene (Hemingfordian NALMA): Thomas Farm Local Fauna, Gilchrist County, Florida.

Miocene (early Barstovian NALMA): Trinity River Local Fauna, San Jacinto County, Texas.

Miocene (medial Barstovian NALMA): Carrot Top Quarry, Brown County, Nebraska. Egelhoff Local Fauna, Brown County, Nebraska. Hottell Ranch rhino quarries, Banner County, Nebraska. Kuhre Quarry, Brown County, Nebraska. Norden Bridge Local Fauna, Brown County, Nebraska.

Miocene (late Barstovian NALMA): Glad Tidings Quarry, Knox County, Nebraska.

Miocene (Clarendonian NALMA): WaKeeney Local Fauna, Trego County, Kansas.

Miocene (Hemphillian NALMA): Haile locality VIA fauna, Alachua County, Florida. Lemoyne Quarry, Keith County, Nebraska. Pipe Creek Sinkhole Biota, Grant County, Indiana [this biota could represent the earliest Pliocene]. Mailbox Prospect, Antelope County, Nebraska. Santee Local Fauna, Knox County, Nebraska. White Cone Local Fauna, Navajo County, Arizona.

Pliocene (Blancan): Beck Ranch Local Fauna, Scurry County, Texas. Big Springs Local Fauna, Antelope County, Nebraska. Hagerman Fauna, Owyhee and Twin Falls counties, Idaho. Hornets Nest Quarry, Knox County, Nebraska. Lisco C quarries, Garden County, Nebraska. Sand Draw Local Fauna, Brown County, Nebraska. White Rock Local Fauna, Republic County, Kansas.

Pleistocene (Irvingtonian NALMA): Courland Canal–Hall Ash Assemblage, Jewell County, Kansas. Cumberland Cave, Allegany County,

Maryland. Gilliland Fauna, Knox County, Texas. Hamilton Cave, Pendleton County, West Virginia. Hanover Quarry No. 1 fissure, Adams County, Pennsylvania. Hansen Bluff Chronofauna, Alamosa County, Colorado. Java Local Fauna, Walworth County, South Dakota. Trout Cave, Pendleton County, West Virginia.

Pleistocene (Rancholabrean NALMA): Bell Cave, Colbert County, Alabama. Cueva de Abra, Tamaulipas, Mexico. Bootlegger Sink, York County, Pennsylvania. Christensen Bog, Hancock County, Indiana. Clark's Cave, Bath County, Virginia. Clear Creek Fauna, Denton County, Texas. Dark Canyon Cave, Eddy County, New Mexico. Dry Cave Fauna, Eddy County, New Mexico. Duck Creek Local Fauna, Ellis County, Kansas. Easley Ranch Local Fauna, Foard County, Texas. East Milford Mammoth site, Halifax County, Nova Scotia, Canada. Fowlkes Cave, Culberson County, Texas. Frankstown Cave, Blair County, Pennsylvania. Guy Wilson Cave, Sullivan County, Tennessee. Gypsum Cave, Clark County, Nevada. Howard Ranch Local Fauna, Hardeman County, Texas. Howell's Ridge Cave, Grant County, New Mexico. Jinglebob Local Fauna, Meade County, Kansas. Kanapolis Local Fauna, Ellsworth County, Kansas. Kingston Saltpeter Cave, Bartow County, Georgia. Kolarik mastodon site, Starke County, Indiana. Ladds Quarry site, Bartow County, Georgia. Lubbock Lake site, Lubbock County, Texas. Miller's Cave Fauna, Llano County, Texas. Mount Scott Local Fauna, Meade County, Kansas. Natural Chimneys site, Augusta County, Virginia. New Paris 4 site, Bedford County, Pennsylvania. New Trout Cave, Pendleton County, West Virginia. Peccary Cave, Newton County, Arkansas. Prairie Creek D Fauna, Daviess County, Indiana. Rancho la Brisca, Sonora, Mexico. Saltville site, Saltville Valley, Virginia. Sandahl Local Fauna, McPherson County, Kansas. Sheriden Pit Cave, Wyandot County, Ohio. Sims Bayou Fauna, Harris County, Texas. Slaton Local Fauna, Lubbock County, Texas. Strait Canyon Fissure, Highland County, Virginia. Williams Local Fauna, Rice County, Kansas. Zoo Cave, Taney County, Missouri.

GENUS *RANA* LINNAEUS, 1758

TRUE FROGS

About 222 species were recognized in the family in 1992 (Duellman, 1993). The range for the genus *Rana* is the same as for the family.

RANA AREOLATA BAIRD AND GIRARD, 1852

CRAWFISH FROG

(FIG. 88)

Fossil Localities. **Miocene (Clarendonian NALMA):** WaKeeney Local Fauna, Trego County, Kansas—Wilson (1968), Holman (1975), Sanchiz (1998) (noted as *Rana* cf. *Rana areolata* by all authors). **Pliocene (Blancan NALMA):** Beck Ranch Local Fauna, Scurry County, Texas—Rogers (1976), Sanchiz (1998). **Pleistocene (Irvingtonian NALMA):** Inglis IA site, Citrus County, Florida—Meylan (1982), Holman (1995b), Sanchiz (1998). **Pleistocene (Rancholabrean NALMA):** Reddick I site,

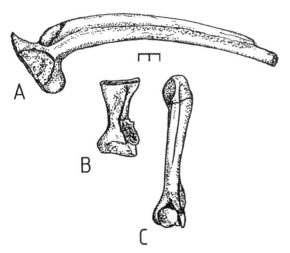

FIGURE 88. Bones of modern *Rana areolata* from Champaign County, Illinois. (A) Right ilium in lateral view. (B) Right scapula in medial view. (C) Right humerus in ventral view. Scale bar = 4 mm and applies to all figures.

Marion County, Florida—Wilson (1975), Holman (1995b), Sanchiz (1998).

The Crayfish Frog, *Rana areolata*, is a moderately large *Rana* that has closely packed dark spots with light borders on its back. This species spends much of its time in crayfish holes, small mammal burrows, or other underground shelters. The range of this species is southern Iowa, Indiana, and Illinois; south along the eastern and western borders of the Ozark Plateau to northern Mississippi, western Louisiana, and eastern Texas; and southeastern Louisiana and North Carolina to peninsular Florida on the coastal plain (Frost, 1985).

Identification of Fossils. In his identification of *Rana areolata* from the Pleistocene (Rancholabrean NALMA) of the Reddick I site in northern Florida, Wilson (1975, p. 10) made the following comments: "*R. areolata* ilia can be distinguished from those of *pipiens* [*Rana pipiens* complex] by the more pronounced craniad expansion of the ventral acetabular expansion that occurs in the latter. . . . This difference was found to be constant in the nine skeletons of *areolata* and seventy skeletons of *pipiens* [*R. pipiens* complex] available for study."

General Remarks. The four ilia tentatively assigned to *Rana areolata* from the Miocene (Clarendonian) WaKeeney Local Fauna from Kansas by Holman (1975) were clearly separable from 210 ilia assigned to the *Rana pipiens* complex.

RANA AURORA BAIRD AND GIRARD, 1852
RED-LEGGED FROG
(FIG. 89)

Fossil Localities. **Pliocene (Blancan NALMA):** Hagerman Fauna, Elmore, Owyhee, and Twin Falls counties, Idaho—Chantell (1970), Sanchiz (1998) (recorded as *Rana* cf. *Rana aurora* by both authors). **Pleistocene (Rancholabrean NALMA):** Newport Beach Mesa site, Orange County, California—Hudson and Brattstrom (1977), Holman (1995b), Sanchiz (1998). Rancho La Brea, Los Angeles County, California—Brattstrom (1953a), Holman (1995b), Sanchiz (1998).

The Red-legged Frog, *Rana aurora*, is a moderate-sized *Rana* that can be identified by the red coloration on the belly and the undersides of the hind legs. Prominent dorsolateral folds are present in this species. This frog is usually found in or near ponds in wet forest, woodland, and grassland habitats, where it is mainly restricted to lowlands and foothills (Stebbins, 1985). The present distribution of this frog is on Vancouver Island, British Columbia, Canada, and south along the Pacific coast of the United States to northern Baja California, Mexico (Frost, 1985).

Identification of Fossils. Chantell (1970) gave a rather rambling account of osteological criteria related to the identification of *Rana* cf. *Rana aurora* from the Hagerman Fauna (Pliocene, Blancan) of Idaho.

Rana cf. *Rana blairi* Mecham, Littlejohn, Oldham, L. E. Brown, and J. R. Brown, 1973
Plains Leopard Frog

Fossil Locality. **Pleistocene (Irvingtonian NALMA):** Albert Ahrens Local Fauna, Nuckolls County, Nebraska—Ford (1992), Holman (1995b), Sanchiz (1998).

At present, this light-colored frog of the *Rana pipiens* complex occurs from western Indiana across the central and southern plains to eastern Colorado and New Mexico and south to central Texas; an isolated population occurs in southeastern Arizona (Frost, 1985).

Identification of Fossils. Ford (1992) identified this frog of the *Rana pipiens* complex as *Rana* cf. *Rana blairi* on zoogeographic grounds.

Rana catesbeiana Shaw, 1802
American Bullfrog
(Fig. 90)

Fossil Localities. **Miocene (Hemphillian NALMA):** Mailbox Prospect site, Antelope County, Nebraska—Parmley

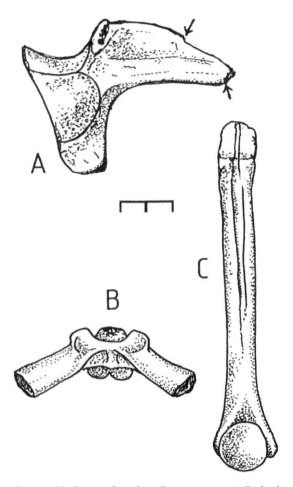

Figure 89. Bones of modern *Rana aurora*. (A) Right ilium in lateral view (anterior end broken; breakage points indicated by arrows). (B) Sacrum in dorsal view. (C) Right humerus in ventral view. Scale bar = 4 mm and applies to all figures.

(1992), Sanchiz (1998) (recorded as *Rana* cf. *Rana catesbeiana* by both authors). Pipe Creek Sinkhole Biota, Grant County, Indiana (this biota possibly represents the earliest Pliocene)—Farlow et al. (2001) (recorded as *Rana* cf. *R. catesbeiana*). Santee site, Knox County, Nebraska—Parmley (1992), Sanchiz (1998) (recorded as *Rana* cf. *R. catesbeiana* by both authors). **Pliocene (Blancan NALMA):** Hornets Nest Quarry, Knox County, Nebraska—Rogers (1984), Sanchiz (1998). Sand Draw Local Fauna, Brown County, Nebraska—Holman (1972a), Sanchiz (1998). White Rock Local Fauna, Republic County, Kansas—Eshelman (1975), Sanchiz (1998) (recorded as *Rana* cf. *R. catesbeiana* by both authors). **Pleistocene (Irvingtonian NALMA):** Albert Ahrens Local Fauna, Nuckolls County, Nebraska—Ford (1992), Holman (1995b), Sanchiz (1998). Courland Canal–Hall Ash Assemblage, Jewell County, Kansas—Rogers

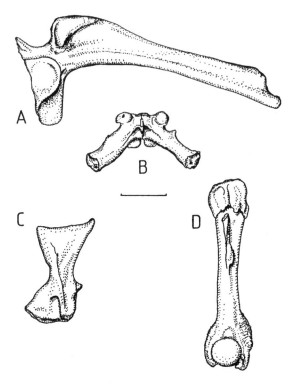

FIGURE 90. Bones of modern *Rana catesbeiana* from Alachua County, Florida. (A) Right ilium in lateral view. (B) Sacrum in dorsal view. (C) Right scapula in medial view. (D) Right humerus in ventral view. Scale bar = 10 mm and applies to all figures.

(1982), Holman (1995b), Sanchiz (1998). Hamilton Cave, Pendleton County, West Virginia—Holman and Grady (1989), Holman (1995b), Sanchiz (1998). Hansen Bluff Chronofauna, Alamosa County, Colorado—Rogers et al. (1985), Holman (1995b), Sanchiz (1998). Inglis IA site, Citrus County, Florida—Meylan (1982), Holman (1995b), Sanchiz (1998). Java Local Fauna, Walworth County, South Dakota—Holman (1977a, 1995b), Sanchiz (1998). **Pleistocene (Rancholabrean NALMA):** Arredondo site, Alachua County, Florida—Lynch (1965), Holman (1995b), Sanchiz (1998). Avenue Local Fauna, Austin, Travis County, Texas—Lundelius (1992), Sanchiz (1998) (recorded as *Rana* cf. *R. catesbeiana* by both authors). Bell Cave, Colbert County, Alabama—Holman et al. (1990), Holman (1995b), Sanchiz (1998). Ben Franklin Local Fauna, Fannin and Delta counties, Texas—Holman (1963b, 1995b), Sanchiz (1998). Boney Spring, Benton County, Missouri—Saunders (1977), Holman (1995b), Sanchiz (1998). Brynjulfson Cave I, Boone County, Missouri—Parmalee and Oesch (1972), Holman (1995b), Sanchiz (1998). Clark's Cave, Bath County, Virginia—Fay (1988), Holman (1995b), Sanchiz (1998). Devil's Den chamber 3, Levy County, Florida—Holman (1978b, 1995b), Sanchiz (1998). Ingleside Fauna, San Patricio County, Texas—Preston (1979), Holman (1995b), Sanchiz (1998). Isle of Hope site, Chatham County, Georgia—Hulbert and Pratt (1998). Jinglebob Local Fauna, Meade County, Kansas—Tihen (1954), Holman (1995b), Sanchiz (1998). Kanapolis Local Fauna, Ellsworth County, Kansas—Holman (1972c, 1995b), Sanchiz (1998). Kingston Saltpeter Cave, Bartow County, Georgia—Fay (1988), Holman (1995b), Sanchiz (1998). Lubbock Lake site, Lubbock County, Texas—Holman (1995b), Johnson (1987), Sanchiz (1998). Miller's Cave Fauna, Llano County, Texas—Holman (1966a, 1995b), Sanchiz (1998). Natural Chimneys site, Augusta County, Virginia—Fay (1988), Holman (1995b), Sanchiz (1998). Peccary Cave, Newton County, Arkansas—Davis (1973), Holman (1995b), Sanchiz (1998). Prairie Creek D Fauna, Daviess County, Indiana—Holman (1992b, 1995b), Holman and Richards (1993), Sanchiz (1998). Strait Canyon Fissure, Highland County, Virginia—Fay (1984), Holman (1995b), Sanchiz (1998). Vero Beach strata 2 and 3, Indian River County, Florida—Weigel (1962), Holman (1995b) (recorded as *Rana* cf. *R. catesbeiana* by both authors); Sanchiz (1998). Williams Local Fauna, Rice County, Kansas—Holman (1984, 1995b), Sanchiz (1998).

The American Bullfrog, *Rana catesbeiana*, the largest North American frog, is mainly green above, with few markings on the body. It lacks dorsolateral ridges on the trunk, as these ridges end posteriorly at about the level of the tympanum. The American Bullfrog is usually found in or at the edge of permanent bodies of still water, ranging from small ponds to large lakes and swamps. At present, the American Bullfrog occurs in eastern North America (except southern Florida); north to Nova Scotia, New Brunswick, southern Quebec, and southern Ontario, Canada; west to the central plains; and south to Veracruz, Mexico (Frost, 1985). This species has been introduced widely in the rest of the world.

Identification of Fossils. The ilium of *Rana catesbeiana* is larger than most *Rana* fossils one comes across and has a very steep slope of the ilial shaft into the dorsal acetabular expansion, which is lacking in frogs of the *Rana pipiens* complex and apparently even in the very large species *Rana grylio*. In young American Bullfrogs, which are as large as many adults of other species of *Rana*, the ilia usually have very porous acetabular areas, especially within the acetabular fossa, whereas ilia of the other species of the same size are usually nonporous.

General Remarks. From a paleoecological standpoint, the presence of *Rana catesbeiana* in a fossil fauna indicates the presence of a permanent body of still water.

RANA CLAMITANS LATREILLE, 1801

GREEN FROG

(FIG. 91)

Fossil Localities. **Miocene (medial Barstovian NALMA):** Hottell Ranch rhino quarries, Banner County, Nebraska—Voorhies et al. (1987), Sanchiz (1998) (recorded as *Rana* cf. *Rana clamitans* by both). **Pleistocene** (Irvingtonian NALMA): Albert Ahrens Local Fauna, Nuckolls County, Nebraska—Ford (1992), Holman (1995b), Sanchiz (1998). Cumberland Cave, Allegany County, Maryland—Holman (1977c, 1995b), Sanchiz (1998). Hamilton Cave, Pendleton County, West Virginia—Holman and Grady (1989), Sanchiz (1998). **Pleistocene (Rancholabrean NALMA):** Clark's Cave, Bath County, Virginia—Fay (1988), Holman (1995b), Sanchiz (1998). Kingston Saltpeter Cave, Bartow County, Georgia—Fay (1988), Holman (1995b), Sanchiz (1998). Natural Chimneys site, Augusta County, Virginia—Fay (1988), Holman (1995b), Sanchiz (1998). New Paris 4 site, Bedford County, Pennsylvania—Fay (1988), Holman (1995b) [recorded as *Rana* (?) *clamitans* by both authors]. New Trout Cave, Pendleton County, West Virginia—Holman and

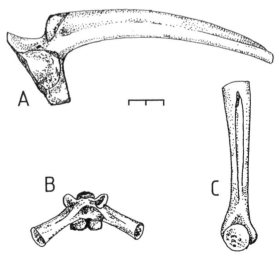

FIGURE 91. Bones of modern *Rana clamitans* from Ingham County, Michigan. (A) Right ilium in lateral view. (B) Sacrum in dorsal view. (C) Right humerus in ventral view. Scale bar = 4 mm and applies to all figures.

Grady (1987), Sanchiz (1998). Prairie Creek D Fauna, Daviess County, Indiana—Holman (1992b, 1995b), Holman and Richards (1993), Sanchiz (1998). Shelton site, Oakland County, Michigan—DeFauw and Shoshani (1991), Holman (1992b, 1995b); Sanchiz (1998) (recorded as *Rana* cf. *R. clamitans*). Sheriden Pit Cave, Wyandot County, Ohio—Holman (1995b, 1997), Sanchiz (1998). Strait Canyon Fissure, Highland County, Virginia—Fay (1984), Holman (1995b), Sanchiz (1998).

The Green Frog (*Rana clamitans*) is greener in the north than in the south, where it is called the Bronze Frog and designated as a different subspecies. *Rana clamitans* is a large *Rana* that can be easily distinguished from the American Bullfrog (*Rana catesbeiana*) on the basis that the Green Frog has a dorsolateral ridge that goes from the tympanum to the end of the body. In the north, *R. clamitans* may be found in or near almost any habitat with shallow, still water. In the south it is more secretive and tends to hide under objects near the water. This species is indigenous to eastern North America, from southern Ontario, Quebec, and Manitoba, Canada, south to central Florida and eastern Texas. Like the American Bullfrog, the Green Frog has been introduced elsewhere, but not as extensively.

Identification of Fossils. Both *Rana clamitans* and *Rana catesbeiana* have the dorsal border of the ilium sloping precipitously into the dorsal acetabular expansion. But the slope is usually even more precipitous in *R. catesbeiana* than in *R. clamitans*, especially in large *R. catesbeiana*. Moreover, ilia of *R. catesbeiana* of the same size as those of adult *R. clamitans* tend to be poorly ossified and porous in the acetabular area, especially within the acetabular fossa itself.

General Remarks. *Rana* fossils from the Shelton Pleistocene (Rancholabrean NALMA) site of Oakland County, Michigan, were identified as both *Rana catesbeiana* and *Rana clamitans* (DeFauw and Shoshani, 1991), when actually only *R. clamitans* is indicated by the fossil elements.

RANA GRYLIO STEJNEGER, 1901
PIG FROG
(FIG. 92)

Fossil Localities. **Pleistocene (Rancholabrean NALMA):** Haile (Rancholabrean complex), Alachua County, Florida—Tihen (1952), Brattstrom (1953b), Holman (1995b), Sanchiz (1998). Reddick I site, Marion County, Florida—Wilson (1975), Holman (1995b), Sanchiz (1998).

The Pig Frog, *Rana grylio*, has a call that sounds like a grunting pig—hence the vernacular name. This distinctive call separates *R. grylio* from *Rana catesbeiana*, which has the "jug-o-rum" call; otherwise, it is difficult to distinguish between the two species, as they are both about the same size and have similar markings. The standard way to tell them apart is to examine the fourth toe: it extends beyond the web in *R. catesbeiana* and ends at the web in *R. grylio*. *Rana grylio* is another very aquatic frog that is found in or near permanent bodies of still water. The natural range of the species is South Carolina and from eastern Texas along the coastal plain to peninsular Florida (Frost, 1985).

Identification of Fossils. Wilson (1975) identified several bones of an individual male frog specimen from the Pleistocene (Rancholabrean NALMA) of the Reddick IA site in Alachua County, Florida, as *Rana grylio*. Wilson determined that the maxilla is lower and longer in *R. grylio* than in *Rana catesbeiana* and *Rana heckscheri*, two other very large species that currently occur in Florida. Turning to the pelvic girdle, we find that the slope of the dorsal border of the ilium into the dorsal acetabular expansion is less steep in *R. grylio* than in *R. catesbeiana* and *Rana clamitans* (see Figs. 90–92).

General Remarks. As in *Rana catesbeiana*, the presence of *Rana gryiio* in a fossil fauna indicates the presence of a permanent body of still water in the vicinity.

RANA PALUSTRIS LeConte, 1825
PICKEREL FROG
(Fig. 93)

Fossil Localities. **Pliocene (Blancan NALMA):** Beck Ranch Local Fauna, Scurry County, Texas—Rogers (1976), Sanchiz (1998). **Pleistocene (Rancholabrean NALMA):** Clark's Cave, Bath County, Virginia—Fay (1988), Holman (1995b), Sanchiz (1998). Lubbock Lake site (Clovis-period level), Lubbock County, Texas—Johnson (1987), Holman (1995b), Sanchiz (1998). Natural Chimneys site, Augusta County, Virginia—Fay (1988), Holman (1995b) [recorded as *Rana* (?) *palustris* by both authors]; Sanchiz (1998) (recorded as *Rana* cf. *Rana palustris*). New Paris 4 site, Bedford County, Pennsylvania—Fay (1988), Holman (1995b) [recorded as *Rana* (?) *palustris* by both authors]; Sanchiz (1998) (recorded as *Rana* cf.*Rana pipiens*). Peccary Cave, Newton County, Arkansas—Davis (1973), Holman (1995b), Sanchiz (1998) (recorded as *Rana* cf. *R. palustris* by all authors).

The Pickerel Frog, *Rana palustris*, is a medium-sized *Rana* with a double row of spots down the back. The upper body coloration is often a greenish brown. It has bright yellow or orange coloration on the

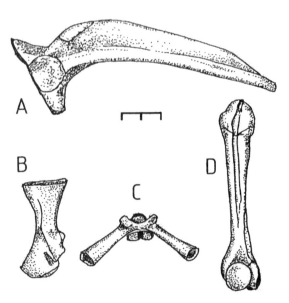

FIGURE 92. Bones of modern *Rana grylio* from Florida. (A) Right ilium in lateral view. (B) Right scapula in medial view. (C) Sacrum in dorsal view. (D) Right humerus in ventral view. Scale bar = 4 mm and applies to all figures.

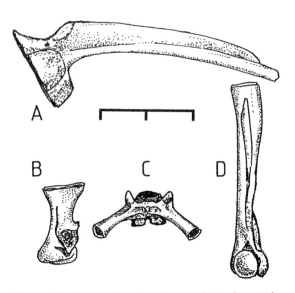

FIGURE 93. Bones of modern *Rana palustris* from Kalamazoo County, Michigan. (A) Right ilium in lateral view. (B) Right scapula in medial view. (C) Sacrum in dorsal view. (D) Right humerus in ventral view. Scale bar = 4 mm and applies to all figures.

bottom of the legs. In the north this species prefers cool, clear water, but in the south it is found in more turbid aquatic habitats. This species wanders fairly far from water out into grassy and shrubby areas from time to time. At present, the Pickerel Frog occurs in eastern North America from Quebec through Ontario, Canada, to Minnesota and south to northern South Carolina, Georgia, and Alabama and eastern Texas (Frost, 1985).

Identification of Fossils. Lynch (1965) reported that the ilium of *Rana palustris* may be separated from that of *Rana pipiens* on the basis of these criteria: the ilial blade is thinner and more delicate, the posterior portion of the ilial blade is more angular, and the ilial prominence is less pronounced in *R. palustris.*

General Remarks. The above identification of *Rana palustris* in the Pliocene (Blancan NALMA) of Scurry County, Texas, and in the Pleistocene (Rancholabrean NALMA) of Lubbock County, Texas, both well west of the modern range of the species (see Conant and Collins, 1998, map, p. 570), is noteworthy.

RANA PIPIENS SCHREBER, 1782
NORTHERN LEOPARD FROG
(FIG. 94)

Fossil Locality. **Pleistocene (Rancholabrean NALMA):** Eardley locality 2, Quebec, Canada—Holman et al. (1997). Sheriden Pit Cave, Wyandot County, Ohio—Holman (1995b, 1997).

This species is often very green in the northern part of its range and browner toward the south. It has two or three rows of dark spots that lie between its prominent dorsolateral ridges. This frog wanders far into grassy meadows and other moist, rather open areas during the summer months and may be found quite far from water from time to time. *Rana pipiens* has a complicated natural range: from southern Quebec west to the extreme southern District of Mackenzie; south to Pennsylvania and Kentucky in the east (isolated records occurring in Maryland and West Virginia); west to the Pacific states; and south in the western United States to Nevada, Arizona, and New Mexico (Conant and Collins, 1998).

Identification of Fossils. Species in the *Rana pipiens* complex have such similar osteology that isolated fossils are difficult to identify. The ilia of this complex, however, tend to have the dorsal border of the ilium sloping much more gently into the dorsal acetabular expansion than in the large frogs outside this complex (e.g., *Rana catesbeiana* and *Rana clamitans*), which have the dorsal border of the ilium sloping

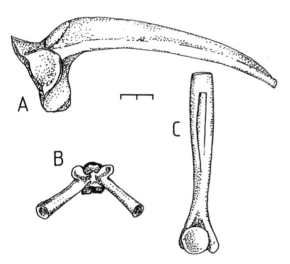

FIGURE 94. Bones of modern *Rana pipiens* from Minnesota. (A) Right ilium in lateral view. (B) Sacrum in dorsal view. (C) Right humerus in ventral view. Scale bar = 4 mm and applies to all figures.

more precipitously into the dorsal acetabular expansion. The above identification of R. pipiens from the Pleistocene (Rancholabrean NALMA) of
northwestern Ohio is based on the fact that the identified ilium is identical to that of modern R. pipiens, in addition to zoogeographic grounds
(see Conant and Collins, 1998, maps, pp. 567, 569).

The Rana pipiens fossil from the Pleistocene (Rancholabrean
NALMA) Eardley, Quebec, location was identified by Holman et al.
(1997) by going through the following process. The ilia, which were the
best-preserved diagnostic elements on this skeleton, were compared with
a series of ilia of modern Rana from northeastern North America. Ilia of
R. pipiens differ strongly from those of Rana catesbeiana, Rana clamitans,
and Rana septentrionalis in having the posterior border of the dorsal ilial
crest sloping much less precipitously into the dorsal acetabular expansion.
Rana pipiens differs markedly from Rana sylvatica in having the dorsal
ilial prominence flattened rather than somewhat swollen and knoblike.
The dorsal ilial prominence of R. sylvatica approaches the condition
found in North American Hyla rather than in Rana. Rana pipiens appears
to be more similar to Rana palustris than to any other species of Rana
in northeastern North America, but it differs in having the dorsal ilial
prominence more distinct from the shaft, and usually, in having the posterior border of the ilial crest sloping more precipitously into the dorsal
acetabular expansion.

General Remarks. The Rana pipiens fossil from the Pleistocene (Rancholabrean NALMA) Eardley, Quebec, locality in Canada is in the form
of a complete skeleton preserved in a lifelike pose. It was found in a
calcareous nodule. The frog lived in a lowland area with grassy meadows
near the so-called Champlain Sea that inundated the area about 10 ka
BP, at the end of the Pleistocene. The Eardley fossil is the most complete
and well-preserved fossil frog that I am aware of in North America.

* RANA PLIOCENICA ZWEIFEL, 1954
(FIG. 95)

Holotype. The posterior part of a skeleton, including the fifth vertebra
and excluding only some phalanges and
the pubis; fragments of anterior elements
present (University of California Museum
of Paleontology: UCMP 1037/34564).

Etymology. The specific name reflects
the original belief that the specimen was
from the Pliocene.

Locality and Horizon. Rodeo locality,
Contra Costa County, California: Miocene (Hemphillian NALMA). Pinole Tuff.

Other Material. No other material is
known.

Diagnosis. The diagnosis is from
Zweifel (1954, p. 85): "A species of Rana
characterized by a relatively deep medial
groove on the ventral surface of the sacral

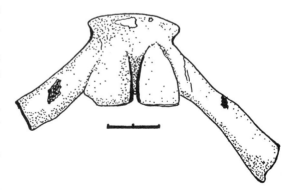

FIGURE 95. Ventral view of the sacral vertebra from the
holotype of Rana pliocenica from the Late Miocene
(Hemphillian) of California. Scale bar = 2 mm.

vertebra (Fig. 2) [see Fig. 95, this volume] and a relatively deep ilial blade."

Description. Excerpts from Zweifel (1954) are slightly modified as follows. The length of dorsal ilial crest is 22.0 mm, and the greatest depth of the crest is 4.1 mm. The length of the tibia is 39.2 mm, and the lengths of vertebrae VI and VII are 6.0 mm. The length of the astragalus is 17.4 mm. The sacral vertebra has the typical ranid characteristic of cylindrical diapophyses and paired condyles that articulate with the urostyle. The first presacral vertebra is amphicoelous. *Rana pliocenica* is characterized by a relatively deep ilium. Of the living western species, *R. pliocenica* most nearly resembles *Rana aurora* both in ilial proportions and in general body size. *Rana pliocenica* may have been the predecessor of *R. aurora.* Paleobotanical studies have shown that the types of plants with which *R. aurora* is associated were well developed in California at the time the extinct frog occurred there.

Rana sphenocephala Cope, 1886
Southern Leopard Frog
(Fig. 96)

Fossil Localities. **Pleistocene (Rancholabrean NALMA):** Arredondo site, Alachua County, Florida—Lynch (1965), Holman (1995b). Devil's Den chamber 3, Levy County, Florida—Holman (1978b, 1995b). Ladds Quarry site, Bartow County, Georgia—Holman (1967b, 1985a, b, 1995b), Wilson (1975). Reddick I site, Marion County, Florida—Gut and Ray (1963), Wilson (1975), Holman (1995b). Sabertooth Cave, Citrus County, Florida—Holman (1958, 1995b). Vero Beach strata 2 and 3, Indian River County, Florida—Weigel (1962), Holman (1995b). Williston IIIA site, Levy County, Florida—Holman (1959a, 1995b, 1996c).

The southern equivalent of *Rana pipiens*, the Southern Leopard Frog (*Rana sphenocephala*), long had the scientific name *Rana utricularia*. *Rana sphenocephala* usually differs from *R. pipiens* in having a more pointy head, fewer spots on the sides of the body, and a light spot in the center of the eardrum. This usually abundant frog can be found near almost any quiet aquatic habitat, and like its northern counterpart, wanders away from the water when the breeding season is over. The modern natural range of the species is southern New York, south along the Atlantic coastal plain to southern Florida, west to eastern Texas, and north to eastern Kansas, southern Illinois, Indiana, and Ohio; most of the Appalachian Mountains are excluded.

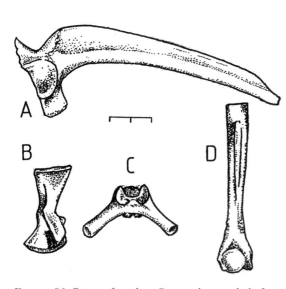

Figure 96. Bones of modern *Rana sphenocephala* from Florida. (A) Right ilium in lateral view. (B) Right scapula in medial view. (C) Sacrum in dorsal view. (D) Right humerus in ventral view. Scale bar = 4 mm and applies to all figures.

Identification of Fossils. Fossils identified as "*Rana utricularia*" (now *Rana sphenocephala*) were almost always originally identified as *Rana pipiens*. The references here are based on zoogeographic grounds (see Conant and Collins, 1998, maps, pp. 567, 569).

RANA SYLVATICA LeCONTE, 1825
WOOD FROG
(FIG. 97)

Fossil Localities. **Pliocene (Blancan NALMA):** Hornets Nest Quarry, Knox County, Nebraska—Rogers (1984), Sanchiz (1998). **Pleistocene (Irvingtonian NALMA):** Albert Ahrens Local Fauna, Nuckolls County, Nebraska—Ford (1992), Holman (1995b), Sanchiz (1998). Cumberland Cave, Allegany County, Maryland—Holman (1977c, 1995b), Sanchiz (1998). Hamilton Cave, Pendleton County, West Virginia—Holman and Grady (1989), Holman (1995b), Sanchiz (1998). Trout Cave, Pendleton County, West Virginia—Holman (1982b, 1995b), Sanchiz (1998). **Pleistocene (Rancholabrean NALMA):** Baker Bluff Cave, Sullivan County, Tennessee—Fay (1988), Holman (1995b), Sanchiz (1998). Clark's Cave, Bath County, Virginia—Fay (1988), Holman (1995b), Sanchiz (1998). Duck Creek Local Fauna, Ellis County, Kansas—Holman (1984, 1995b), Sanchiz (1998). Frankstown Cave, Blair County, Pennsylvania—Fay (1988), Holman (1995b), Sanchiz (1998). Natural Chimneys site, Augusta County, Virginia—Fay (1988), Holman (1995b), Sanchiz (1998). New Paris 4 site, Bedford County, Pennsylvania—Fay (1988), Holman (1995b), Sanchiz (1998). New Trout Cave, Pendleton County, West Virginia—Holman and Grady (1987), Holman (1995b), Sanchiz (1998). Prairie Creek D Fauna, Daviess County, Indiana—Holman (1992b, 1995b), Holman and Richards (1993), Sanchiz (1998). Worm Hole Cave, Pendleton County, West Virginia—Holman and Grady (1994), Holman (1995b), Sanchiz (1998).

The Wood Frog, *Rana sylvatica*, a fairly small *Rana*, is one of the most distinct species of the genus in North America. A broad, dark patch, extending through the eye and back to the shoulder, is a distinguishing feature. This frog extends farther north than any other North American amphibian or reptile and is able to freeze solid during the winter and thaw out without any harmful effects. As its range extends northward, its legs get shorter and it takes on a rather toadlike appearance. The Wood Frog occurs mainly in woodland habitats, where it breeds in temporary ponds early in the spring. The present distribution of this species includes Alaska and Labrador, Canada, and south to northern Idaho, Minnesota, northern

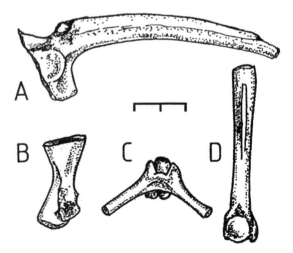

FIGURE 97. Bones of modern *Rana sylvatica* from Ingham County, Michigan. (A) Right ilium in lateral view. (B) Right scapula in medial view. (C) Sacrum in dorsal view. (D) Right humerus in ventral view. Scale bar = 4 mm and applies to all figures.

Georgia, and Maryland; isolated populations occur in southern Missouri and adjacent Arkansas and in northern Colorado (Frost, 1985).

Identification of Fossils. The ilium of *Rana sylvatica* is the most easily identified of all ilia of the species of *Rana* that I am familiar with. The dorsal prominence usually has a rounded tubercle on it, and this is sometimes so rounded that it resembles that of a *Hyla* more than it does that of a *Rana*. Sometimes the tubercle is a little roughened, and it can be less distinct in some specimens than in others.

General Remarks. The fact that the ilium of *Rana sylvatica* is fairly easy to identify probably accounts for the large number of fossil records of this taxon. This frog is not related to any of the established species groups (Frost, 1985). I have implied that it might be related to some of the "brown frogs" of Europe (Holman, 1998).

FAMILY MICROHYLIDAE GÜNTHER, 1859 "1858" (1843)

The large family Microhylidae occurs in North and South America, in Africa south of the Sahara, and from India and the Koreas to northern Australia (Frost, 1985). Seventy genera and about 313 species were recognized in 1992 (Duellman, 1993).

Osteological characters used to define the family are as follows. Eight procoelous, holochordal presacral vertebrae are present, and they can have either the imbricate or non-imbricate condition. All the vertebrae are procoelous in the cophylines and genyophrynines, but the eighth presacral is biconcave and the sacrum is biconvex in the other subfamilies. The first and second presacrals are not fused, and the cervical condyles of the first presacral are widely separated. Ribs are absent in all taxa of the family. The sacral vertebra has broadly dilated diapophyses and a bicondylar articulation with the urostyle. This articulation is fused in some brevicipines. The urostyle lacks transverse processes. The pectoral girdle is firmisternal. The omosternum is absent in most taxa but occurs in brevicipines, cophylines, and some dyscophines and melanobatrachines. The sternum is cartilaginous. If present, the clavicles do not overlie the scapulae anteriorly. Clavicles are reduced or absent in all taxa other than the brevicipines. In most taxa the palatines are reduced or absent. The parahyoid is absent, and the cricoid ring is complete. Both the premaxilla and the maxilla lack teeth, except in the dyscophines and some cophylines. The calcaneum and astragalus are only fused proximally and distally. There are two tarsalia, and the phalangeal formula is normal in most taxa. The phalangeal formula is reduced in the melanobatrachines, but it is increased by the addition of short, cartilaginous intercalary elements between the penultimate and terminal phalanges in the phrynomerines.

GENUS *GASTROPHRYNE* FITZINGER, 1843

NORTH AMERICAN NARROW-MOUTHED TOADS

Living species of *Gastrophryne* occur from the southern United States south to Costa Rica (Frost, 1985). Five species were recognized in the genus both in 1985 and in 1992 (Duellman, 1993). *Gastrophryne* sp. was identified from the Pleistocene (Irvingtonian NALMA) of the Inglis IA site, Citrus County, Florida, by Meylan (1982). The other North American fossil records are of the extant species *Gastrophryne carolinensis* and *Gastrophryne olivacea*. Van Devender et al. (1985, p. 35) gave osteological criteria for the identification of the ilium and humerus of *Gastrophryne*: "The ilia of *Gastrophryne* are very distinctive and differ from other small anurans including *Acris*, *Bufo*, *Hyla*, and *Pseudacris* in the following combination of characters: (1) the head [acetabular area] of the ilium is at a sharper angle with the shaft; (2) dorsal acetabular expansion is greatly reduced; (3) ventral acetabular expansion is moderately deep but very broad. The humeri are also distinctive and exhibit the following characters: (1) well-ossified at a very small size; (2) the shaft is relatively straight; (3) the lateral epicondyle is very poorly developed; (4) the lateral crests poorly developed or absent; (5) the posterior surface opposite the ulnar condyle is rounded rather than flattened."

Gastrophryne carolinensis (Holbrook, 1836)
Eastern Narrow-mouthed Toad
(Fig. 98)

Fossil Localities. **Miocene (Hemingfordian NALMA):** Thomas Farm Local Fauna, Gilchrist County, Florida—Auffenberg (1956), Holman (1965), Sanchiz (1998) (recorded as *Gastrophryne* cf. *Gastrophryne carolinensis* by all authors). **Pleistocene (Rancholabrean NALMA):** Arredondo site, Alachua County, Florida—Holman (1962a, 1995b), Lynch (1965), Sanchiz (1998). Kingston Saltpeter Cave, Bartow County, Georgia—Fay (1988), Holman (1995b), Sanchiz (1998). Ladds Quarry site, Bartow County, Georgia—Holman (1985a, b, 1995b), Sanchiz (1998). Reddick I site, Marion County, Florida—Holman (1962a, 1995b), Wilson (1975), Sanchiz (1998). Sabertooth Cave, Citrus County, Florida—Holman (1958, 1995b), Sanchiz (1998). Williston IIIA site, Levy County, Florida—Holman (1959a, 1995b, 1996c), Sanchiz (1998).

The Eastern Narrow-mouthed Toad, *Gastrophryne carolinensis*, is a very small, plump, toadlike anuran with a brownish or reddish brown coloration and a pointy head that is indistinct from its body. The living animal has a lightly spotted, highly pigmented belly, distinguishing it from its western equivalent, *Gastrophryne olivacea*, which has a plain belly. *Gastrophryne carolinensis* is rather ubiquitous in its choice of habitats within its range as long as moisture and places to hide are available. I once watched an adult in northern Florida at dusk sitting by an anthill, eating one ant after another as they emerged from or re-

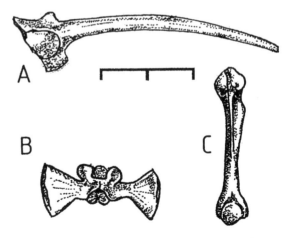

Figure 98. Bones of modern *Gastrophryne carolinensis* from Harris County, Texas. (A) Right ilium in lateral view. (B) Sacrum in dorsal view. (C) Right humerus in ventral view. Scale bar = 4 mm and applies to all figures.

treated to the opening of the hill. The natural range of the modern species is in the southeastern United States, from southern Florida and eastern Maryland, to eastern Kansas and eastern Texas, to Iowa, where a disjunct population occurs (Frost, 1985).

Identification of Fossils. The ilium of *Gastrophryne carolinensis* differs from that of *Gastrophryne olivacea* in that it has a well-developed, triangular dorsal prominence. The dorsal prominence is lower and less triangular in *G. olivacea*, and the dorsal part of the ventral acetabular expansion is more deeply undercut anteriorly than in *G. carolinensis.*

General Remarks. The Miocene (Hemingfordian NALMA) ilium from Florida listed above is so similar to that of *Gastrophryne carolinensis* and differs from that of *Gastrophryne olivacea* to such an extent that it appears that the two species could have differentiated by early Miocene times.

Gastrophryne olivacea (Hallowell, 1857 "1856")
Great Plains Narrow-mouthed Toad
(Fig. 99)

Fossil Localities. **Pleistocene (Rancholabrean NALMA):** Clear Creek Fauna, Denton County, Texas—Holman (1963b, 1995b), Sanchiz (1998). Rancho la Brisca locality, Sonora, Mexico—Van Devender et al. (1985), Sanchiz (1998) (recorded as *Gastrophryne* cf. *Gastrophryne olivacea* by both).

This species was differentiated from modern *Gastrophryne carolinensis* in the preceding section. This is also an ant-eating species that prefers moist areas where it can find shelter during the daylight hours. The modern range of this species is from southern Arizona to Nayarit, Mexico, and from Kansas and Missouri to Chihuahua, Durango, Tamaulipas, and San Luis Potosi, Mexico (Frost, 1985).

Identification of Fossils. See the generic section and the preceding account.

General Remarks. The small size of both species of *Gastrophryne* probably accounts for their rather rare occurrence as fossils.

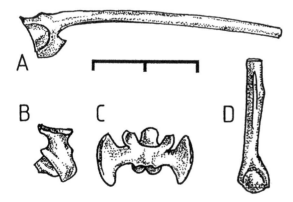

FIGURE 99. Bones of *Gastrophryne olivacea* from Maverick County, Texas. (A) Right ilium in lateral view. (B) Left scapula in medial view. (C) Sacrum in dorsal view. (D) Right humerus in ventral view. Scale bar = 4 mm and applies to all figures.

NOTES

1. Hanover Quarry, No. 1 fissure, Adams County, Pennsylvania, early (?) Irvingtonian NALMA.

2. Guy Wilson Cave, Sullivan County, Tennessee, Woodfordian (full glacial): 19.7 ± 0.6 ka BP (I-4163).

3. Guy Wilson Cave, Sullivan County, Tennessee, Woodfordian (full glacial): 19.7 ± 0.6 ka BP (I-4163).

3

Chronological Accounts

North American anurans are represented by poorly known or primitive families from the Jurassic until the Late Eocene, when three modern North American families appear. The Oligocene has a diminished anuran fauna, but by the end of the Miocene all the living modern families, genera, and even some living species are represented. The Pliocene and Pleistocene constitute a remarkable interval of evolutionary stasis in North American anurans, as almost all of them represent currently living species.

MESOZOIC ERA

JURASSIC ANURANS

The Jurassic period lasted for 62 Ma (206–144 Ma BP) and has been divided into the Early, Middle, and Late Jurassic epochs. Here I shall use international geologic age (IUGS) units (International Union of Geological Sciences, 2002) for the divisions (ages) within the Jurassic and Cretaceous epochs. The Jurassic saw the expansion of dinosaur communities dominated by megaherbivores (giant herbivores). Frogs are poorly represented in the period.

Although the proanuran fossils *Triadobatrachus* and *Czatkobatrachus* indicate that salientian evolution toward the anuran state was well under way in the Late Triassic, the first true jumping frog was not known until *Prosalirus bitis* Shubin and Jenkins, 1995 was described from the Sinemurian and Pliensbachian (Lower Jurassic) strata of the Kayenta Formation, Coconino County, northeastern Arizona (Jenkins and Shubin, 1998). Studies of this material (see Jenkins and Shubin, 1998) show that *Prosalirus* had essentially the same jumping mechanism as in modern frogs. The Kayenta vertebrate fauna is of further interest because it has the earliest known caecilian, which was found to have small legs. Today, caecilians are completely legless lissamphibians. Occurring in the same faunal assemblage with *Eocaecilia* and *Prosalirus* were lungfishes, turtles, lizards, crocodilians, pterosaurs, dinosaurs, and early mammals.

The second oldest true jumping frog, *Vieraella herbsti*, dating from later in the Early Jurassic, was found in southern Patagonia (Argentina and Chile). The familial relationships of *Prosalirus* remain uncertain, but Estes and Reig (1973) suggested that *Vieraella* could be considered at least a structural ancestor to the primitive living families Leiopelmatidae and Discoglossidae because it shows many characteristics of the living genus *Leiopelma*. Sanchiz (1998), however, placed both of these genera in the category Anura incertae sedis. In the Early Jurassic, the two earliest frogs were thousands of miles apart, as indicated by the relationship of the continents to one another during the Pliensbachian age, when *Prosalirus* was hopping about in what is now Arizona (see Roček, 2000, p. 1296). Because of the dearth of anuran remains from the Early Jurassic, it would be guesswork to try to pinpoint the geographic location of the origin of true jumping frogs.

As far as I am aware, no anuran remains are known from the Middle Jurassic of North America. In fact it appears that the only recorded anuran remains in the world from this time are those of *Eodiscoglossus oxoniensis*, of the living but primitive family Discoglossidae, from the medial Middle Jurassic strata of Great Britain; and *Notobatrachus degiustoi*, of the living but primitive family Leiopelmatidae, from the Middle and Upper Jurassic strata of Patagonia. Nevertheless, we can at least say that living anuran families first appear in the Middle Jurassic of Europe and South America.

Five anuran occurrences from Late Jurassic times are known in North America. *Rhadinosteus parvus* is known from the Morrison Formation, at Dinosaur National Monument, Utah. The strata at the site of discovery are from the Kimmeridgian age, which lasted from about 154 to 151 Ma BP. *Rhadinosteus* is questionably assigned to the living but primitive family Rhinophrynidae, mainly on the basis of having ecto-chordal vertebrae (Henrici, 1998). At present, this family has a restricted occurrence, ranging from southern Texas and Michoacan, Mexico, south to Costa Rica.

A second anuran genus, *Enneabatrachus hechti*, is assigned to the living but primitive family Discoglossidae. *Enneabatrachus* is thought to be from the Tithonian age (151–154 Ma BP) (Evans and Milner, 1993), of the Late Jurassic. The site of discovery is the Como Bluff Quarry Nine locality, Wyoming. At present, the family Discoglossidae has a wide distribution in the Old World, occurring in Europe, North Africa, Israel, the western and eastern parts of the former Soviet Union, China, North and South Korea, Borneo, and the Philippines. A second anuran ilium from the Como Bluff Quarry Nine locality is said to represent the family Pelobatidae (Evans and Milner, 1993). Rage (personal communication, in Roček, 2000, p. 1302), however, pointed out that the elongate ventral acetabular expansion and very small dorsal prominence of this specimen are similar to those of the Discoglossidae, not those of the Pelobatidae.

Two problematical Late Jurassic North American anurans, *Eobatrachus agilis* Marsh, 1887 and *Comobatrachus aenigmatis* Hecht and Estes, 1960, are from the Morrison Formation of Como Bluff, Wyoming. Each name was designated as a nomen vanum by Sanchiz (1998). This means they are names that cannot be defined from the available specimens or from their diagnoses, and it is likely that the names will never again be formally recognized by zoologists. As both were based on skeletal elements that are not particularly diagnostic, I did not include either of them in Chapter 2. It has been suggested that the humerus of *Eobatrachus* resembles that of the Pipidae most closely but that it cannot be assigned to this family and that the humerus of *Comobatrachus* is somewhat similar to that of the Discoglossidae (see discussion in Roček, 2000, p. 1302). Evans and Milner (1993) suggested the differences between the type humeri of *Eobatrachus* and *Comobatrachus* may be ontogenetic and that these genera cannot be placed in any recognized family.

In summary, only one anuran — *Prosalirus bitis* — is known from the Early Jurassic of North America. This taxon is the earliest known true jumping frog in the world. The relationship of *Prosalirus* to other anuran groups is still unknown. According to the published literature, the Middle Jurassic has yielded no North American frogs. Although five anuran occurrences are reported from the Late Jurassic of North America, only one, *Enneabatrachus hechti*, is unquestionably assigned to a family, and that family is the Discoglossidae. *Rhadinosteus parvus* has been questionably referred to the Rhinophrynidae.

It has been suggested that two problematical genera, *Eobatrachus* and *Comobatrachus*, have affinities with the Pipidae and Discoglossidae, respectively, but both of these suggestions are probably ill-founded. Finally, it has been suggested that an unnamed frog ilium is from a member of the Pelobatidae, but this has also been disputed.

CRETACEOUS ANURANS

On a global scale, the Cretaceous is divided into only two epochs: the Early Cretaceous (144–99 Ma BP) and the Late Cretaceous (99–65 Ma BP). The dates for the ages within these epochs can be found in Table 1 (p. 35). Although the Early Cretaceous interval lasted 45 Ma, anurans are poorly known from this epoch in North America; in fact, all the remains are from sites of Comanchean age (equivalent to Berriasian–Albian ages) in central Texas, and none has been identified to the extent that it could be included in Chapter 2.

Nevertheless, this material is worth noting here, as it has the potential to be useful in the future, especially in the zoogeographic sense. Winkler et al. (1990) reported that at least some of this Texas material belongs to the family Discoglossidae. Beyond that, other Comanchean Texas frog remains that were not referred to any family are especially noteworthy because their maxillae have a distinct, pitted ornamentation and their vertebrae are amphicoelous.

Pitted ornamentation and amphicoelous vertebrae together are characteristic of the gobiatid discoglossids that occur in the central Asian Cretaceous, and Roček (2000) and Roček and Nessov (1993) suggested that some specimens described by Winkler et al. (1990) might belong to this distinctive group, which is considered a family rather than a subfamily by Roček (2000).

Turning to the Late Cretaceous epoch, we find that fossil anuran material is a little better known than in the earlier part of the period, yet the relationships of most of these taxa remain somewhat doubtful. Fossil frogs are known from only three of the six ages of the Late Cretaceous: the Cenomanian (99.0–93.5 Ma BP), the Campanian (83.5–71.3 Ma BP), and the Maastrichtian (71.3–65 Ma BP). Cenomanian frogs of the Kaiparowits Plateau in southern Utah were discussed in a preliminary way by Eaton and Cifelli (1988) and listed as Utah anurans by Roček (2000). Other anurans listed as "Montana discoglossids I, II, and III" by Roček (2000) are from Campanian strata of that state.

A problematical form, *Nezpercius dodsoni*, was described on the basis of two ilia from the Campanian Judith River Formation type area, in north-central Montana. The fossil elements of this taxa are questionably anuran or even salientian, as discussed in Chapter 2.

A left ilium from the Maastrichtian of the Lance Formation, Wyoming, appears to be identical to two ilia of the modern *Leiopelma* of the primitive family Leiopelmatidae (Estes, 1964). This ilium was not discussed by Sanchiz (1998), who does not recognize this family in the fossil record of North America. Actually, this ilium is as clearly assignable to the Leiopelmatidae as other Mesozoic and Paleogene frogs are to their respective families, and the herpetological community should be aware of its existence.

Theatonius lancensis of the late Maastrichtian of the Lance Formation, Wyoming, is a taxon of uncertain family, but it is distinctive in that it is the only taxon in the Mesozoic that undoubtedly lacks teeth on the maxilla, and it has a unique pattern of sculpturing on the dermal cranial bones (Gardner, 2000).

Scotiophryne pustulosa, known from the Campanian through the Maastrichtian and possibly the Paleocene (localities in California, Montana, New Mexico, and Wyoming), was included with the Discoglossidae in Chapter 2, but Gardner (2000) considered this taxon to be familia incertae sedis. Estes and Sanchiz (1982b) made a good case for the inclusion of *Paradiscoglossus americanus* from the Maastrichtian of the Lance Formation, Wyoming, with the Discoglossidae (see Chapter 2). This taxon was included with the Discoglossidae without comment by Sanchiz (1998), but Gardner (2000) included it with his indeterminate families.

Palaeobatrachus occidentalis from the late Maastrichtian of the Lance Formation, Wyoming, although questionably referred to the genus by both Sanchiz (1998) and Gardner (2000), certainly represents the family Palaeobatrachidae, the only extinct anuran family that is currently unequivocally recognized. All the other numerous records of this family are known only from the Late Cretaceous to the middle Pleistocene of Europe. *Eopelobates* sp. represents the family Pelobatidae but is of uncertain generic status in North America (see Chapter 2), where it is known from Campanian–Maastrichtian strata in Montana, New Mexico, and Wyoming (Lance, Hell Creek, and Fruitland formations).

In summary, frogs from the Early Cretaceous epoch are all from central Texas and poorly known. Some of the Texas material represents the family Discoglossidae, including some with pitted cranial elements and amphicoelous vertebrae such as occur in the gobiatid discoglossids of the Cretaceous of central Asia. Anurans of the Late Cretaceous epoch include three unnamed discoglossids from the Campanian of Utah; a newly named Campanian genus that probably is not an anuran and may not even be a salientian; and an ilium from the Maastrichtian that appears to be identical to that of the modern genus *Leiopelma* of the family Leiopelmatidae.

The distinct genus *Theatonius* of the Maastrichtian lacks teeth on the maxilla and has a unique cranial sculpturing pattern but cannot be identified with any anuran family. The genera *Scotiophryne* (Campanian–Maastrichtian) and *Paradiscoglossus* (Maastrichtian) are probable discoglossids. *Palaeobatrachus occidentalis* of the late Maastrichtian is the only known representative of the extinct family Palaeobatrachidae in the North American fossil record. *Eopelobates* sp. of the Maastrichtian is a member of the Pelobatidae, but it is uncertain whether it is the same genus as *Eopelobates* of the Old World. We can say, though, that the Discoglossidae, Palaeobatrachidae, and Pelobatidae were present in North America by the Late Cretaceous.

CENOZOIC ERA

PALEOCENE ANURANS

The Paleocene, the first epoch of the Cenozoic era, lasted about 10 Ma and was in a sense a lag time when both marine and terrestrial communities reorganized in the wake of the extinction of the dinosaurs and great marine reptiles. The Paleocene is divided into only Early and Late Paleocene ages. In the Cenozoic era, I shall refer to North American land-mammal ages (NALMAs; see Fig. 10), because the NALMA terms reflect recent North American stratigraphic studies well (e.g., Woodburne, 1987; Prothero and Emry, 1996) and are widely used.

There are few Paleocene sites in North America, and frogs are particularly rare. A probable record of *Scotiophryne*, a Cretaceous genus referred by some to the Discoglossidae and considered by others to be of uncertain family, has been reported from the Early Paleocene (Torrejonian NALMA) of the Tongue River Formation, Montana (Estes, 1969). *Scotiophryne* was also reported from the latest Cretaceous (Maastrichtian age) to Early Paleocene (Puercan or Torrejonian NALMA) of the Tullock Formation, Montana (Bryant, 1989). A Late Paleocene (Tiffanian NALMA) discoglossid-like frog was reported by Estes (1975) on the basis of a single humerus from the Princeton Quarry, Park County, Wyoming. If any or all of these records do represent the Discoglossidae, they would represent the last known surviving members of the family in North America.

A record of *Eorhinophrynus* sp. from the Late Paleocene (Tiffanian NALMA) Princeton and Fritz quarries in the Fort Union Formation, Wyoming (Estes, 1975), is the first record of the Rhinophrynidae in North America. At present the family occurs from extreme southern Texas to Costa Rica.

In summary, the Paleocene probably represents the last occurrence of the currently Old World family Discoglossidae and the first unquestionable occurrence of the currently New World subtropical and tropical family Rhinophrynidae in North America.

EOCENE ANURANS

In general, the Eocene was a time of reorganization of terrestrial vertebrate communities that were dominated by large mammalian species.
The world climate was relatively equable until, toward the end of the
epoch, it began to deteriorate. Many families of modern mammals originated in Eocene times. Anurans of the Eocene are much better known
than those of the Paleocene, but they are not greatly diverse. The Eocene
lasted from 54.8 to 33.7 Ma BP; thus, this epoch was about twice as long
as the Paleocene. The Eocene is divided into Early, Middle, and Late
ages.

Turning to the Early Eocene (Wasatchian NALMA), we find records
of two frog taxa. *Eorubeta nevadensis*, of the Sheep Pass Formation, Nevada, was once considered to belong to the family Leptodactylidae
(Hecht, 1960), but Lynch (1971) demonstrated that this taxon does not
resemble any known frog genus and is of uncertain family. *Scaphiopus
(Scaphiopus) guthriei*, however, not only represents a living genus of the
primitive living family Pelobatidae, but represents the subgenus *Scaphiopus* as well. *Scaphiopus guthriei* was unearthed from the Lysite Member
of the Wind River Formation, Wyoming.

Middle Eocene frogs have been found in strata of Bridgerian
NALMA and probably Uintan NALMA. *Eorhinophrynus septentrionalis*
is known from the Bridgerian Tabernacle Butte area of the Bridger Formation, Wyoming. This species represents the family Rhinophrynidae but
is osteologically separable from the sole living genus, *Rhinophrynus* (Henrici, 1991). The enigmatic pelobatid taxon *Eopelobates grandis* has been
tentatively reported from the Bridgerian Tabernacle Butte area of the
Bridger Formation, Wyoming. The well-studied rhinophrynid genus *Chelomophrynus bayi* is known from a locality of probable Uintan NALMA
in the Wagon Bed Formation, Wyoming. This genus is well differentiated
from both the modern genus *Rhinophrynus* and the extinct genus *Eorhinophrynus*.

Three modern genera—*Rhinophrynus*, *Scaphiopus*, and *Hyla*—are
known from the Late Eocene (Chadronian NALMA), along with the
enigmatic genus *Eopelobates*. *Rhinophrynus canadensis* is known from a
great deal of material from the Calf Creek Local Fauna, near East End,
Saskatchewan, Canada. *Rhinophrynus canadensis* is remarkably similar to
the modern species *Rhinophrynus dorsalis* in a number of ways but can
be separated from *R. dorsalis* on the basis of characters of the ilium and
humerus. Both of these species are burrowing toads.

Scaphiopus (Scaphiopus) skinneri, a large species, is also known from
the Calf Creek Local Fauna, Saskatchewan, and is thought to be a basal
member of the so-called eastern clade of the subgenus *Scaphiopus*. The
problematical pelobatid *Eopelobates grandis* was originally described from
the Chadronian of the Indian Creek locality, South Dakota. Finally, the
first record in North America of the family Hylidae—in the form of *Hyla
swanstoni*—is from the Chadronian Calf Creek Local Fauna, Saskatchewan. Oddly, *H. swanstoni* appears to be most similar to *Hyla miofloridana* of the Early Miocene of Texas.

In summary, the Eocene is much richer in anuran taxa than the
Paleocene. The Early Eocene marks the last occurrence of the Old World

family Discoglossidae in North America and the first appearance of a currently living North American genus and subgenus, *Scaphiopus (Scaphiopus) guthriei*. The odd Early Eocene form *Eorubeta nevadensis* is of unknown family. The Middle Eocene marks the appearance of two extinct genera of the family Rhinophrynidae: *Chelomophrynus bayi* and *Eorhinophrynus septentrionalis*. The problematical pelobatid *Eopelobates grandis* was reported from the Middle Eocene of Wyoming.

The Late Eocene marks the appearance of three modern genera. The living genus *Rhinophrynus* is represented for the first time by *Rhinophrynus canadensis*, of Saskatchewan, Canada; and the living genus *Scaphiopus*, continuing on from previous times, is represented by *Scaphiopus (Scaphiopus) skinneri*, also of Saskatchewan. Finally, *Hyla swanstoni*, of the Late Eocene of Saskatchewan, represents the first appearance of the family Hylidae and of the living genus *Hyla* in North America. The controversial pelobatid *Eopelobates grandis* is known from the Late Eocene of South Dakota.

OLIGOCENE ANURANS

The Oligocene, like the Paleocene, represents a relatively short interval of about 10 Ma (33.7–23.8 Ma BP). Also, like the Paleocene, the Oligocene is divided into Early and Late ages only. The climate began to deteriorate in the Late Eocene (e.g., Prothero and Emry, 1996; Holman, 2000a), and this continued throughout the Oligocene. Only two anuran genera are known from the North American Oligocene, and one of them occurs very near the end of the epoch.

Scaphiopus (Scaphiopus) skinneri, a pelobatid also known from the Late Eocene (Chadronian NALMA) of Saskatchewan, Canada, was described from the Early Oligocene (Orellan NALMA) of the White River Group, North Dakota. As previously mentioned, *Scaphiopus skinneri* is considered to be the basal member of the "eastern clade" of the subgenus *Scaphiopus*. *Scaphiopus* sp., with no subgenus designation, is known from the Early Oligocene (Whitneyan NALMA) I-75 Fauna, of northern Florida.

In the Late Oligocene (early Arikareean NALMA) the first *Scaphiopus* species of the subgenus *Spea* is known in the form of *Scaphiopus (Spea) neuter* from the Wounded Knee area of the Sharps Formation, South Dakota. This species is considered to be the basal member of the subgenus *Spea* clade. Finally, the first member of the currently Old World family Pelodytidae, in the form of *Tephrodytes brassicarvalis*, is known from the Late Oligocene–Early Miocene (Arikareean NALMA) Cabbage Patch beds of the Flint Creek basin, Montana.

In summary, two basal forms in the evolution of the subgenera *Scaphiopus* and *Spea* of the genus *Scaphiopus* are known from the Oligocene — *Scaphiopus (Scaphiopus) skinneri* from the Early Oligocene (Orellan NALMA) of North Dakota; and *Scaphiopus (Spea) neuter* of the Late Oligocene (Arikareean NALMA) of South Dakota. *Tephrodytes brassicarvalis* of the Late Oligocene–Early Miocene (Arikareean NALMA) transition in Montana represents the first North American occurrence of the currently Old World family Pelodytidae.

Miocene Anurans

The Miocene saw the return of a more equable climate and the rapid spread of grassland communities throughout much of the world. Frogs became modern at the generic level (and even at the specific level in a few cases), and most of the present North American families are known from this epoch. The Miocene is one of the long (18.5 Ma) epochs of the Cenozoic era, occurring between 23.8 and 5.3 Ma BP. The Miocene is divided into Early, Middle, and Late ages. Five NALMAs occur in the Miocene. The first one, the Arikareean, overlaps the Late Oligocene; and the last one, the Hemphillian, overlaps the Early Pliocene. The Miocene has been so well studied that stratigraphic units within the various NALMA units are often used in faunal reports.

The earliest part of the Early Miocene (late Arikareean NALMA) is poorly represented by anurans, as only *Scaphiopus (Spea) neuter*, from the Harrison Formation, Nebraska, is known. This taxon is considered to be a basal member of the subgenus *Spea* clade. The late part of the Early Miocene (Hemingfordian NALMA) is represented in Colorado and Florida by *Acris* sp. of the family Hylidae from the Martin Canyon Local Fauna (Colorado) and the Thomas Farm Local Fauna (Florida).

The Hemingfordian Thomas Farm Local Fauna, of north-central Florida, has yielded an especially interesting Early Miocene anuran assemblage. The extinct bufonid species *Bufo praevius* is thought to be a member of the *Bufo valliceps* group of Tihen (1962a, b). The odd, extinct hylid genus and species, *Proacris mintoni*, has an ilial morphology that thus far does not compare well with that of any other North American hylid genus. The extinct hylid species *Acris barbouri* is related to the living species *Acris crepitans* and *Acris gryllus*, but it is osteologically more distinct from these extant species than they are from each other. The extinct hylid species *Hyla miofloridana* appears to be most similar to the living species *Hyla gratiosa* in the southeastern United States. *Hyla* sp. indet. was also identified at the Thomas Farm Local Fauna, Florida. The modern family Ranidae is represented by many fragmentary fossil elements of the *Rana pipiens* complex, but it cannot be determined what living species of this group is most closely related to these Florida fossils. Finally, the family Microhylidae is represented by *Gastrophryne* cf. *Gastrophryne carolinensis*, a living species. This is the earliest North American record of the Microhylidae.

Middle Miocene anurans are especially well represented in the Barstovian NALMA, where frog remains have been identified from Nebraska, Saskatchewan, South Dakota, and Texas. The family Pelobatidae is represented by material referred to *Scaphiopus* cf. *Scaphiopus (Scaphiopus) alexanderi* (an extinct species) from the Wood Mountain Formation, Saskatchewan, Canada; two extinct species, *Scaphiopus (Scaphiopus) wardorum* and *Scaphiopus (Scaphiopus) hardeni*, from the Valentine Formation, Nebraska; and the extant species *Scaphiopus (Spea) bombifrons*, also from the Valentine Formation, Nebraska.

The family Bufonidae is represented in the Barstovian NALMA by three extinct *Bufo* species from the Valentine Formation, Nebraska, with one of these also known from the Wood Mountain Formation, Saskatchewan. These are *Bufo hibbardi*, *Bufo kuhrei* (a large species), and *Bufo valentinensis* (also known from Saskatchewan). The family Hylidae is represented in the Barstovian by the genera *Acris*, *Hyla*, and *Pseudacris*. *Acris* sp. indet. is known from the Valentine Formation, Nebraska. *Hyla* sp. indet. is known from the Hottell Ranch rhino quarries and Valentine Formation localities, Nebraska, and from the Bijou Hills Local Fauna and Glenn Olson quarries, South Dakota. The extinct species *Hyla miocenica* is known from the early Barstovian of the Fleming Formation of Texas. *Pseudacris* sp. indet. has been reported from the Glad Tidings Prospect locality and Valentine Formation, Nebraska, as well as the Bijou Hills Local Fauna, South Dakota. The odd extinct form *Pseudacris nordensis* (which might represent some other hylid genus, or even a new hylid genus) has been reported from the medial Barstovian NALMA Valentine Formation, Nebraska.

The most numerous Barstovian NALMA frog fossils at any given locality tend to be those of the *Rana pipiens* complex, none of which has been assigned to species. These records are from the early Barstovian of Texas and medial and late Barstovian localities in Nebraska.

A tentative record, *Rana* cf. *Rana clamitans*, an extant species, is known from the medial Barstovian NALMA Hottell Ranch rhino quarries, Nebraska. The overriding abundance of *Rana pipiens* complex, frogs of moderate size, in contrast to the single record of the large form, *Rana* cf. *R. clamitans*, may be due to the temporary nature of ponds in the vicinity of the fossil sites, as frogs of the *R. pipiens* group are far less aquatic than *Rana clamitans* and many other large *Rana* species.

The extinct *Miopelodytes gilmorei* from the Middle Miocene (Barstovian or Clarendonian NALMA) Elko shales, Nevada, is a rather surprising addition to the North American Miocene fauna, as it represents the family Pelodytidae, which currently occurs in the Old World. As far as I can determine (if the identification is correct), this is the last appearance of this currently Old World frog family in North America.

The Clarendonian NALMA, spanning the early part of the Late Miocene, is represented by frogs of the families Pelobatidae, Bufonidae, and Hylidae and by a strange anuran that defies family assignment. This frog, *Tregobatrachus hibbardi*, from the Ogallala Formation, Kansas, is unquestionably an anuran, but it has a morphology unlike that of any known anuran. This is unusual for such a comparatively late time in anuran history. Other frogs of this NALMA are rather modern in nature.

Except for a record of the small extinct toad *Bufo pliocompactilis* from the Mission Local Fauna, South Dakota, all of the Clarendonian records below (and the *Tregobatrachus* record above) are from the WaKeeney Local Fauna of the Ogallala Formation, Kansas. Here the family Pelobatidae is represented by the extinct species *Scaphiopus (Scaphiopus) hardeni*. Five species of bufonids, all in the genus *Bufo*, have been identified from the WaKeeney Local Fauna, including *Bufo* cf. *Bufo cognatus*, an extant species; *Bufo hibbardi*, an extinct species; *Bufo marinus*, representing an extant giant toad that occurs in extreme southern Texas today; the small extinct toad *Bufo pliocompactilis*; and *Bufo valentinensis*, also an extinct form. Hylid families are represented by *Acris* sp. indet., *Hyla* sp. indet., and *Pseudacris* sp. indet. Ranid families are represented in the WaKeeney Local Fauna by numerous elements from the

Rana pipiens complex and by *Rana* cf. *Rana areolata*, a modern species that hides in crayfish burrows.

The late part of the Late Miocene is represented by the Hemphillian NALMA, which overlaps the Early Pliocene somewhat. Miocene Hemphillian anuran records are known from Chihuahua, Mexico, Florida, Kansas, and Nebraska. The Pelobatidae is represented by the extinct *Scaphiopus (Scaphiopus) wardorum* from the Santee site, Nebraska, and by *Scaphiopus (Scaphiopus)* sp. of the White Cone Local Fauna, Arizona. Many species of *Bufo* of the family Bufonidae are known from the Miocene part of the Hemphillian, including the following eight extinct species: *Bufo alienus*, of Kansas; *Bufo campi*, of Chihuahua, Mexico; *Bufo hibbardi*, of Kansas and Nebraska (Nebraska record listed as *Bufo* cf. *Bufo hibbardi*); *Bufo holmani*, of Nebraska; *Bufo pliocompactilis*, of Nebraska and Arizona; *Bufo spongifrons*, of Kansas; *Bufo tiheni*, of Florida; and *Bufo valentinensis*, of Nebraska. The modern species *Bufo woodhousii* makes its first appearance in the early part of the late Hemphillian. *Bufo* species tentatively referred to modern species include *Bufo* cf. *Bufo cognatus*, of Kansas and Nebraska; and *Bufo speciosus* or *cognatus*, of Nebraska.

The family Hylidae is represented in the Miocene portion of the Hemphillian NALMA by *Hyla* sp. of the Lemoyne Quarry site, Nebraska; and the White Cone Local Fauna, Arizona. The Ranidae are represented by *Rana pipiens* complex fossils from the Haile VIA locality, Florida; the Mailbox Prospect and Santee sites, Nebraska; and the White Cone Local Fauna, Arizona. The first record of the American Bullfrog, *Rana catesbeiana*, is provided by the tentative identification of the species at the Mailbox Prospect and Santee sites, Nebraska. The extinct *Rana pliocenica* is known from the Miocene Hemphillian of the Rodeo locality, California. Finally, the extinct pelobatid, *Scaphiopus (Scaphiopus) alexanderi*, is known from a questionable Hemphillian site at Fish Lake, Nevada.

In summary, the Miocene anuran fauna was mainly dominated by modern families and genera, although most species are considered extinct. The currently Old World family Pelodytidae occurred for the last time in North America. The few modern species that were recognized (some tentatively) were in the families Pelobatidae, Bufonidae, and Ranidae. A large number of extinct species of *Bufo* were recognized in the Miocene, at least indicating that the group had become quite morphologically diverse. The extinct genus *Tregobatrachus* cannot be assigned to any known anuran family, and *Proacris mintoni* and *Pseudacris nordensis* of the Hylidae cannot be allied with any known hylid genus.

Turning to a more detailed summary featuring Early, Middle, and Late Miocene anuran events, we find that the Early Miocene was represented in North America by the families Pelobatidae, Bufonidae, Hylidae, Ranidae, and Microhylidae. This marked the first appearance of the Bufonidae, Ranidae, and Microhylidae in North America. The modern genera *Scaphiopus* (Pelobatidae), *Bufo* (Bufonidae), *Rana* (Ranidae), and *Gastrophryne* (Microhylidae) are recorded. The odd genus *Proacris* also represents the Hylidae. All of the modern genera are represented by extinct species, except the tentatively identified extant taxon *Gastrophryne carolinensis*.

In the Middle Miocene the extant North American families Pelobatidae, Bufonidae, Hylidae, and Ranidae were present, along with the extinct genus *Miopelodytes*, representing the last appearance of the currently Old World family Pelodytidae. All the pelobatids are represented by the extant genus *Scaphiopus*, of which all species are extinct except *Scaphiopus (Spea) bombifrons*, which currently occupies the Great Plains area of the United States.

Three extinct *Bufo* species are known. The family Hylidae occurs in the Middle Miocene in the form of *Acris*, *Hyla*, and *Pseudacris*, but these have not been given specific names in the modern literature. The exact generic placement within the Hylidae of the odd species *Pseudacris nordensis* is difficult. The family Ranidae is represented by abundant remains of frogs of the *Rana pipiens* complex, as well as by the tentatively identified modern ranid species *Rana clamitans*.

The Late Miocene is represented by the Pelobatidae, Bufonidae, Hylidae, and Ranidae and by the strange genus *Tregobatrachus* (Clarendonian NALMA), which cannot be assigned to any known anuran family. Pelobatids are represented by a single extinct species, *Scaphiopus (Scaphiopus) wardorum*. The Bufonidae are represented by the extant species *Bufo* cf. *Bufo cognatus*, *Bufo* cf. *Bufo speciosus* or *Bufo cognatus*, *Bufo marinus*, and *Bufo woodhousii* and eight extinct species. Whether all these extinct species are valid or not, there is no question that the genus *Bufo* underwent a lot of morphological change in the Late Miocene, as giant, medium-sized, and very small species occurred. The Hylidae in the Late Miocene, too, are represented by the modern genera *Acris*, *Hyla*, and *Pseudacris*, but again, these have not been given specific names in the modern literature. The family Ranidae was again well represented by *Rana pipiens* complex species, as well as by the extinct species *Rana pliocenica* and material tentatively identified as being from the modern species *Rana areolata* and *Rana catesbeiana*.

PLIOCENE ANURANS

The modern anuran fauna of North America was essentially in place in the Pliocene and was almost completely modern in the Pleistocene. The major events that occurred in the anuran community in the Pliocene–Pleistocene interval were range changes in some species, brought about by climatic changes in the Pleistocene. But other than species that were displaced by advancing ice sheets and that reoccupied habitats once they became stabilized after the ice retreated, most frogs stayed put. Probably the most remarkable aspects of the Pliocene and Pleistocene anuran fauna were the remarkable evolutionary stasis and lack of extinctions during the interval.

Next to the Pleistocene, the Pliocene is the shortest epoch of the Cenozoic, occurring from 5.3 to 1.8 Ma BP, a duration of 3.5 Ma. The Pliocene is divided into Early and Late units. Most of the Pliocene occurs within the Blancan NALMA, but the Hemphillian overlaps the beginning of the Pliocene somewhat, and a very small part of the Irvingtonian NALMA overlaps the end of the Pliocene. Compared with the rest of the world, few vertebrate sites of Pliocene age are known in North America, but enough sites are available to provide a fairly good estimate of the status of the anuran fauna. Most North American Pliocene anuran species are living today. The three extinct species recognized are all in the genus *Bufo*.

The Blancan NALMA takes up most of the Pliocene epoch. All of the anurans of the Blancan have been identified as currently living species, with the exception of three somewhat questionable extinct species of *Bufo*. The family Pelobatidae is represented by *Scaphiopus* cf. *Scaphiopus (Spea) bombifrons*, from the San Draw Local Fauna, Nebraska, and *Scaphiopus* cf. *Scaphiopus (Spea) hammondii*, from the Hagerman Local Fauna, Idaho. The three extinct *Bufo* species of the Bufonidae are *Bufo rexroadensis*, of the Fox Canyon locality, Kansas, the Big Springs Local Fauna, Nebraska, and the Beck Ranch Local Fauna, Texas; *Bufo suspectus* (thought to be the Pliocene equivalent of the Miocene *Bufo valentinensis*), of the Rexroad Formation, Kansas; and *Bufo* cf. *Bufo spongifrons*, of the Big Springs Local Fauna, Nebraska.

Bufo rexroadensis may have been derived from *Bufo hibbardi* of the Late Miocene (Hemphillian NALMA) of Kansas and, in turn, may have given rise to the modern species *Bufo woodhousii*. But there is a possibility that *Bufo rexroadensis* itself is indeed *B. woodhousii*, considering the fact that the latter species has been reported from Pliocene (Blancan NALMA) localities in Arizona, Kansas, and Texas. *Bufo* cf. *Bufo spongifrons* is included with the so-called western section of the *Bufo americanus* group but is said to not be closely related to any species within this group (Tihen, 1962b). There is some question about whether the identification of *Bufo* cf. *B. spongifrons* from the Blancan Pliocene of Nebraska is correct.

Currently living *Bufo* species from the Blancan Pliocene include the following. *Bufo* cf. *Bufo alvarius* is known from the Benson locality, Arizona. *Bufo cognatus* has been reported from the Rexroad and White Rock local faunas, Kansas; the Big Springs and Hornets Nest quarries, Nebraska; and the Beck Ranch Local Fauna, Texas. *Bufo* cf. *Bufo speciosus* has been reported from the Fox Canyon, Rexroad, and Wendell Fox Pasture localities, Kansas. *Bufo* cf. *Bufo valliceps* has been identified from the Beck Ranch Local Fauna, Texas. Finally, *Bufo woodhousii* has been reported from the Benson Fauna, Arizona; the Borchers Fauna and White Rock Local Fauna, Kansas; and the Beck Ranch Local Fauna, Texas.

Turning to the Pliocene (Blancan NALMA) Hylidae, we find two genera, *Acris* and *Pseudacris*, represented. *Acris* sp. indet. was reported from the Rexroad Local Fauna, Kansas; and the Beck Ranch Local Fauna, Texas. *Pseudacris* sp. indet. was reported from the Saunders Local Fauna, Kansas; and the Hagerman Fauna, Idaho.

The family Ranidae is represented by numerous remains of the *Rana pipiens* complex from the Hagerman Fauna, Idaho; the White Rock Local Fauna, Kansas; the Big Springs Local Fauna, Hornets Nest Quarry, Lisco C rhino quarries, and San Draw Local Fauna, Nebraska; and the Beck Ranch Local Fauna, Texas. Currently living *Rana* species include *Rana areolata*, from the Beck Ranch Local Fauna, Texas; *Rana* cf. *Rana aurora*, from the Hagerman Fauna, Idaho; *Rana catesbeiana*, from the White Rock Local Fauna, Kansas (referred to *Rana* cf. *Rana catesbeiana*), and the Hornets Nest Quarry and Sand Draw Local Fauna, Nebraska; *Rana palustris*, from the Beck Ranch Local Fauna, Texas; and *Rana sylvatica*, from the Hornets Nest Quarry, Nebraska.

In summary, the anuran fauna of the Pliocene is essentially modern, as all five genera and 13 of 16 frogs referred to the specific level are taxa that are currently living. All three extinct species are in the genus *Bufo*, but at least one of these, *Bufo rexroadensis*, may be the living species *Bufo woodhousii.*

PLEISTOCENE ANURANS

The Pleistocene is by far the shortest epoch of the Cenozoic (1.8–0.01 Ma BP), yet it has far more anuran fossil records than any other Cenozoic epoch, simply because there are so many Pleistocene sites. The Pleistocene consists of two NALMA units: (1) most of the Irvingtonian, which overlaps slightly with the Pliocene and extends in the Pleistocene from 1.8 to 0.15 Ma BP; and (2) the Rancholabrean, which extends from 0.15 to 0.01 Ma BP and marks the beginning of the Holocene.

From a human standpoint, the Pleistocene, which ended only about 140 human lifetimes ago, is the most important geologic interval. The Ice Age, as it is popularly known, is characterized by gigantic, moving ice sheets that changed the face of the Earth in many areas, including North America, and saw a massive, worldwide extinction of large mammalian taxa. This was coupled with the rise of humans, whose livestock eventually replaced the large, extinct mammalian herbivores; whose cereal-producing grass species replaced native grassland communities; and whose agricultural practices and factories are now polluting the environment.

Irvingtonian Land Mammal Age. Although the Irvingtonian lasted much longer than the Rancholabrean, it has far fewer records of frog species. This is because there are far fewer Irvingtonian fossil sites. Five families of anurans are known, and no extinct species occur in Irvingtonian sites. The family Pelobatidae is represented by only a single species, *Scaphiopus (Spea) bombifrons*, which is known from the Hansen Bluff Chronofauna, Colorado; the Nash Local Fauna, Kansas; and the Albert Ahrens Local Fauna, Nebraska. The family Leptodactylidae is also represented by a single species, *Eleutherodactylus* cf. *Eleutherodactylus marnockii*, identified from the Vera and Gilliland faunas, Texas.

Bufonids are represented by seven living species. *Bufo alvarius* has been reported from the El Golfo locality, Sonora, Mexico. *Bufo americanus* is known from the Cumberland Cave Fauna, Maryland; the Albert Ahrens Local Fauna, Nebraska; the Hanover Quarry, Pennsylvania; and Trout Cave, West Virginia. *Bufo cognatus* has been reported from the Hansen Bluff Chronofauna, Colorado, and the Albert Ahrens Local Fauna, Nebraska. *Bufo fowleri* is known from the Conard Fissure, Arkansas; Cumberland Cave, Maryland; and Hamilton Cave, West Virginia. *Bufo* cf. *Bufo hemiophrys* has been recorded from the Courland Canal, Cudahy, and Hall Ash faunas, Kansas; and the Vera Local Fauna, Texas. *Bufo terrestris* is known from the Leisey Shell Pit Fauna, Florida. *Bufo woodhousii* is known from the Hansen Bluff locality, Colorado; and the Gilliland Fauna, Texas.

Species of three genera of hylid frogs—*Acris, Hyla,* and *Pseudacris*—have been reported from the Irvingtonian NALMA: *Acris* cf. *Acris crepitans,* from the Vera Local Fauna, Texas; "*Hyla chrysoscelis* or *H. versicolor,*" from the Albert Ahrens Local Fauna, Nebraska, and Hamilton Cave, West Virginia; *Hyla cinerea,* from the Inglis IA site, Florida; *Pseudacris* cf. *Pseudacris clarkii,* from the Albert Ahrens Local Fauna, Nebraska; *Pseudacris crucifer,* from Cumberland Cave, Maryland, and Hamilton Cave, West Virginia; and *Pseudacris triseriata,* from the Hansen Bluff Chronofauna, Colorado, the Cudahy Fauna, Kansas, and Cumberland Cave, Maryland.

The family Ranidae is represented by at least six species, all in the genus *Rana.* No extinct species are recognized within this assemblage. Fossil elements referred to the *Rana pipiens* complex but not to any particular species in the complex are abundant in the Irvingtonian and have been recorded from the Hansen Bluff Chronofauna, Colorado; the Courtland Canal and Hall Ash assemblages, Kansas; Cumberland Cave, Maryland; Hanover Quarry, Pennsylvania; the Java Local Fauna, South Dakota; Hamilton and Trout caves, West Virginia; and the Gilliland Fauna, Texas.

Rana areolata is known from the Inglis IA site, Florida; and *Rana* cf. *Rana blairi* has been recorded from the Albert Ahrens Local Fauna, Nebraska. *Rana catesbeiana* is a widespread frog in the Irvingtonian of North America: it has been reported from the Hansen Bluff Chronofauna, Colorado; the Inglis IA site, Florida; the Courland Canal and Hall Ash assemblages, Kansas; the Albert Ahrens Local Fauna, Nebraska; the Java Local Fauna, South Dakota; and Hamilton Cave, West Virginia. *Rana clamitans* is known from Cumberland Cave, Maryland; the Albert Ahrens Local Fauna, Nebraska; and Hamilton Cave, West Virginia. Finally, *Rana sylvatica* has been reported from Cumberland Cave, Maryland; the Albert Ahrens Local Fauna, Nebraska; and Hamilton and Trout caves, West Virginia.

In summary, five families—Pelobatidae, Leptodactylidae, Bufonidae, Hylidae, and Ranidae—have been reported from the Irvingtonian NALMA of the Pleistocene epoch. Seven genera—*Scaphiopus* (Pelobatidae); *Eleutherodactylus* (Leptodactylidae); *Bufo* (Bufonidae); *Acris, Hyla,* and *Pseudacris* (Hylidae); and *Rana* (Ranidae)—have been identified. Twenty-one species among these genera have been identified, and none of these represents an extinct species.

Rancholabrean Land Mammal Age. Although the Rancholabrean NALMA lasted only 140 ka, seven anuran families—Rhinophrynidae, Pelobatidae, Leptodactylidae, Bufonidae, Hylidae, Ranidae, and Microhylidae—and a plethora of genera and species have been recovered from sites ranging from caves, tar pits, river terraces, and quaking bogs to ancient seaways (Holman, 2001b). Only one species is regarded as extinct, and even its status is somewhat questionable. The remarkable evolutionary stasis of anurans in the face of Pliocene and Pleistocene climatic changes will be part of a special discussion in the following chapter.

Turning to the taxonomic content of Rancholabrean anurans, we find that the family Rhinophrynidae is represented by the single living species, *Rhinophrynus dorsalis,* from the Cueva de Abra locality, Tamaulipas,

Mexico. The family Pelobatidae is represented by both subgenera of the genus *Scaphiopus*. These taxa include *Scaphiopus (Scaphiopus) couchii*, from Deadman Cave, Arizona; Rancho la Brisca, Sonora, Mexico; Dark Canyon, Howell's Ridge, and Shelter Cave, New Mexico; and Friesenhahn and Schulze caves, Texas. *Scaphiopus (Scaphiopus) holbrookii* is known from Peccary Cave, Arkansas; Arredondo, Devil's Den, Orange Lake, Reddick I, Sabertooth Cave, Vero Beach strata 2 and 3, and Williston IIIA sites, Florida; Kingston Saltpeter Cave, Georgia; Bootlegger Sink, Pennsylvania; and Clark's Cave and the Natural Chimneys site, Virginia.

In the subgenus *Spea* of *Scaphiopus*, *Scaphiopus (Spea) bombifrons* is known from Cragin Quarry and the Jinglebob Fauna of Kansas, as well as the Dry Cave Fauna of New Mexico. *Scaphiopus (Spea) hammondii* has been recorded from Deadman Cave, Arizona; Dry Cave, New Mexico; and Smith Creek Cave, Nevada. *Scaphiopus (Spea) intermontana* is known from Hidden and Smith Creek caves, Nevada, as well as from Bechan Cave, Utah; and *Scaphiopus (Spea) multiplicata* has been been reported from Dark Canyon, New Mexico.

The family Leptodactylidae is represented in New Mexico and Texas; and Sonora and Tamaulipas, Mexico. *Eleutherodactylus augusti* is known from Dark Canyon, New Mexico; Rancho la Brisca, Sonora, Mexico; and Friesenhahn and Schulze caves, Texas. *Eleutherodactylus* cf. *Eleutherodactylus cystignathoides* has been recorded from the Cueva de Abra locality, Tamaulipas, Mexico; and *Eleutherodactylus marnockii* is known from the Easley Ranch and Schulze Cave faunas, Texas. Turning to the genus *Leptodactylus*, we find that *Leptodactylus* cf. *Leptodactylus labialis* has been reported from the Cueva de Abra locality, Tamaulipas, Mexico; and *Leptodactylus melanonotus* is known from the Rancho la Brisca site, Sonora, Mexico.

The family Bufonidae is well represented from coast to coast in the Rancholabrean NALMA. *Bufo alvarius* is known from the Rancho la Brisca locality, Sonora, Mexico. *Bufo americanus* is widely known as a fossil in eastern North America; it has been found at Bell Cave, Alabama; Peccary Cave, Arkansas; Kingston Saltpeter Cave and Ladds Quarry, Georgia; the Meskill Road site, Michigan; Sheriden Pit Cave, Ohio; Kelso Cave, Ontario, Canada; Bootlegger Sink, Frankstown Cave, and the New Paris 4 site, Pennsylvania; Baker Bluff, Cheek Bend, Guy Wilson, and Robinson caves, Tennessee; Clark's Cave, Natural Chimneys, and Strait Canyon Fissure, Virginia; New Trout Cave, West Virginia; and Moscow Fissure, Wisconsin.

Bufo boreas is known from Costeau Pit, Newport Beach Mesa, Potter Creek Cave, and Rancho La Brea sites, California; as well as from Smith Creek Cave in Nevada. *Bufo cognatus* is known from Kansas, Sonora, Mexico, and Texas: it has been reported from Butler Spring, Cragin Quarry, Jinglebob, and Sandahl local faunas, Kansas; the Rancho la Brisca locality, Sonora, Mexico; and Friesenhahn Cave, Jones Fauna, and Lubbock Lake site, Texas.

Bufo fowleri is rather widely known in Rancholabrean localities in eastern North America. It is known from the Bradenton, Devil's Den Chamber 3, Haile (Rancholabrean complex), Ichetucknee River, Kanapaha I, Kendrick IA, and Reddick I sites, Florida. It has also been recorded from Kingston Saltpeter Cave, Georgia; Zoo Cave, Missouri; the Bootlegger Sink and New Paris 4 sites, Pennsylvania; Baker Bluff, Cheek Bend, and Guy Wilson caves, Tennessee; and the Clark's Cave, Natural Chimneys, Saltville, and Strait Canyon Fissure sites, Virginia.

Bufo hemiophrys has been recorded from the Medicine Hat Fauna, Alberta, Canada. *Bufo* cf. *Bufo kelloggi* and *Bufo mazatlanensis* have been identified from the Rancho la Brisca locality, Sonora, Mexico. Another western form, *Bufo punctatus*, is known from the Deadman Cave and Wolcott Peak sites, Arizona; the Redtail Peaks and Tunnel Ridge sites, California; Gypsum Cave, Nevada; and the Dry Cave Fauna, New Mexico. *Bufo quercicus* is known only from the Reddick I site, Florida. Remains identified as *Bufo speciosus* or *Bufo cognatus* are recorded from the Howard Ranch Local Fauna, Texas.

Bufo terrestris has been identified from a rather large number of localities in Florida, including the Arredondo, Bradenton, Devil's Den Chamber 3, Haile (Rancholabrean complex), Hornsby Spring, Ichetucknee River, Kanapaha I, Orange Lake, Reddick I, Williston IIIA, and Winter Beach sites; *B. terrestris* has also been identified from the Ladds Quarry site, Georgia, and has been tentatively identified from Cheek Bend Cave, Tennessee. *Bufo valliceps* is known only from Fowlkes Cave, Texas.

Another western form, *Bufo woodhousii*, is known from Deadman Cave, Arizona; Duck Creek, Jinglebob, Jones, Sandahl, and Williams local faunas, Kansas; Smith Creek Cave, Nevada; Dark Canyon and Dry caves, New Mexico; and the Clear Creek, Easley Ranch, Fowlkes Cave, Lubbock Lake, Schulze Cave, and Slaton faunas, Texas. A putative extinct subspecies of *B. woodhousii, Bufo woodhousii bexarensis*, is known from Friesenhahn Cave, Texas.

The family Hylidae is represented in the Rancholabrean NALMA by four genera: *Acris, Hyla, Pseudacris,* and *Pternohyla. Hyla baderi* represents the only extinct North American Pleistocene anuran recognized in this book; and even *H. baderi* is somewhat questionable, as it is known from only one site among all the rich Pleistocene faunas in Florida.

Acris crepitans is known from the Butler Spring, Kanapolis, and Williams local faunas, Kansas; and from the Clear Creek, Easley Ranch, Howard Ranch, Lubbock Lake, and Slaton faunas, Texas. *Acris gryllus* has been reported from the Arredondo site, Florida.

The genus *Hyla* is well represented in the Rancholabrean of eastern North America. The western species, *Hyla arenicolor,* has been tentatively reported from Howell's Ridge Cave, New Mexico, and identified from the Rancho la Brisca locality, Sonora, Mexico. The extinct species *Hyla baderi* is known only from the Arredondo site (pit II) of Florida. The two sibling, cryptic species, *Hyla chrysoscelis* and *Hyla versicolor,* have been identified as "*Hyla chrysoscelis* or *H. versicolor*" from the following: Ladds Quarry, Georgia; the Kanapolis Local Fauna, Kansas; Frankstown Cave, Pennsylvania; and the Easley Ranch, Howard Ranch, and Schulze Cave faunas, Texas.

Four currently southeastern United States species of *Hyla* have been identified from southeastern Rancholabrean localities. *Hyla cinerea* has been reported from the Arredondo and Reddick I sites, Florida. *Hyla femoralis* is known from the Haile (Rancholabrean complex) and Williston IIIA sites, Florida, and from Robinson Cave, Tennessee. *Hyla gratiosa* has been reported from Bell Cave unit I, Alabama, and from the Arredondo site, Florida. Finally, *Hyla squirella* has been reported from the Haile (Rancholabrean complex), Florida.

Pseudacris species are known in the Rancholabrean from coast to coast. *Pseudacris clarkii* is known from the Easley Ranch Local Fauna, Texas. *Pseudacris crucifer* is rather widely spread in the eastern United States: it has been reported from Kingston Saltpeter Cave and Ladds Quarry, Georgia; Frankstown Cave and the New Paris 4 site, Pennsylvania; Cheek Bend Cave, Tennessee; Clark's Cave, Virginia; and New Trout Cave, West Virginia. The closely related species *Pseudacris nigrita* is known from the Reddick I site, Florida. *Pseudacris ornata* is known from the Arredondo site, Florida.

The far western species *Pseudacris regilla* is known from the Newport Beach Mesa site, California. *Pseudacris streckeri* has been reported from the Howard Ranch Local Fauna and Miller's Cave Fauna in Texas. *Pseudacris triseriata* occurs widely in the eastern United States, as well as more western localities, but oddly has been reported (sometimes tentatively) from only the following: the Cragin Quarry, Cudahy, Jinglebob, Jones, and Sandahl local faunas, Kansas; the Dark Canyon Fauna and Dry Cave Fauna unit II, New Mexico; the Doby Springs Local Fauna, Oklahoma; and Sheriden Pit Cave, Ohio. Finally, the genus *Pternohyla*, in the form of *Pternohyla fodiens*, is known only from the Rancho la Brisca locality, Sonora, Mexico.

The family Ranidae, represented in the Rancholabrean NALMA by the genus *Rana*, occurs from coast to coast and from Nova Scotia and Quebec, Canada, south to Sonora and Tamaulipas, Mexico. A problem with the genus *Rana* is that paleontologists have not been able to identify species within the *Rana pipiens* complex because of the osteological similarity among these taxa. Thus, many Rancholabrean records are listed as "*Rana pipiens* complex" in the literature. A list of localities for which this designation has been used can be found on pp. 172–173. Turning to other records of *Rana*, we find that *Rana areolata* is known from the Reddick I site, Florida, and that *Rana aurora* has been recorded at the Newport Beach Mesa and Rancho La Brea sites, California.

Rana catesbeiana is known from a relatively large number of localities, from eastern North America to Kansas and Texas, and these are as follows: Bell Cave, Alabama; Peccary Cave, Arkansas; Kingston Saltpeter Cave, Georgia; Arredondo, Devil's Den Chamber 3, and Vero Beach strata 2 and 3, Florida; the Prairie Creek D Fauna, Indiana; the Jinglebob, Kanapolis, and Williams local faunas, Kansas; Boney Spring and Brynjulfson Cave I, Missouri; the Avenue (tentatively), Ben Franklin, Ingleside, Lubbock Lake, and Miller's Cave faunas, Texas; and the Clark's Cave, Natural Chimneys, and Strait Canyon Fissure sites, Virginia.

Rana clamitans has been reported from a relatively large number of Rancholabrean localities in eastern North America: Kingston Saltpeter Cave, Alabama; the Prairie Creek D Fauna, Indiana; the Shelton site, Michigan; Sheriden Pit Cave, Ohio; the New Paris 4 site, Pennsylvania (tentatively); the Clark's Cave, Natural Chimneys, and Strait Canyon Fissure sites, Virginia; and New Trout Cave, West Virginia. *Rana grylio*, on the other hand, has been reported only from the Haile (Rancholabrean complex) and Reddick I sites, Florida.

Rana palustris has been tentatively reported from Peccary Cave, Arkansas, and the New Paris 4 site, Pennsylvania. It has been identified from the Lubbock Lake (Clovis period) site, Texas; Clark's Cave, Virginia; and the Natural Chimneys site, Virginia (tentatively). *Rana pipiens*, a related species, has been reported from Eardley locality 2, Quebec, Canada, and the Sheriden Pit Cave, Ohio. *Rana sphenocephala*, a species related to the previous two, is known from the Ladds Quarry site, Georgia, as well as from several sites in Florida, including the Arredondo, Devil's Den Chamber 3, Reddick I, Sabertooth Cave, Vero Beach strata 2 and 3, and Williston IIIA sites.

Finally, *Rana sylvatica* is known from the Prairie Creek D Fauna, Indiana; Duck Creek Local Fauna, Kansas; the Frankstown Cave and New Paris 4 sites, Pennsylvania; Baker Bluff Cave, Tennessee; the Clark's Cave and Natural Chimney sites, Virginia; and New Trout and Worm Hole caves, West Virginia.

The family Microhylidae is represented in the Rancholabrean by only two species, both in the genus *Gastrophryne*. *Gastrophryne carolinensis* is known from the Kingston Saltpeter Cave and Ladds Quarry sites, Georgia; and the Arredondo, Reddick I, Sabertooth Cave, and Williston IIIA sites, Florida. *Gastrophryne olivacea* has been tentatively identified from the Rancho la Brisca locality, Sonora, Mexico. It has been positively identified from the Clear Creek Fauna, Texas.

In summary, six families — Rhinophrynidae, Pelobatidae, Leptodactylidae, Hylidae, Ranidae, and Microhylidae — are known from the Rancholabrean NALMA. These families are together represented by 11 living genera. At least 49 species are recorded, only one of which is recognized as extinct.

OVERVIEW OF FROGS IN THE PLEISTOCENE

The Pleistocene was a time of stress for vertebrate populations of all kinds. Anurans were certainly affected by these stresses, especially in areas where massive ice sheets advanced over and withdrew from frog ranges in a cyclic fashion. In North America, the ice sheet advanced farthest in the central Great Lakes region (Fig. 100). The most southern penetration occurred during the Illinoian glacial age, when the Laurentide Ice Sheet extended deep into southern Illinois, Indiana, and Ohio. In the Wisconsinan, the last glacial age of the Pleistocene epoch, the ice did not extend quite as far south in these states (see Fig. 100).

The effects of the ice sheets were numerous. Topographic changes are still evident from Pleistocene times, as ice sheets drastically changed the landscapes over which they passed. It is estimated that the average thickness of the ice was about 2.5 km and that in places it was 3.2 km or more. Hills, valleys, ridges, lakes, streams, swamps, and bogs — features produced by the activity of the ice — obviously influenced the ecology and distribution of Pleistocene frogs.

The movements of the ice sheets effected dramatic climatic changes.

FIGURE 100. Maximum extent of the ice sheet in the Pleistocene of North America in Illinoian and Wisconsinan times: IL, Illinois; IN, Indiana; MI, Michigan; OH, Ohio; ON, Ontario; NO ON, northern Ontario; UP MI, Upper Peninsula of Michigan; WI, Wisconsin.

But the classic idea of alternating purely cold glacial and warm interglacial climates is oversimplified. Pleistocene vertebrate faunas in North America indicate that climates were cold near the ice-sheet borders but that in the central and southern United States the climate was more equable than it is at present, with warmer winters and cooler summers (e.g., Lundelius et al., 1983; Holman, 2001b). This phenomenon has been referred to as the Pleistocene climatic equability model (Holman, 1995b, 2001b).

Vegetational communities were either altered or eliminated during the Pleistocene. For years it was postulated that the major vegetational associations withdrew southward in bandlike units as the ice sheet advanced and then moved northward again in the same way as the ice sheet retreated. Thus, it was envisioned that during glacial times, a barren tundra association existed in a deep band south of the glacial front and that this was followed by a deep band of coniferous forest that graded into temperate deciduous forest, which penetrated far into the southeastern United States. This classic concept has been referred to as the stripe hypothesis (Holman, 2001b).

The modern theory has been referred to as the plaid hypothesis (Holman, 2001b). The concept here is that during a large part of the Pleistocene, a cold climate and coniferous or even tundra vegetation existed near the edge of the ice sheet but that plant communities in the central and southern United States were a mixture of the original plants of the area and invading northern forms. This concept arose from the suggestion that plant and animal species reacted individually, rather than as groups, to Pleistocene changes and that the mixed communities in the area would have been able to coexist quite well in the equable climate of the time (see Holman, 1995b).

Amphibian populations also faced alternating flooding and re-emergence of habitats caused by rising and falling lake and sea levels. This was due to the fact that global water was bound up in the ice sheets during much of the Pleistocene: during full glacial events, lake and sea levels dropped, but during interglacials the reverse occurred. Florida, for instance, was inundated by the sea during interglacial times to the extent that it was cut off from the mainland, with only its high, central part, the "Ocala Island," remaining above sea level (Cooke, 1945). On the other hand, in southeastern Canada, the St. Lawrence River system, which is currently freshwater, became a marine system called the Champlain Sea (Harington, 1988; Holman, 2001b).

In both North America and Europe the glacial periods lasted much longer than the interglacial ones, and this must have had an impact on the amphibian populations of the Pleistocene. North American frogs somehow coped with the fluctuations in the climate, and this may be reflected in some of their physiological adaptations to cold climates. Three anuran species, *Hyla versicolor*, *Pseudacris triseriata*, and *Rana sylvatica*, currently occur in the central Great Lakes region where, during the Pleistocene, the ice obliterated their habitats several times. These three species have the amazing ability to freeze solidly in the winter and thaw out in the spring with no harm done (Layne and Lee, 1995). *Rana sylvatica* couples in amplexus have been experimentally frozen solid, and when thawed out, resumed their copulating activities! In these forms, glycogen in the liver is converted to an "antifreeze" derivative that keeps the cells from breaking apart during the freezing process. Obviously, there must be other physiological adaptations to cold climates in anurans that have not yet been discovered.

A certain lag time in community development and organization must have affected anuran populations in the Pleistocene. Retreating ice sheets of the epoch left a sterile mass of mud, silt, sand, and gravel behind. Obviously, there must have been a series of floras that developed before stable vegetational communities became established. Unfortunately, there is little information available about how much time it took for such communities to develop. If this were known, we could better estimate the time of re-entry of the less cold-tolerant anuran species into deglaciated areas.

Evolutionary Stasis in Pleistocene Anurans. Despite all of these Pleistocene stresses, the frog fauna, at least in continental North America (also the British Isles and the European continent; Holman, 1998) existed in a state of evolutionary stasis during the epoch (Holman, 2000b). Although mammals and birds had a devastating extinction episode at the family, generic, and specific levels in the Pleistocene of North America (e.g., Kurtén and Anderson, 1980; Martin and Klein, 1984), only one frog species, *Hyla baderi*, is regarded as extinct in the Pleistocene. Actually, there is some question about the validity of *H. baderi*, as it has been found in only one Rancholabrean NALMA site in northern Florida (see Chapter 2).

Rather than accepting without question the concept of the evolutionary stasis of anurans through the Pleistocene, we should consider some possible modifications of the concept. Loss of information occurs during the death and preservation of most vertebrate fossils, and frogs are especially prone to postmortem information loss because of the fragility of their skeletons. Thus, specific identifications of Pleistocene anurans are usually made on the basis of the relatively dense bones, such as the humerus, ilium, and sacral vertebra that survive the fossilization process. Obviously, one of the most acute problems is the probability that extinct species, and possibly even genera, may have gone unrecognized. A case in point concerns the sibling cryptic species *Hyla chrysoscelis* and *Hyla versicolor*, which have identical skeletons as far as anyone has been able to tell. It is possible that extinct cryptic species could have gone undetected in Pleistocene specimens.

A big problem concerns the *Rana pipiens* complex, which contains several modern species that cannot be distinguished osteologically. It is quite possible that now extinct frogs of this evolutionarily plastic group could have existed in the Pleistocene. Modern toads of the genus *Bufo*, even in different species groups, are often difficult or impossible to distinguish. In fact, one of the main problems in the study of Pleistocene anurans in both North America and Europe is that some people have tried to carry the identification of fragmentary frog bones too far.

A frustrating problem in the study of Pleistocene frogs is that when they do occur as relatively complete specimens, they are usually so embedded in matrix that they are difficult to prepare and study. Some fossil frogs occur merely as impressions in rock rather than as true bones. This leaves the articular surfaces and muscle scars obscure in the postcranial skeleton, and the skull bones may be indistinguishable from one another. Fortunately, these kinds of specimens are not common in North American Pleistocene studies. Often, a frog sacrum or ilium, unbroken and free of matrix, is much more useful for the identification of fossil anuran taxa than a so-called complete frog skeleton.

An important psychological problem among many paleoherpetologists is the ingrained idea that all Pleistocene amphibian and reptile species, other than large tortoises or unique island taxa, must belong to modern species. This bias undoubtedly emerged from the frustration felt by the "new" generation of paleoherpetologists in the 1950s in response to the zealous naming of Pleistocene species by their predecessors. In the present paleoherpetological community, anyone who attempts to describe an extinct Pleistocene frog will immediately be taken to task by his or her colleagues; but someone who identifies a modern species on the basis of a fragmentary ilium or sacrum is usually not subject to such scrutiny.

As has been argued here before and will be again, you need a comparative anuran skeleton collection to be able to make reliable identifications of fossil frogs because of the osteological variations that occur within species. But limitations exist, as seldom do you get an adequate sample of fossil amphibian species. Moreover, because of declining amphibian populations, getting a comparative modern skeletal collection together that is large enough for variational studies often becomes a legal or at least an ethical problem. Finally, in the Pleistocene of North America, Rancholabrean NALMA sites are numerous, but Irvingtonian NALMA sites are comparatively rare.

Nevertheless, all of the above having been stated, the Pleistocene herpetofaunas that have been documented at numerous sites in continental North America have been strikingly more stable than the mammalian and avian faunas.

So, what about this frog stasis in the Pleistocene? To begin with, ectothermic animals such as anurans, which have low metabolic rates and are able to aestivate or brumate during inclement climatic conditions, would have many advantages over the endothermic birds and mammals during times of climatic fluctuation. The ability of several anurans to freeze solid in the winter and to thaw out in the spring with no harmful effects (Layne and Lee, 1995) shows that they must have been highly adaptive during the Pleistocene cold intervals. Continued studies of aestivation and brumation patterns in frogs should be useful in establishing parameters for estimating their Pleistocene tolerances.

The small body mass of most North American frogs would appear to be advantageous in the wake of the Pleistocene habitat shrinkage that must have occurred during ice-sheet fluctuations (see Holman, 1991, 1995b, 1998); in contrast, the larger endotherms, especially the megaherbivores, would have required large tracts of available habitat (Owen-Smith, 1987).

In addition, it is likely that few frogs were directly dependent, either as predators, scavengers, or commensals, upon the large mammals that became extinct during the Pleistocene. The well-documented fauna of the Pleistocene at Rancho La Brea, California, showed that the dung beetles and scavenging or commensal birds that were dependent on the mammalian megaherbivores became extinct along with these mammals (Harris and Jefferson, 1985). All of the amphibian and reptilian species, however, survived to the present.

In contrast to this, in Australia the giant varanid lizard, *Megalania*, and the giant constricting snake, *Wonambi* (Murray, 1984), were top predators and became extinct during the Pleistocene, along with many large marsupial herbivores. Actually, few suggestions have been made in Pleistocene studies about the possible ecological relationships of frogs to the large extinct mammals, if any existed. It seems likely that frogs existed well outside the web of interactions that took place between the large endotherms and their predators and scavengers.

The parenting process in endothermic birds and mammals is obviously a part of the reproductive stress syndrome and would appear to be a considerable drain on their energy budget during times of climatic change. Of course, parenting in the avian or mammalian sense is lacking in North American frogs. Moreover, the high reproductive potential of frogs, in contrast to the low fecundity of larger endotherms (especially such megaherbivores as mastodons and mammoths), might heavily favor the anurans during times of climatic change.

For more than three decades, the hypothesis of human overkill has been suggested as an important, if not the major, factor in the Pleistocene extinction of the large herbivorous mammals and their predators and scavengers (e.g., Martin and Wright, 1967; Martin and Klein, 1984). Holman (1959a) compared extinction percentages of amphibians and reptiles with those of mammals in the Williston IIIA Fauna of the Pleistocene (Rancholabrean NALMA) in Florida and suggested that the rise of humans was an important factor in the differential between extinctions in the herpetological and mammalian components of the fauna. This might reflect the lesser desirability of most amphibians and reptiles as a food resource, compared with large mammalian herbivores. Obviously, the huge land tortoises of the genus *Hesperotestudo* would have been an easily gathered source of food for humans (e.g., at the Little Salt Spring site in Florida; Clausen et al., 1979), and this was probably one of the major causes of the extinction of those tortoises at the end of the Pleistocene.

In summary, studies of many aspects of physiology, cryobiology, and ecology of living frogs may bear directly on the question of why there was a general evolutionary stasis among amphibians and reptiles during the Pleistocene, when climatic oscillations and great environmental changes took place. Further studies of frogs in the Pleistocene food web, as well as modern studies in this subject, especially in megaherbivore-dominated communities in Africa, might provide more insights. Finally, studies of human feeding activities in the late Pleistocene of North America might provide data on the importance of frogs as a Pleistocene food resource for humans.

Range Adjustments of Pleistocene Frogs. Amphibians and reptiles made far fewer range adjustments in the Pleistocene than did mammals (Holman, 1991, 1995b, 1998, 2000b) or birds. In fact, in North America, the tendency was for both salamanders and frogs to stay in place, south of the periglacial areas, during the Pleistocene (Fay, 1988; Holman, 1995b). This conclusion is based on the study of more than 150 North American amphibian faunas (Holman, 1995b). Some would argue that rapid evolution of new ecological tolerances in the face of climatic selective pressures in amphibians and reptiles in the Pleistocene did occur. Nevertheless, the equable climate (with warmer winters and cooler summers) south of the glaciated areas during the Pleistocene in North America was probably quite compatible with the existing physiological adaptations of frogs, as well as other amphibians and reptiles.

The fact that frogs tended to stay put is especially well documented in the late Pleistocene of the Appalachians, where several large herpetofaunas were studied by Fay (1988). Here the few amphibian invading species were mainly from the south, rather than the north! Nevertheless, Fay pointed out that it would be difficult to detect any northern amphibian invaders, since most amphibian species that occur in Appalachia today also occur much farther north. One can only invoke again the Pleistocene climatic equability hypothesis to explain the presence of the few southern forms.

The Ozark region of Missouri and Arkansas is an ancient, isolated uplift with numerous and interesting modern amphibian taxa. Here the amphibian situation is even more striking than in Appalachia, as all of the Pleistocene species are not only extant, but currently living in the vicinity of their respective fossil sites (Holman, 1995b). Other than California, the Pacific coastal region has practically no Pleistocene records of amphibians; and in California, where several Pleistocene herpetofaunas have been reported (Holman, 1995b), few anurans have been recorded. Nevertheless, the few frogs that have been identified represent extant species that still occur in the vicinity of their respective fossil sites.

A few extralimital records of Pleistocene frogs are known. Some of these are from the northern plains. The early Pleistocene (Irvingtonian NALMA) Java Local Fauna in north-central South Dakota yielded *Rana catesbeiana*, which currently ranges north in the region only to southern South Dakota (Conant and Collins, 1998, map, p. 555). The Pleistocene (Irvingtonian) Albert Ahrens Local Fauna is considered to represent a full glacial period. Here the northeastern extralimital form *Rana sylvatica* occurs with the eastern extralimital taxa *Bufo americanus*, "*Hyla chrysoscelis* or *H. versicolor*," and *Rana clamitans* and the southern extralimital *Pseudacris* cf. *Pseudacris clarkii*. The nearest area where all these frog species live today is extreme northwestern Missouri. Ford (1992) suggested that these animals lived in a more equable climate than occurs in the area at present. In southwestern Kansas, the Jinglebob Local Fauna (Rancholabrean NALMA) yielded *Pseudacris* cf. *Pseudacris triseriata*, which is really more extralocal than extralimital (see Conant and Collins, 1998, map, p. 543). Probably the area was a little less arid at the time.

Turning to the southern plains in North America, we find the late Pleistocene (Rancholabrean NALMA) Easley Ranch Local Fauna (Lynch, 1964), in north-central Texas, which yielded the extralimital southern leptodactylid frog *Eleutherodactylus marnockii* (see Conant and Collins, 1998, map, p. 510). The Pleistocene (Rancholabrean) Howard Ranch Local Fauna, also in north-central Texas, produced the remains of "*Hyla chrysoscelis* or *H. versicolor*" and *Pseudacris streckeri*, both of which currently occur east of the Texas locality (see Conant and Collins, 1998, maps, pp. 536, 550). The Pleistocene (Rancholabrean) Fowlkes Cave site, in trans-Pecos Texas, yielded *Bufo valliceps*, an extralimital southeastern form (see Conant and Collins, 1998, map, p. 521).

Turning to the Rocky Mountain area of the southwestern United States, we find in the late Pleistocene (Rancholabrean NALMA) Dry Cave site, near Carlsbad, in southeastern New Mexico, *Pseudacris triseriata*, an extralimital northern species (see Conant and Collins, 1998, map, p. 543) that occurs north of the vicinity in other parts of the general area. In the Pleistocene (Rancholabrean) Smith Creek Canyon Fauna, *Bufo boreas* and *Bufo* cf. *Bufo woodhousii* have been reported from the Snake Range in White Pine County, Nevada, near the Utah border, but neither of these species occurs in the Snake Range today (Mead et al., 1982).

In the southeastern United States, several extralimital Pleistocene (Rancholabrean NALMA) records of *Bufo fowleri* (see Chapter 2) are known from peninsular Florida, whereas at present this species occurs only in the Florida panhandle (Conant and Collins, 1998, map, p. 519).

The reason for the contraction of the range of this species is poorly understood.

Re-invasion of Previously Glaciated Areas. By far the most extensive movements of Pleistocene frog populations were re-invasions of previously glaciated areas. Although the present distributions of some populations of amphibians and reptiles in Michigan are fragmented (see Holman, 2001a), Holman (1992c) developed a simple model of the postglacial herpetological recolonization in Michigan. This model was based mainly on the postglacial vegetational development documented by paleobotanical and palynological data; the ecological tolerances of the amphibians and reptiles living in the state at present; and the Michigan fossil record. Routes of entry for glacially displaced amphibians and reptiles were suggested (Fig. 101).

The postglacial frog fauna of the Upper Peninsula of Michigan probably was largely derived from a Wisconsin corridor, because the width and depth of the Straits of Mackinac constrained movement. This invasion must have been some time after 9.9 ka BP, since at that time the Upper Peninsula had glacial ice along its northern edge (Fig. 102). The

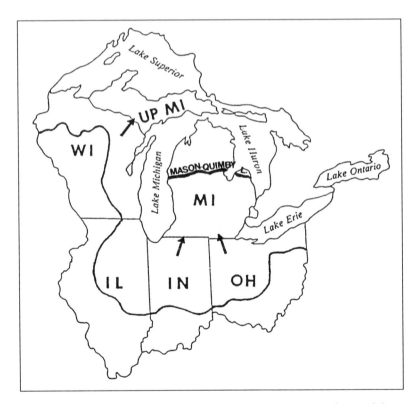

FIGURE 101. Arrows indicate the most probable re-entry routes for amphibians and reptiles into deglaciated Michigan at the end of the Pleistocene. No undisputable Pleistocene vertebrate fossils and very few Paleo-Indian artifacts have been found above the Mason–Quimby Line in Michigan. The dark line in Wisconsin, Illinois, Indiana, and Ohio indicates the maximum extent of the ice sheet in the Wisconsinan glacial age. IL, Illinois; IN, Indiana; MI, Michigan; OH, Ohio; UP MI, Upper Peninsula of Michigan; WI, Wisconsin.

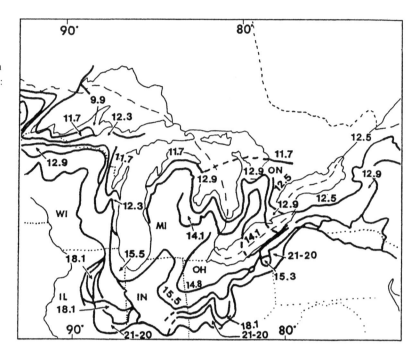

FIGURE 102. Ice margins at various dates (in thousands of years before present [BP]) in the Great Lakes region: IL, Illinois; IN, Indiana; MI, Michigan; OH, Ohio; ON, Ontario; WI, Wisconsin.

Lower Peninsula probably derived its postglacial amphibian faunas from Ohio and Indiana. The paleobotanical record and the time necessary for stable plant communities to develop suggest that the tundra-tolerant, freeze-tolerant species *Rana sylvatica* could have reached southern Michigan by about 13 ka BP.

The Holman (1992c) model suggested three categories (Table 3) of invading herpetological species: primary invaders, whose ecological tolerances include coniferous forest and in one case (*Rana sylvatica*), tundra; secondary invaders, adapted to mixed coniferous–broadleaf forests; and tertiary invaders, adapted to broadleaf forests. Members of the tertiary group are all currently restricted to the southern part of the Lower Peninsula of Michigan, where many of them exist as peripheral or isolated populations and are listed as threatened or endangered (Holman, 1992c).

Able to tolerate tundra conditions at present, *Rana sylvatica* has the ability to freeze during winter hibernation with no damaging effects and seems to regularly do this, at least in some parts of its range. This species may have been the first primary amphibian invader of postglacial Michigan as the great mass of sterile, glacially derived mud and sand gave way to tundra between about 14.8 and 13 ka BP. The other primary forms (Table 3) probably entered southern Michigan about 12.5 ka BP, when the first spruce forests developed, and then later (about 11.8 ka BP), when jack pine and red pine communities began to develop. Primary species probably did not reach the Upper Peninsula of Michigan until at least 9.9 ka BP, because glacial ice rimmed the northern border of the area until that time.

Secondary invaders (Table 3) probably began to enter southern Michigan about 10 ka BP as mixed coniferous–broadleaf forests began to de-

TABLE 3. Amphibian taxa considered to be primary, secondary, and tertiary invaders of postglacial Michigan.

Primary invaders	Secondary invaders	Tertiary invaders
Salamanders	Anurans	Salamanders
Ambystoma laterale	Acris crepitans	Ambystoma opacum
Ambystoma opacum	Bufo woodhousii	Ambystoma texanum
Ambystoma tigrinum	Hyla chrysoscelis	Siren intermedia
Hemidactylium scutatum		
Necturus maculosus		
Notopthalmus viridescens		
Plethodon cinereus		
Anurans		
Bufo americanus		
Hyla versicolor		
Pseudacris crucifer		
Pseudacris triseriata		
Rana catesbeiana		
Rana clamitans		
Rana palustris		
Rana pipiens		
Rana septentrionalis		
Rana sylvatica		

Note: No frogs are in the tertiary invader category. The present ranges of primary invading frog species appear to be less fragmented than those for any other similar grouping of Michigan amphibians.

Source: Holman (1992c).

velop in the southern part of the state. The entrance of the tertiary invaders probably began with the establishment of mixed hardwood forests in southern Michigan, about 9.9 ka BP, and probably persisted through the general warming period that lasted until about 2.5 ka BP. The present ranges of primary invading frog species appear to be the least fragmented of any other similar grouping of Michigan amphibians and reptiles (see county records of amphibians and reptiles on file at the Michigan State University Museum, East Lansing).

Great Lakes Region Pleistocene Anuran Record. The Michigan model, of course, can only be verified by the fossil record, which is unfortunately rather weak in the Great Lakes area.

Pre-Wisconsinan evidence that could set the stage for the interpretation of the late Wisconsinan frog fauna is lacking. But the Hopwood Farm locality in south-central Illinois (King and Saunders, 1986) indicates that both amphibian and reptilian faunas were flourishing in a mild, equable climate in Sangamonian times. Here, the giant land tortoise, *Hesperotestudo crassiscutata* (formerly known as *Geochelone crassiscutata*) is known from the Sangamonian zone of a locality that transverses Illinoian to Sangamonian times. This giant tortoise indicates above-freezing temperatures throughout the year in south-central Illinois, at least during the portion of the Sangamonian age represented by the Hopwood Farm locality (see Holman, 2001b).

Late Wisconsinan anurans, however, are known from several sites in the Great Lakes region. The Prairie Creek D Fauna site, in Daviess County, southwestern Indiana, has been dated at about 14.5 ka BP, when the ice margin could have been as close as northeastern Indiana (see Fig.

102); yet the herpetofauna was quite diverse (Holman and Richards, 1993). Some of the amphibians must have recolonized southwestern Indiana rather rapidly from more southern areas, for during the 21–20 ka BP advance, and even as late as the 14.8 ka BP advance (see Fig. 102), the ice front was as close as 95 km from the Prairie Creek D Fauna site. One could certainly reason that the vegetation at the time of the 21–20 ka BP advance, and even as late as the 14.8 ka BP advance, was boreal forest (Delcourt et al., 1983), an environment that would have suited mainly cold-adapted amphibians and reptiles. The Prairie Creek D frog fauna consisted of *Bufo* sp., *Rana catesbeiana*, *Rana clamitans*, *Rana pipiens* complex, and *Rana sylvatica*. Those frogs identified to the specific level were all considered primary invaders of postglacial Michigan by Holman (1992c) and may be found today in areas with a significant amount of coniferous vegetation.

In Hancock County, east-central Indiana, the Christensen Bog, a kettle bog that formed from 13 to 11.7 ka BP, yielded the remains of an amphibian, turtles, birds, and mammals. This site lies just west of the Union City Moraine, which is correlated with the 14.8 ka BP ice advance. For this reason, we might suspect that Hancock County, Indiana, would have been tundra-like at 14.8 ka BP and practically uninhabitable for most herpetological species. Nevertheless, when the ice front was probably situated in southern Ontario and the northern part of the Lower Peninsula of Michigan, a frog and three species of turtles inhabited the site. The frog, *Rana pipiens* complex, today occurs northward into areas with significant amounts of coniferous forest and was considered by Holman (1992c) one of the primary invaders of Michigan.

The Meskill Road Water Well site, in St. Clair County, southeastern Michigan, yielded remains of *Bufo americanus* from sediments that are thought to be those of Glacial Lake Whittlesey and thus to represent a date of about 13 ka BP (Holman, 1988). *Bufo americanus* is considered a primary invader of Michigan (Holman, 1992c). Given the location of the 12.9 ka BP ice margin, if the 13 ka BP date is correct the toad must have lived almost in sight of the glacier! *Bufo americanus* has also been found at the Moscow Fissure site in extreme southwestern Wisconsin (Foley, 1984). A radiocarbon date of 17.05 ± 1.5 ka BP is available for the site. At this time the western extent of the ice margin was in southeastern Wisconsin and continued southward into central Illinois.

The Shelton mastodon site in Oakland County, southeastern Lower Peninsula of Michigan, has yielded radiocarbon dates of 12.3–11.7 ka BP. *Rana clamitans* was identified from the site (DeFauw and Shoshani, 1991). *Rana clamitans* was considered a primary invader of postglacial Michigan. Ice covered the Shelton site at 14.8 ka BP, and as recently as 12.9 ka BP the ice was as close as Tuscola, Huron, and Sanilac counties, in the "thumb" area of Michigan. This certainly indicates a quite rapid re-invasion of a formerly glaciated area.

The Sheriden Pit Cave site in Wyandot County, northwestern Ohio, has yielded a radiocarbon date of 11.7 ka BP. Anurans reported from the site by Holman (1995b, 1997) include *Bufo americanus*, *Bufo fowleri*, *Bufo* sp., *Pseudacris triseriata*, *Rana clamitans*, *Rana pipiens*, *Rana sylvatica*, and *Rana* sp. Six of the seven anurans were considered primary invaders of postglacial Michigan by Holman (1992c). On the other hand, *B. fowleri* was considered a secondary invader, adapted to both mixed coniferous–broadleaf and broadleaf forests in rather sandy situations. At 14.8 ka BP ice still covered northwestern Ohio, but by 11.7 ka BP the ice margin was in the upper part of the Lower Peninsula of Michigan and in southern Ontario. Thus, the invasion of this rather large number of amphibian species must have taken place within at most 3.1 ka.

Remarks. Although the sample size is rather small, the fossil record in the Great Lakes region supports the Michigan model of postglacial re-invasion by amphibian species. In fact, all the amphibians that have been identified from the late Pleistocene of Michigan, Ohio, Indiana, and Wisconsin are considered primary invaders, with the exception of one proposed secondary invader, *Bufo fowleri*, recorded from the Sheriden Pit Cave Fauna, northwestern Ohio.

One might ask if there are any aspects of these studies that pertain to the present or to the future. If future amphibian populations were to be extirpated from large geographic areas, at what rate might one expect recolonization to occur? A consideration of Pleistocene re-invasion patterns as analogs might have some bearing on this question, as the following discussion suggests.

Four species of anurans (*Rana catesbeiana*, *Rana clamitans*, *Rana pipiens* complex, and *Rana sylvatica*) co-occur in the Prairie Creek D Fauna (dated at about 14.5 ka BP), southwestern Indiana, and at the Sheriden Pit Cave site (dated at about 11.7 ka BP), northwestern Ohio. The latitudinal distance between the two sites is approximately 280 km. The time difference between the two sites is about 2.8 ka. If we assume that these populations represent the earliest amphibian re-invaders in the area, the mean rate of movement would be about 0.1 km per year. This would indicate that it took an average of about 1 ka for the four frog species to move 100 km northward during postglacial time. Although this estimate is admittedly rough, it does indicate that it might take groups of amphibians a considerable amount of time to naturally recolonize a large area — in fact, several human lifetimes. On the other hand, fossil localities in Michigan do indicate that relatively rapid recolonization of individual frog species (e.g., *Bufo americanus* and *R. clamitans*) may have occurred.

Epilogue

This is a long epilogue, as epilogues go, because the classification and the phylogeny of anurans are in an embryonic state, owing to a poor fossil record. Thus, I feel the subject needs to be discussed in a special section that people will look at, perhaps to find out what an epilogue really is. Heatwole and Carroll (2000, p. 1462), in the latest state-of-the-art work on amphibian paleontology, stated in the prefix to their appendix II: "Because of major gaps in the knowledge of the fossil record, understanding of the interrelationships of the principal groups of amphibians remains incomplete. Any cladogram would give an erroneous impression of the precision of the comprehension of their phylogenetic history. All that is practical at the present times is an informal classification."

Again, to this end, Carroll (2000b, p. 977) stated the following: "One of the most important goals of Willi Hennig (1966) and the current practitioners of phylogenetic systematics [cladistics] is to establish a system of classification that directly reflects evolutionary history. Unfortunately, this is not yet possible for the Amphibia because of the major gaps in the fossil record. It remains extremely difficult to establish the pattern of interrelationships among the well established orders. Without direct knowledge of the occurrence and specific combination of characters in fossils, it is difficult to determine the polarity of character transformation in particular lineages, or the homology of similar derived features in different lineages. Until this has been established, major groupings of amphibian orders can only be recognized on the basis of broad similarities. It is an important advance in understanding simply to recognize that the interrelationships among previously identified taxa cannot yet be either refuted or firmly established. Only additional information from fossils will enable advancement from this time of uncertainty to one in which well substantiated phylogenies can be proposed."

Obviously, to rectify this situation, we must build a much better foundation in baseline anuran paleontology. To this end, we first need to discuss the naming of fossil anuran taxa. Various iconoclasts have said that paleontological species names are, at best, just good scientific guesses. For this reason, paleontological generic names have been used more than specific ones in studies of broad evolutionary patterns. In terms of classification and phylogenetic studies, however, the description of invalid species and the improper identification of some already existing species cloud very much the issues at hand. If characters used to describe fossil frogs species are individually variable, the cladograms mean nothing. For this reason, anuran characters taken from the literature may be valid or invalid, depending on the number of specimens of each species, fossil or recent, that were available for study.

Many times in vertebrate paleontological history, incorrect naming as two fossil species, one representing the male and the other the female, has occurred. This can happen in naming fossil frogs, especially if the species was named on the basis of bones associated with the reproductive process (e.g., the nuptial flange on the humerus of some male anurans). Fortunately, students of anuran osteology and paleontology can, with an adequate series of modern skeletons of known sex, ascertain what characters are sexually dimorphic, thus avoiding postpublication embarrassment.

Intracolumnar variation, or lack of it, in presacral frog vertebrae can be a problem. For instance, presacral frog vertebrae may be so similar to one another that the position in the vertebral column (e.g., II, IV, or VII) of isolated fossil vertebrae cannot be determined; yet subtle differences in each presacral vertebra could be taxonomically useful. The cervical and sacral vertebrae, however, because of their complexity, are easily separated from the other vertebrae in the column and have been widely used in fossil studies. Ultramicroscopic studies of foramina for the entrance and exit of blood vessels or specific spinal nerves (as was done by Edwards, 1976, for salamanders) in presacral vertebrae, as well as developmental studies, could provide phylogenetic information for both fossil and modern anurans.

Ontogenetic variation can be problematic in anuran paleontology. As far as I am aware, all frog bones change in structure both externally and internally as the animals go from newly metamorphosed individuals or hatchlings to the mature adult and finally to the aging stage. A few of the ontogenetic variations that may occur are (1) changes in proportions of large units, such as the body and skull; (2) changes in the proportions of individual elements of the skeleton; (3) changes in the bony or cartilaginous matrix of the skeleton; (4) fusions of various bones in the skull, vertebral column, and limb girdles; (5) changes in the external features of individual bones; (6) changes in the internal structure of individual bones; and (7) degeneration of bone or pathological conditions in old individuals.

Ontogenetic changes that I have noticed during the study of series of skeletons of modern anuran species are as follows. In general, the head is proportionally larger in newly metamorphosed individuals than in adults, but often not as much as one would expect. The humerus, ilium, scapula, and sacrum tend to change in general shape and proportions during ontogeny, but also, at times, not as much as one would expect.

The bony skeleton tends to become more ossified and the cartilaginous units tend to become more calcified with age; the lack of ossification is often regional in individual bones such as the ilium, where the shaft may become completely ossified earlier than the acetabular region. Calcification of the cartilaginous articular ends of long bones, as well as cartilaginous elements of the pectoral girdle, often occurs during the aging process.

Fusion of bones tends to occur during the growth and aging process, including "natural" fusion of paired frontoparietals in some taxa; fusion of the scapula to other elements of the pectoral girdle; occasional fusion of the sacral vertebra to the vertebra in front of it; and various irregular fusions in old individuals that may be a reflection of poor diet in captivity or various pathologies.

In general, structures such as the dorsal prominence of the ilium in individual bones of anurans tend to become more robust and roughened with age. But again, this increase robustness and rugosity in the dorsal prominence of the ilium with age is not as marked as one might expect.

Variation caused by environmental conditions can often be seen in frog skeletons. It is hardly necessary to discuss the skeletal aberrations in modern frogs that are caused or enhanced by the present deterioration of environmental conditions, as a rather vast literature is now available. Most of the environmentally produced variations I have seen in both fossil and recent frog skeletons, however, have been the healed broken legs in ranid frogs, which I suppose, should not be surprising. Sometimes the healed area is ugly, with one bone overlapping the other, and other times a smooth healing has occurred. But occasionally, ridges or flared areas associated with a nicely healed limb fracture on long bones of the pelvic girdle have been interpreted by one paleontologist or another as a unique specific character. I have not seen other pathologies on fossil anuran bones, but I am sure they must exist.

I am aware that many amphibians can regenerate severed fingers and toes. I observed that this happened frequently in a *Scaphiopus holbrookii* population in northern Florida that had been marked by toe-clipping for recapture. The regenerated toes lacked pigment and appeared to either lack supporting bones or have a different bony structure than the normal toes.

Bones of the same frog species may be different sizes and shapes in different geographic areas. The fact that *Rana sylvatica* is much more short-legged in boreal and tundra situations than in more southern areas is ingrained in the herpetological literature.

Genome variations within frog species have probably caused the most trouble in the identification and naming of anuran species. Anyone studying populations of amphibians and reptiles in the field soon learns that different individuals can be identified as easily as an instructor can learn to recognize the names of a hundred or more students in a classroom. A lot of this is possible, I am sure, because of the variability of gene complexes that cause individuals in both the classroom and the field populations of frogs to look different from one another.

Looking at a bone such as a humerus or an ilium from individuals of a single species of frog from the same pond can illustrate the osteological variation that occurs in what are assumed to be closely related, con-specific individuals. This can be a frustrating exercise if one is attempting to find a character on a bone that separates frog species A from frog species B, but it needs to be done if anuran paleontology is to get any better than it is.

More fossil frogs are needed! From the frantic fossil wars of Marsh and Cope until shortly before World War II, the vertebrate paleontological community was mainly engaged in the collection, exhibition, and study of large fossil vertebrates. The cry seemed to be "the bigger the better" during this interval. But after World War II, several vertebrate paleontologists (briefly discussed in Chapter 1) began purposely collecting small vertebrate fossils (called microvertebrate fossils by some) by simple washing and sieving techniques. This led to the study of small vertebrate fossil bones, such as those of fishes, amphibians, reptiles, bats, mice, shrews, and chipmunks. The fact that the myriad of small vertebrates from paleontological sites could be combined with the big ones quickly led to faunal and paleoecological studies as an adjunct to studies of the lineages of large vertebrates or to studies using large vertebrates as stratigraphic index fossils. It would be helpful to the cause of "a fossil frog phylogeny" if washing and sieving techniques were routinely done on all vertebrate fossil digs.

REFERENCES

Agassiz, L. 1835. Résumé des travaux de la section d'Histoire naturelle et de celle des Sciences médicales pendant l'année 1833. Mémoires de la Société des sciences naturelles de Neuchâtel 1:17–28.

Andreae, J. G. R. 1776. Briefe aus der Schweiz nach Hannover geschrieben in dem Jahre 1763. Zürich und Winterthur: Zweiter Abdruck.

Applegarth, J. S. 1980. Herpetofauna (anurans and lizards) of Eddy County, New Mexico: Quaternary changes and environmental implications. Dissertation Abstracts International, Section B: The Sciences and Engineering 40(8): 3613.

Armstrong-Ziegler, J. G. 1980. Amphibia and Reptilia from the Campanian of New Mexico. Fieldiana: Geology (New Series), No. 4, 39 pp.

Auffenberg, W. 1956. Remarks on some Miocene anurans from Florida, with a description of a new species of *Hyla*. Breviora, No. 52, 11 pp.

Auffenberg, W. 1957. A new species of *Bufo* from the Pliocene of Florida. Florida Academy of Sciences, Quarterly Journal 20:14–20.

Auffenberg, W. 1958. A small fossil herpetofauna from Barbuda, Leeward Islands, with the description of a new species of *Hyla*. Florida Academy of Sciences, Quarterly Journal 21:248–254.

Bayrock, L. A. 1964. Fossil *Scaphiopus* and *Bufo* in Alberta. Journal of Paleontology 38:1111–1112.

Bell, C. J. 2000. Biochronology of North American microtine rodents. *In* J. S. Noller, J. M. Sowers, and W. R. Lettis, eds., Quaternary geology, methods and applications. Washington, D.C.: American Geophysical Union, pp. 379–405.

Benton, M. J. 2000. Stems, nodes, crown clades, and rank-free lists: is Linnaeus dead? Biological Reviews 75:633–648.

Blob, R. W., Carrano, M. T., Rogers, R. R., Forster, C. A., and Espinoza, N. R. 2001. A new fossil frog from the Upper Cretaceous Judith River Formation of Montana. Journal of Vertebrate Paleontology 21:190–194.

Bolt, J. R. 1977. Dissorophoid relations and ontogeny, and the origin of the Lissamphibia. Journal of Paleontology 51:235–249.

Bolt, J. R. 1979. *Amphibamus grandiceps* as a juvenile dissorophid: evidence and implications. *In* M. H. Nitecki, ed., Mazon Creek fossils. New York: Academic Press, pp. 529–563.

Brattstrom, B. H. 1953a. The amphibians and reptiles from Rancho La Brea. Transactions of the San Diego Society of Natural History 11:365–392.

Brattstrom, B. H. 1953b. Records of Pleistocene reptiles and amphibians from Florida. Florida Academy of Sciences, Quarterly Journal 16:243–248.

Brattstrom, B. H. 1955a. Pliocene and Pleistocene amphibians and reptiles from southeastern Arizona. Journal of Paleontology 29:150–154.

Brattstrom, B. H. 1955b. Records of some Pliocene and Pleistocene reptiles and amphibians from Mexico. Bulletin, Southern California Academy of Sciences 54:1–4.

Brattstrom, B. H. 1958a. Additions to the Pleistocene herpetofauna of Nevada. Herpetologica 14:36.

Brattstrom, B. H. 1958b. New records of Cenozoic amphibians and reptiles from California. Bulletin, Southern California Academy of Sciences 57:5–12.

Brattstrom, B. H. 1976. A Pleistocene herpetofauna from Smith Creek Cave, Nevada. Bulletin, Southern California Academy of Sciences 75:283–284.

Bryant, L. J. 1989. Non-dinosaurian lower vertebrates across the Cretaceous–Tertiary boundary in northeastern Montana. University of California Publications in Geological Sciences, No. 134, 107 pp.

Camp, C. L. 1917. An extinct toad from Rancho La Brea. University of California Publications, Bulletin of the Department of Geological Sciences 10:287–292.

Cannatella, D. C. 1985. A phylogeny of primitive frogs (Archaeobatrachians). Ph.D. dissertation, University of Kansas, Lawrence.

Cannatella, D. C., and Hillis, D. M. 1993. Amphibian relationships: phylogenetic analysis of morphology and molecules. Herpetological Monographs 7:1–7.

Carroll, R. L. 2000a. *Eocaecilia* and the origin of caecilians. *In* H. Heatwole and R. L. Carroll, eds., Amphibian biology. Vol. 4: Palaeontology, the evolutionary history of amphibians. Chipping Norton, Australia: Surrey Beatty & Sons, pp. 1402–1411.

Carroll, R. L. 2000b. The fossil record and large-scale patterns of amphibian evolution. *In* H. Heatwole and R. L. Carroll, eds., Amphibian biology. Vol. 4: Palaeontology, the evolutionary history of amphibians. Chipping Norton, Australia: Surrey Beatty & Sons, pp. 973–978.

Chantell, C. J. 1964. Some Mio-Pliocene hylids from the Valentine Formation of Nebraska. American Midland Naturalist 72:211–225.

Chantell, C. J. 1965. A Lower Miocene *Acris* (Amphibia: Hylidae) from Colorado. Journal of Paleontology 39:507–508.

Chantell, C. J. 1966. Late Cenozoic hylids from the Great Plains. Herpetologica 22:259–264.

Chantell, C. J. 1968a. The osteology of *Acris* and *Limnaoedus* (Amphibia: Hylidae). American Midland Naturalist 79:169–182.

Chantell, C. J. 1968b. The osteology of *Pseudacris* (Amphibia: Hylidae). American Midland Naturalist 80:381–391.

Chantell, C. J. 1970. Upper Pliocene frogs from Idaho. Copeia 1970:654–664.

Chantell, C. J. 1971. Fossil amphibians from the Egelhoff Local Fauna of north-central Nebraska. Contributions from the Museum of Paleontology, University of Michigan 23:239–246.

Churcher, C. S., and Dods, R. R. 1979. *Ochotona* and other vertebrates of possible Illinoian age from the Kelso Cave, Halton County, Ontario, Canada. Canadian Journal of Earth Sciences 14:326–331.

Clarke, B. T. 1988. Evolutionary relationships of the discoglossid frogs: osteological evidence. Ph.D. thesis, City of London Polytechnic, London.

Clausen, C. J., Cohen, A. D., Emiliani, C., Holman, J. A., and Stipp, J. J. 1979. Little Salt Spring, Florida: a unique underwater site. Science 203:609–614.

Clemens, W. A., Lillegraven, J. A., Lindsay, E. H., and Simpson, G. G. 1979. Where, when, and what: a survey of known Mesozoic mammal distribution. *In* J. A. Lillegraven, Z. Kielen-Jaworowska, and W. A. Clemens, eds., Mesozoic mammals: The first two-thirds of mammalian history. Berkeley: University of California Press, pp. 7–58.

Collins, J. T., and Taggart, T. W. 2002. Standard common and current scientific names for North American amphibians, turtles, reptiles, and crocodilians. 5th ed. Lawrence, Kan.: Center for North American Herpetology.

Conant, R. C., and Collins, J. T. 1998. Field guide to reptiles and amphibians, eastern and central North America. 3rd ed. Boston: Houghton Mifflin.

Cooke, C. M. 1945. Geology of Florida. Bulletin, Florida Geological Survey, No. 29, 339 pp.

Cope, E. D. 1864. On the limits and relations of the Raniformes. Proceedings of the Academy of Natural Sciences of Philadelphia 16:181–184.

Cope, E. D. 1865. Sketch of the primary groups of the Batrachia Salientia. Natural History Review, a Quarterly Review of Biological Science 5:97–120.

Cope, E. D. 1866. On the structure and distribution of the genera of the arciferous Anura. Journal of the Academy of Natural Sciences of Philadelphia 6: 67–97.

Cope, E. D. 1884, "1883." The Vertebrata of the Tertiary formations of the west. Book 1. *In* F. V. Hayden, ed., United States Geological Survey of the Territories. Vol. 3. Washington, D.C.: US Department of the Interior, 1009 pp.

Crother, B. I., ed. 2001. Scientific and standard English names of amphibians and reptiles of North America north of Mexico with comments regarding confidence in our understanding. Society for the Study of Amphibians and Reptiles, Herpetological Circular No. 29, 82 pp.

Davis, L. C. 1973. The herpetofauna of Peccary Cave, Arkansas. M.S. thesis, University of Arkansas, Fayetteville.

DeFauw, S. L., and Shoshani, J. 1991. *Rana catesbeiana* and *R. clamitans* from the late Pleistocene of Michigan. Journal of Herpetology 25:95–99.

Delcourt, H. R., Delcourt, P. A., and Webb, T., III. 1983. Dynamic plant ecology: the spectrum of vegetational change in space and time. Quaternary Science Reviews 1:153–175.

Dole, J. W. 1968. Homing in leopard frogs, *Rana pipiens*. Ecology 49:386–399.

Duellman, W. E. 1993. Amphibian species of the world: additions and corrections. University of Kansas Museum of Natural History, Special Publication 21, 372 pp.

Duellman, W. E., and Trueb, L. 1994. Biology of the amphibians. Baltimore and London: Johns Hopkins University Press.

Eaton, J. G., and Cifelli, R. L. 1988. Preliminary report on Cretaceous mammals of the Kaiparowits Plateau, southern Utah. Contributions to Geology, University of Wyoming 26(2):45–55.

Edwards, J. L. 1976. Spinal nerves and their bearing on salamander phylogeny. Journal of Morphology 148:305–328.

Emslie, S. D., and Morgan, G. S. 1995. Taphonomy of a late Pleistocene carnivore den, Dade County, Florida. *In* D. W. Steadman and J. I. Mead, eds., Late Quaternary environments and deep history: A tribute to Paul S. Martin. Hot Springs: The Mammoth Site of Hot Springs, South Dakota, Inc. Scientific Papers, No. 3, pp. 65–83.

Eshelman, R. E. 1975. Geology and paleontology of the early Pleistocene (late Blancan) White Rock Fauna from north-central Kansas. University of Michigan, Papers on Paleontology, No. 13, 60 pp.

Esteban, M., Castanet, J., and Sanchiz, B. 1995. Size inference based on skeletal fragments of the common European frog (*Rana temporaria* L.). Herpetological Journal 5:229–235.

Estes, R. 1964. Fossil vertebrates from the Late Cretaceous Lance Formation, eastern Wyoming. University of California Publications in Geological Sciences, No. 49, 187 pp.

Estes, R. 1969. A new fossil discoglossid frog from Montana and Wyoming. Breviora, No. 328, 7 pp.

Estes, R. 1970. New fossil pelobatid frogs and a review of the genus *Eopelobates*. Bulletin of the Museum of Comparative Zoology, Harvard University 139: 293–339.

Estes, R. 1975. Lower vertebrates from the Fort Union Formation, Late Paleocene, Big Horn Basin, Wyoming. Herpetologica 31:365–385.

Estes, R., and Báez, A. 1985. Herpetofauna of North and South America during the Late Cretaceous and Cenozoic: evidence for interchange? *In* F. G. Stehli and S. D. Webb, eds., The great American biotic interchange. New York: Plenum, pp. 139–196.

Estes, R., and Reig, O. A. 1973. The early fossil record of frogs: a review of the evidence. *In* J. L. Vial, ed., Evolutionary biology of the anurans: Contemporary research on major problems. Columbia: University of Missouri Press, pp. 11–63.

Estes, R., and Sanchiz, B. 1982a. Early Cretaceous lower vertebrates from Galve (Teruel), Spain. Journal of Vertebrate Paleontology 2:21–39.

Estes, R., and Sanchiz, B. 1982b. New discoglossid and palaeobatrachid frogs from the Late Cretaceous of Wyoming and Montana, and a review of other frogs from the Lance and Hell Creek formations. Journal of Vertebrate Paleontology 2:9–20.

Estes, R., and Tihen, J. A. 1964. Lower vertebrates from the Valentine Formation of Nebraska. American Midland Naturalist 72:453–472.

Estes, R., Špinar, Z. V., and Nevo, E. 1978. Early Cretaceous pipid tadpoles from Israel (Amphibia:Anura). Herpetologica 34:374–393.

Evans, S. E., and Borsuk-Białynicka, M. 1998. A stem-group frog from the Early Triassic of Poland. Acta Palaeontologica Polonica 43:573–580.

Evans, S. E., and Milner, A. R. 1993. Frogs and salamanders from the Upper Jurassic Morrison Formation (Quarry Nine, Como Bluff) of North America. Journal of Vertebrate Paleontology 13:24–30.

Farlow, J. A., Sunderman, J. A., Havens, J. J., Swinehart, A. L., Holman, J. A., Richards, R. L., Miller, N. G., Martin, R. A., Hunt, R. M., Jr., Storrs, G. W., Curry, B. B., Fluegman, R. H., Dawson, M. R., and Flint, M. E. T. 2001. The Pipe Creek Sinkhole Biota, a diverse late Tertiary continental fossil assemblage from Grant County, Indiana. American Midland Naturalist 145: 367–378.

Fay, L. P. 1984. Mid-Wisconsinan and mid-Holocene herpetofaunas of eastern North America: a study in minimal contrast. Special Publication of Carnegie Museum of Natural History, No. 8, pp. 14–19.

Fay, L. P. 1986. Wisconsinan herpetofaunas of the central Appalachians. Virginia Division of Mineral Resources Publication, No. 75, pp. 126–128.

Fay, L. P. 1988. Late Wisconsinan Appalachian herpetofaunas: relative stability in the midst of change. Annals of Carnegie Museum 57:189–220.

Ferguson, J. H., and Lowe, C. H., Jr. 1969. Evolutionary relationships in the *Bufo punctatus* group. American Midland Naturalist 81:435–466.

Foley, R. L. 1984. Late Pleistocene (Woodfordian) vertebrates from the driftless area of southwestern Wisconsin, the Moscow Fissure Local Fauna. Reports of Investigations, Illinois State Museum, No. 39, 50 pp.

Ford, K. M., III. 1992. Herpetofauna of the Albert Ahrens Local Fauna (Pleistocene: Irvingtonian), Nebraska. M.S. thesis, Michigan State University, East Lansing.

Ford, L. S., and Cannatella, D. C. 1993. The major clades of frogs. Herpetological Monographs 7:94–117.

Fox, R. C. 1976. An edentulous frog (*Theatonius lancensis*, new genus and species) from the Upper Cretaceous Lance Formation of Wyoming. Canadian Journal of Earth Sciences 13:1486–1490.

Frost, D. R., ed. 1985. Amphibian species of the world, a taxonomic and geographical reference. Lawrence, Kansas: Allen Press and Association of Systematics Collections.

Gao, K., and Wang, Y. 2001. Mesozoic anurans from Liaoning Province, China, and phylogenetic relationships of archaeobatrachian anuran clades. Journal of Vertebrate Paleontology 21:460–476.

Gardner, J. D. 2000. Systematics of albanerpetonids and other lissamphibians from the Late Cretaceous of western North America. Ph.D. dissertation, University of Alberta, Edmonton.

Gehlbach, F. R. 1965. Amphibians and reptiles from the Pliocene and Pleistocene of North America: a chronological summary and selected bibliography. Texas Journal of Science 27:56–70.

Gervais, P. 1859. Zoologie et paléontologie françaises. Nouvelles recherches sur les animaux vertébrés dont on trouve les ossements enfouis dans le sol de la France. Paris: Arthus Bertrand.

Gilmore, C. W. 1928. Fossil lizards of North America. Memoirs of the National Academy of Sciences, No. 22, 201 pp.

Goldfuss, A. 1831. Beiträge zur Kenntniss verschiedener Reptilien der Vorwelt. Nova Acta Physico-Medica Academiae Caesareae Leopoldino-Carolinae Naturae Curiosorum 15:61–128.

Green, M., and Holman, J. A. 1977. A late Tertiary stream channel fauna from South Bijou Hill, South Dakota. Journal of Paleontology 51:543–547.

Guilday, J. E. 1962. The Pleistocene local fauna of the Natural Chimneys, Augusta County, Virginia. Annals of Carnegie Museum 36:87–122.

Guilday, J. E., Hamilton, H. W., and McCrady, A. D. 1966. The bone breccia of Bootlegger Sink, York County, PA. Annals of Carnegie Museum 38:145–163.

Guilday, J. E., Hamilton, H. W., and McCrady, A. D. 1969. The Pleistocene vertebrate fauna of Robinson Cave, Overton County, Tennessee. Palaeovertebrata 2:25–75.

Gut, H. J., and Ray, C. E. 1963. The Pleistocene vertebrate fauna of Reddick, Florida. Florida Academy of Sciences, Quarterly Journal 26:315–328.

Harington, C. R. 1978. Quaternary vertebrate faunas of Canada and Alaska and their suggested chronological sequence. Syllogeus, National Museum of Natural Sciences, No. 15, 105 pp.

Harington, C. R. 1988. Marine mammals of the Champlain Sea, and the problem of whales in Michigan. In N. R. Gadd, ed., The late Quaternary development of the Champlain Sea basin. Geological Association of Canada, Special Paper 35, pp. 225–240.

Harris, A. H. 1987. Reconstruction of mid-Wisconsinan environments in southern New Mexico. National Geographic Research 3:142–151.

Harris, J. M., and Jefferson, G. T., eds. 1985. Rancho La Brea: treasures of the tar pits. Los Angeles: Natural History Museum of Los Angeles County.

Heatwole, H., and Carroll, R. L., eds. 2000. Amphibian biology. Vol. 4: Palaeontology, the evolutionary history of amphibians. Chipping Norton, Australia: Surrey Beatty & Sons.

Hecht, M. K. 1959. Amphibians and reptiles. In P. O. McGrew, ed., The geology and paleontology of the Elk Mountain and Tabernacle Butte areas, Wyoming. Bulletin of the American Museum of Natural History 117:130–146.

Hecht, M. K. 1960. A new frog from an Eocene oil-well core in Nevada. American Museum Novitates, No. 2006, 14 pp.

Hecht, M. K. 1962. A reevaluation of the early history of the frogs, part I. Systematic Zoology 11:39–44.

Hecht, M. K. 1963. A reevaluation of the early history of the frogs, part II. Systematic Zoology 12:20–35.

Hecht, M. K. 1969. The living lower tetrapods: their interrelationships and phylogenetic position. Annals of the New York Academy of Sciences 167:74–79.

Hecht, M. K. 1970. The morphology of *Eodiscoglossus*, a complete Jurassic frog. American Museum Novitates, No. 2424, 17 pp.

Hecht, M. K., and Estes, R. 1960. Fossil amphibians from Quarry Nine. Postilla, No. 46, 19 pp.

Henrici, A. C. 1991. *Chelomophrynus bayi* (Amphibia, Anura, Rhinophrynidae), a new genus from the Middle Eocene of Wyoming: ontogeny and relationships. Annals of Carnegie Museum 60:94–144.

Henrici, A. C. 1994. *Tephrodytes brassicarvalis*, new genus and species (Anura: Pelodytidae), from the Arikareean Cabbage Patch beds of Montana, USA, and pelodytid–pelobatid relationships. Annals of Carnegie Museum 63:155–183.

Henrici, A. C. 1998. A new pipoid anuran from the Late Jurassic Morrison Formation at Dinosaur National Monument, Utah. Journal of Vertebrate Paleontology 18:321–332.

Henrici, A. C. 2000. Reassessment of the North American pelobatid anuran *Eopelobates guthriei*. Annals of Carnegie Museum 69:145–156.

Henrici, A. C., and Fiorillo, A. R. 1993. Catastrophic death assemblage of *Chelomophrynus bayi* (Anura, Rhinophrynidae) from the Middle Eocene Wagon Bed Formation of central Wyoming. Journal of Paleontology 67:1016–1026.

Heusser, H. 1960. Über die Beziehungen der Erdkröte (*Bufo bufo*) zu ihrem Laichplatz II. Behaviour 16:93–109.

Heyer, W. R. 1970. Studies on the frogs of the genus *Leptodactylus* (Amphibia: Leptodactylidae). VI. Biosystematics of the *melanonotus* group. Los Angeles County Museum, Contributions in Science, No. 191, 48 pp.

Hibbard, C. W. 1949. Techniques of collecting microvertebrate fossils. Contributions from the Museum of Paleontology, University of Michigan 8:7–19.

Hibbard, C. W. 1960. An interpretation of Pliocene and Pleistocene climates in North America. President's Address, 62nd Annual Report of the Michigan Academy of Science, Arts, and Letters, pp. 5–30.

Holman, J. A. 1958. The herpetofauna of Saber-tooth Cave, Citrus County, Florida. Copeia 1958:276–280.

Holman, J. A. 1959a. Amphibians and reptiles from the Pleistocene (Illinoian) of Williston, Florida. Copeia 1959:96–102.

Holman, J. A. 1959b. A Pleistocene herpetofauna near Orange Lake, Florida. Herpetologica 15:121–125.

Holman, J. A. 1962a. Additional records of Florida Pleistocene amphibians and reptiles. Herpetologica 15:121–125.

Holman, J. A. 1962b. A Texas Pleistocene herpetofauna. Copeia 1962:255–261.

Holman, J. A. 1963a. A new rhinophrynid frog from the Early Oligocene of Canada. Copeia 1963:706–708.

Holman, J. A. 1963b. Late Pleistocene amphibians and reptiles of the Clear Creek and Ben Franklin local faunas of Texas. Journal of the Graduate Research Center 31:152–167.

Holman, J. A. 1964. Pleistocene amphibians and reptiles from Texas. Herpetologica 20:73–83.

Holman, J. A. 1965. Early Miocene anurans from Florida. Florida Academy of Sciences, Quarterly Journal 28:68–82.

Holman, J. A. 1966a. The Pleistocene herpetofauna of Miller's Cave, Texas. Texas Journal of Science 28:372–377.

Holman, J. A. 1966b. A small Miocene herpetofauna from Texas. Florida Academy of Sciences, Quarterly Journal 29:267–275.

Holman, J. A. 1967a. Additional Miocene anurans from Florida. Florida Academy of Sciences, Quarterly Journal 30:121–140.

Holman, J. A. 1967b. A Pleistocene herpetofauna from Ladds, Georgia. Bulletin of the Georgia Academy of Science 25:154–166.

Holman, J. A. 1968. Lower Oligocene amphibians from Saskatchewan. Florida Academy of Sciences, Quarterly Journal 31:273–289.

Holman, J. A. 1969a. Herpetofauna of the Pleistocene Slaton Local Fauna of Texas. Southwestern Naturalist 14:203–212.

Holman, J. A. 1969b. Pleistocene amphibians from a cave in Edwards County, Texas. Texas Journal of Science 31:63–67.

Holman, J. A. 1969c. The Pleistocene amphibians and reptiles of Texas. Publications of the Museum, Michigan State University, Biological Series 4: 163–192.

Holman, J. A. 1970a. Herpetofauna of the Wood Mountain Formation (Upper Miocene) of Saskatchewan. Canadian Journal of Earth Sciences 7:1317–1325.

Holman, J. A. 1970b. A Pleistocene herpetofauna from Eddy County, New Mexico. Texas Journal of Science 21:29–39.

Holman, J. A. 1970c. A small Pleistocene herpetofauna from Tamaulipas. Florida Academy of Sciences, Quarterly Journal 32:153–158.

Holman, J. A. 1971. Herpetofauna of the Sandahl Local Fauna (Pleistocene: Illinoian) of Kansas. Contributions from the Museum of Paleontology, University of Michigan 23:349–355.

Holman, J. A. 1972a. Amphibians and reptiles. *In* M. F. Skinner and C. W. Hibbard, eds., Early Pleistocene preglacial and glacial rocks and faunas of north-central Nebraska. Bulletin of the American Museum of Natural History 148: 55–71.

Holman, J. A. 1972b. Herpetofauna of the Calf Creek Local Fauna (Lower Oligocene: Cypress Hills Formation) of Saskatchewan. Canadian Journal of Earth Sciences 9:1612–1631.

Holman, J. A. 1972c. Herpetofauna of the Kanapolis Local Fauna (Pleistocene: Yarmouth) of Kansas. Michigan Academician 5:87–98.

Holman, J. A. 1973a. Herpetofauna of the Mission Local Fauna (Lower Pliocene) of South Dakota. Journal of Paleontology 47:462–464.

Holman, J. A. 1973b. New amphibians and reptiles from the Norden Bridge Fauna (Upper Miocene) of Nebraska. Michigan Academician 6:149–163.

Holman, J. A. 1974. A Late Pleistocene herpetofauna from southwestern Missouri. Journal of Herpetology 8:343–346.

Holman, J. A. 1975. Herpetofauna of the WaKeeney Local Fauna (Lower Pliocene: Clarendonian) of Trego County, Kansas. *In* G. R. Smith and N. E. Friedland, eds., Studies on Cenozoic paleontology and stratigraphy in honour of Claude W. Hibbard. University of Michigan, Papers on Paleontology, No. 12, pp. 49–66.

Holman, J. A. 1976. The herpetofauna of the lower Valentine Formation, north-central Nebraska. Herpetologica 32:262–268.

Holman, J. A. 1977a. America's northernmost Pleistocene herpetofauna (Java, north-central South Dakota). Copeia 1977:191–193.

Holman, J. A. 1977b. Amphibians and reptiles of the Gulf Coast Miocene of Texas. Herpetologica 33:391–403.

Holman, J. A. 1977c. The Pleistocene (Kansan) herpetofauna of Cumberland Cave, Maryland. Annals of Carnegie Museum 46:157–172.

Holman, J. A. 1978a. Herpetofauna of the Bijou Hills Local Fauna (Late Miocene: Batstovian) of South Dakota. Herpetologica 34:253–257.

Holman, J. A. 1978b. The late Pleistocene herpetofauna of Devil's Den Sinkhole, Levy County, Florida. Herpetologica 34:228–237.

Holman, J. A. 1979. Herpetofauna of the Nash Local Fauna (Pleistocene: Aftonian) of Kansas. Copeia 1979:747–749.

Holman, J. A. 1980. Paleoclimatic implications of Pleistocene herpetofaunas of eastern and central North America. Transactions of the Nebraska Academy of Sciences 8:131–140.

Holman, J. A. 1981. A herpetofauna from the eastern extension of the Harrison Formation (Early Miocene: Arikareean), Cherry County, Nebraska. Journal of Vertebrate Paleontology 1:49–56.

Holman, J. A. 1982a. New herpetological species and records from the Norden Bridge Fauna (Miocene: late Barstovian) of Nebraska. Transactions of the Nebraska Academy of Sciences 10:31–36.

Holman, J. A. 1982b. The Pleistocene (Kansan) herpetofauna of Trout Cave, West Virginia. Annals of Carnegie Museum 51:391–404.

Holman, J. A. 1984. Herpetofauna of the Duck Creek and Williams local faunas (Pleistocene: Illinoian) of Kansas. Special Publication of Carnegie Museum of Natural History, No. 8, pp. 20–38.

Holman, J. A. 1985a. Herpetofauna of Ladds Quarry. National Geographic Research 1:423–436.

Holman, J. A. 1985b. New evidence of the status of Ladds Quarry. National Geographic Research 1:569–570.

Holman, J. A. 1986a. Butler Spring herpetofauna of Kansas (Pleistocene: Illinoian) and its climatic significance. Journal of Herpetology 20:568–570.

Holman, J. A. 1986b. The known herpetofauna of the late Quaternary of Virginia poses a dilemma. In J. N. McDonald and S. O. Bird, eds., The Quaternary of Virginia. Virginia Division of Mineral Resources Publication, No. 75, pp. 36–42.

Holman, J. A. 1987. Herpetofauna of the Egelhoff site (Miocene: Barstovian) of north-central Nebraska. Journal of Vertebrate Paleontology 7:109–120.

Holman, J. A. 1988. The status of Michigan's Pleistocene herpetofauna. Michigan Academician 20:125–132.

Holman, J. A. 1991. North American Pleistocene herpetofaunal stability and its impact on the interpretations of modern herpetofaunas: an overview. Scientific Papers, Illinois State Museum 23: 227–235.

Holman, J. A. 1992a. Amphibians from a second century Roman well at Tiddington, Warwickshire. Bulletin, British Herpetological Society 39:5–7.

Holman, J. A. 1992b. Late Quaternary herpetofauna of the central Great Lakes region, U.S.A.: zoogeographical and paleoecological implications. Quaternary Science Reviews 11:345–351.

Holman, J. A. 1992c. Patterns of herpetological re-occupation of post-glacial Michigan: amphibians and reptiles come home. Michigan Academician 24: 453–466.

Holman, J. A. 1995a. A late medial Miocene herpetofauna from northeastern Nebraska. Herpetological Natural History 3:143–150.

Holman, J. A. 1995b. Pleistocene amphibians and reptiles in North America. New York: Oxford University Press.

Holman, J. A. 1996a. Glad Tidings, a late Middle Miocene herpetofauna from northeastern Nebraska. Journal of Herpetology 30:430–432.

Holman, J. A. 1996b. Herpetofauna of the Trinity River Local Fauna (Miocene: early Barstovian), San Jacinto County, Texas, USA. Tertiary Research 17:5–10.

Holman, J. A. 1996c. The large Pleistocene (Sangamonian) herpetofauna of the Williston IIIA site, north-central Florida. Herpetological Natural History 4: 35–47.

Holman, J. A. 1997. Amphibians and reptiles from the Pleistocene (late Wisconsinan) of Sheriden Pit Cave, northwestern Ohio. Michigan Academician 29: 1–20.

Holman, J. A. 1998. Pleistocene amphibians and reptiles in Britain and Europe. New York: Oxford University Press.

Holman, J. A. 1999. Early Oligocene (Whitneyan) snakes from Florida (USA), the second oldest colubrid snakes in North America. Acta Zoologica Cracoviensia 42:447–454.

Holman, J. A. 2000a. Fossil snakes of North America: Origin, evolution, distribution, paleoecology. Bloomington: Indiana University Press.

Holman, J. A. 2000b. Pleistocene Amphibia: evolutionary stasis, range adjustments, and recolonization patterns. *In* H. Heatwole and R. L. Carroll, eds., Amphibian biology. Vol. 4: Palaeontology, the evolutionary history of amphibians. Chipping Norton, Australia: Surrey Beatty & Sons, pp. 1445–1458.

Holman, J. A. 2001a. Fossil dunes and soils near Saginaw Bay, a unique herpetological habitat. Michigan Academician 33:135–153.

Holman, J. A. 2001b. In quest of Great Lakes Ice Age vertebrates. East Lansing: Michigan State University Press.

Holman, J. A., and Grady, F. 1987. Herpetofauna of New Trout Cave. National Geographic Research 3:305–317.

Holman, J. A., and Grady, F. 1989. The fossil herpetofauna (Pleistocene: Irvingtonian) of Hamilton Cave, Pendleton County, West Virginia. National Speleological Society, Bulletin 51:34–41.

Holman, J. A., and Grady, F. 1994. A Pleistocene herpetofauna from Worm Hole Cave, Pendleton County, West Virginia. National Speleological Society, Bulletin 56:46–49.

Holman, J. A., and Harrison, D. L. 2000. Early Oligocene (Whitneyan) snakes from Florida (USA), a unique booid. Acta Zoologica Cracoviensia 43:127–134.

Holman, J. A., and Harrison, D. L. 2001. Early Oligocene (Whitneyan) snakes from Florida (USA): remaining boids, indeterminate colubroids, summary and discussion of the I-75 Local Fauna snakes. Acta Zoologica Cracoviensia 44:25–36.

Holman, J. A., and McDonald, J. N. 1986. A late Quaternary herpetofauna from Saltville, Virginia. Brimleyana 12:85–100.

Holman, J. A., and Richards, R. L. 1993. Herpetofauna of the Prairie Creek site, Daviess County, Indiana. Proceedings of the Indiana Academy of Science 102:115–131.

Holman, J. A., and Schloeder, M. E. 1991. Fossil herpetofauna of the Lisco C quarries (Pliocene: early Blancan) of Nebraska. Transactions of the Nebraska Academy of Sciences 18:19–29.

Holman, J. A., and Sullivan, R. M. 1981. A small herpetofauna from the type section of the Valentine Formation (Miocene: Barstovian), Cherry County, Nebraska. Journal of Paleontology 55:138–144.

Holman, J. A., and Winkler, A. J. 1987. A mid-Pleistocene (Irvingtonian) herpetofauna from a cave in southcentral Texas. Pearce-Sellards Series: An Occasional Publication of the Texas Memorial Museum, No. 44, 17 pp.

Holman, J. A., Bell, G., and Lamb, J. 1990. A late Pleistocene herpetofauna from Bell Cave, Alabama. Herpetological Journal 1:521–529.

Holman, J. A., Fisher, D. C., and Kapp, R. O. 1986. Recent discoveries of fossil vertebrates in the Lower Peninsula of Michigan. Michigan Academician 18:431–463.

Holman, J. A., Harington, C. R., and Mott, R. J. 1997. Skeleton of a leopard frog (*Rana pipiens*) from Champlain Sea deposits (ca. 10 000 BP) near Eardley, Quebec. Canadian Journal of Earth Sciences 34:1150–1155.

Hood, C. H., and Hawksley, O. 1975. A Pleistocene fauna from Zoo Cave, Taney County, Missouri. Missouri Speleology 15:1–42.

Hudson, D. M., and Brattstrom, B. H. 1977. A small herpetofauna from the late Pleistocene of Newport Beach Mesa, Orange County, California. Bulletin, Southern California Academy of Sciences 76:16–20.

Hulbert, R. C., Jr., ed. 2001. The fossil vertebrates of Florida. Gainesville: University Press of Florida.

Hulbert, R. C., Jr., and Morgan, G. S. 1989. Stratigraphy, paleoecology, and vertebrate fauna of the Leisey Shell Pit Local Fauna, early Pleistocene (Irvingtonian) of southwestern Florida. Papers in Florida Paleontology, No. 2, 19 pp.

Hulbert, R. C., Jr., and Pratt, A. E. 1998. New Pleistocene (Rancholabrean) vertebrate faunas from coastal Georgia. Journal of Vertebrate Paleontology 18: 412–429.

Hulse, A. C. 1977. *Bufo kelloggi.* Catalogue of American Amphibians and Reptiles 200:1–2.

Hunt, A. P., and Lucas, S. G. 1992. Stratigraphy, paleontology, and age of the Fruitland and Kirtland formations (Upper Cretaceous), San Juan Basin, New Mexico. Guidebook, New Mexico Geological Society 43:217–239.

Hunt, A. P., and Lucas, S. G. 1993. Cretaceous vertebrates of New Mexico. Bulletin of the New Mexico Museum of Natural History and Science 2:77–91.

International Union of Geological Sciences. 2002. International stratigraphic chart. Internet: www.iugs/pubs/intrstratchart.htm

Jenkins, F. A., Jr., and Shubin, N. H. 1998. *Prosiluris bitis* and the anuran caudopelvic mechanism. Journal of Vertebrate Paleontology 18:495–510.

Jenkins, F. A., Jr., and Walsh, D. 1993. An Early Jurassic caecilian with limbs. Nature (London) 365:246–250.

Johnson, E. 1987. Vertebrate remains. *In* E. Johnson, ed., Late Quaternary studies on the southwest High Plains. College Station: Texas A & M University Press, pp. 49–89.

King, J. E., and Saunders, J. J. 1986. *Geochelone* in Illinois and the Illinoian–Sangamonian vegetation of the type region. Quaternary Research 25:89–99.

Klippel, W. E., and Parmalee, P. W. 1982. The paleontology of Cheek Bend Cave. Phase II report to the Tennessee Valley Authority [contract No. TVA-TV 49244A], 249 pp.

Kluge, A. G. 1966. A new pelobatine frog from the Lower Miocene of South Dakota with a discussion of the evolution of the *Scaphiopus–Spea* complex. Los Angeles County Museum, Contributions in Science, No. 113, 26 pp.

Kurtén, B., and Anderson, E. 1980. Pleistocene mammals of North America. New York: Columbia University Press.

Lartet, E. 1851. Notice sur la colline de Sansan, suivie d'une récapitulation des diverses espèces d'animaux vértebrés fossiles trouvés soit à Sansan, soit dans d'autres gisements du terrain tertiare miocène dans le bassin Sous-Pyrénéen. Auch, France: J. A. Portes.

Layne, J. R., Jr., and Lee, R. E., Jr. 1995. Adaptions of frogs to freezing. Climatic Research 5:53–59.

Lindsay, E. 1984. Late Cenozoic mammals from northwestern Mexico. Journal of Vertebrate Paleontology 4:208–215.

Linnaeus, C. 1758. Systema naturae per regna tria naturae, secundum classes, ordines, genera, species, cum characteribus, differentiis, synonymis, locis. Editio decimereformata [10th revised ed.] Stockholm: Laurentii Salvii.

Lundelius, E. L. 1992. The Avenue Local Fauna, late Pleistocene vertebrates from terrace deposits at Austin, Texas. Annales Zoologici Fennici 28:329–340.

Lundelius, E. L., Graham, R. W., Anderson, E., Guilday, J., Holman, J. A., Steadman, D. W., and Webb, S. D. 1983. Terrestrial vertebrate faunas. *In* H. E. Wright, ed., Late Quaternary environments of the United States. Vol. 1: The late Pleistocene, S. Porter, ed. Minneapolis: University of Minnesota Press, pp. 311–353.

Lynch, J. D. 1964. Additional hylid and leptodactylid remains from the Pleistocene of Texas and Florida. Herpetologica 20:141–142.

Lynch, J. D. 1965. The Pleistocene amphibians of Pit II, Arredondo, Florida. Copeia 1965:72–77.

Lynch, J. D. 1966. Additional treefrogs (Hylidae) from the North American Pleistocene. Annals of Carnegie Museum 38:265–271.

Lynch, J. D. 1971. Evolutionary relationships, osteology, and zoogeography of leptodactyloid frogs. Miscellaneous Publications, University of Kansas, Museum of Natural History, No. 53, 238 pp.

Lynch, J. D. 1978. A re-assessment of the telmatobiine leptodactylid frogs of Patagonia. Occasional Papers of the Museum of Natural History, University of Kansas, No. 72, 57 pp.

Martin, P. S., and Klein, R. G., eds. 1984. Quaternary extinctions: A prehistoric revolution. Tucson: University of Arizona Press.

Martin, P. S., and Wright, H. E., eds. 1967. Pleistocene extinctions: The search for a cause. New Haven: Yale University Press.

Mead, J. I., and Bell, C. J. 1994. Late Pleistocene and Holocene herpetofaunas of the Great Basin and Colorado Plateau. In K. T. Harper, L. L. St. Clair, K. H. Thorne, and W. M. Hess, eds., Natural History of the Colorado Plateau and Great Basin. Boulder: University Press of Colorado, pp. 255–274.

Mead, J. I., Thompson, R. S., and Van Devender, T. R. 1982. Late Wisconsinan and Holocene fauna from Smith Creek Canyon, Snake Range, Nevada. Transactions of the San Diego Society of Natural History 20:1–26.

Mead, J. I., Roth, E. L., Van Devender, T. R., and Steadman, D. W. 1984. The late Wisconsinan vertebrate fauna from Deadman Cave, southern Arizona. Transactions of the San Diego Society of Natural History 20:247–276.

Mead, J. I., Sankey, J. T., and McDonald, H. G. 1998. Pliocene (Blancan) herpetofaunas from the Glenns Ferry Formation, southern Idaho. In W. A. Akersten, H. G. McDonald, D. J. Meldrum, and M. E. T. Flint, eds., And whereas . . . : Papers on the vertebrate paleontology of Idaho honoring John A. White. Vol. 1. Idaho Museum of Natural History, Occasional Paper 36, pp. 94–109.

Mecham, J. S. 1959. Some Pleistocene amphibians and reptiles from Friesenhahn Cave, Texas. Southwestern Naturalist 3:17–27.

Meszoely, C. A. M., Špinar, Z. V., and Ford, R. L. E. 1984. A new palaeobatrachid frog from the Eocene of the British Isles. Journal of Vertebrate Paleontology 3:143–147.

Meyer, H. von. 1852. III. Beschreibung der fossilen Decapoden, Fische, Batrachier und Säugethiere aus den tertiären Süsswassergebilden des nördlichen Böhmens. In A. E. Reuss and H. von Meyer, Die tertiären Süsswassergebilde des nördlichen Böhmens und ihre fossilen Thierreste (Schluss). Palaeontographica 2:1–73.

Meyer, H. von. 1860. Frösche aus Tertiär-Gebilden Deutschland's. Palaeontographica 7:123–182.

Meylan, P. A. 1982. The squamate reptiles of the Inglis IA Fauna (Irvingtonian: Citrus County, Florida). Bulletin of the Florida State Museum, Biological Sciences 27:1–85.

Meylan, P. A. 1995. Pleistocene amphibians and reptiles from the Leisey Shell Pit, Hillsborough County, Florida. Bulletin of the Florida Museum of Natural History 37(Pt. I):273–297.

Milner, A. R. 1988. The relationships and origin of living amphibians. In M. J. Benton, ed., The phylogeny and classification of tetrapods. Vol. 1: Amphibians, reptiles, birds. Oxford: Clarendon Press, pp. 59–102.

Murray, P. 1984. Extinction down under: a bestiary of extinct Australian late Pleistocene monotremes and marsupials. In P. S. Martin and R. G. Klein, eds., Quaternary extinctions: A prehistoric revolution. Tucson: University of Arizona Press, pp. 600–628.

Nevo, E. 1995. Evolution and extinction. In Encyclopedia of environmental biology. Vol. I. London: Academic Press, pp. 717–745.

Owen-Smith, N. 1987. Pleistocene extinctions: the pivotal role of megaherbivores. Paleobiology 13:351–362.

Palmer, A. R., and Geissman, J., compilers. 1999. 1999 geologic time scale. Washington, D.C.: Geological Society of America.

Parmalee, P. W., and Oesch, R. D. 1972. Pleistocene and Recent faunas from the Brynjulfson Caves, Missouri. Reports of Investigations, Illinois State Museum, No. 25, 52 pp.

Parmley, D. 1986. Herpetofauna of the Rancholabrean Schulze Cave Local Fauna of Texas. Journal of Herpetology 22:82–87.

Parmley, D. 1988a. Additional Pleistocene amphibians and reptiles from the Seymour Formation, Texas. Journal of Herpetology 22:82–87.

Parmley, D. 1988b. Late Pleistocene anurans from Fowlkes Cave, Culberson County, Texas. Texas Journal of Science 40:357–358.

Parmley, D. 1992. Frogs in Hemphillian deposits of Nebraska, with the description of a new species of *Bufo*. Journal of Herpetology 26:274–281.

Parmley, D., and Peck, D. L. 2002. Amphibians and reptiles of the late Hemphillian White Cone Local Fauna, Navajo County, Arizona. Journal of Vertebrate Paleontology 22:175–178.

Parsons, T. S., and Williams, E. E. 1963. The relationships of the modern Amphibia: a re-examination. Quarterly Review of Biology 38:26–53.

Patton, T. H. 1969. An Oligocene land vertebrate fauna from Florida. Journal of Paleontology 43:543–546.

Pomel, A. 1844. Description géologique et paléontologique des collines de la Tour-de-Boulade et du Puy-du-Teiller (Puy-de-Dome). Bulletin de la Société géologique de France 1:579–596.

Pomel, A. 1853. Catalogue méthodique et descriptif des vertébrés fossiles découverts dans le bassin hydrogéographique supérieur de la Loire, surtout dans la vallée de son affluent principal, l'Allier. Paris: J. B. Ballière.

Preston, R. E. 1979. Late Pleistocene cold-blooded vertebrate faunas from the mid-continental United States. I. Reptilia: Testudines, Crocodilia. University of Michigan, Papers on Paleontology, No. 19, 53 pp.

Prothero, D. R., and Emry, R. J., eds. 1996. The Eocene–Oligocene Transition in North America. Cambridge: Cambridge University Press.

Rage, J-C., and Roček, Z. 1989. Redescription of *Triadobatrachus massinoti* (Piveteau, 1936) an anuran amphibian from the Early Triassic. Palaeontographica 206:1–16.

Roček, Z. 1994. Taxonomy and distribution of Tertiary discoglossids (Anura) of the genus *Latonia* v. Meyer, 1843. Geobios 27:717–751.

Roček, Z. 1995. Heterochrony: response of Amphibia to cooling events. Geolines (Prague) 3:55–58

Roček, Z. 2000. Mesozoic anurans. *In* H. Heatwole and R. L. Carroll, eds., Amphibian biology. Vol. 4: Palaeontology, the evolutionary history of amphibians. Chipping Norton, Australia: Surrey Beatty & Sons, pp. 1295–1331.

Roček, Z., and Nessov, L. A. 1993. Cretaceous anurans from central Asia. Palaeontographica Abteilung A: Palaozoolgie–Stratigraphie 226:1–54.

Roček, Z., and Rage, J-C. 2000a. Anatomical transformations in the transition from Temnospondyl to Proanuran stages. *In* H. Heatwole and R. L. Carroll, eds., Amphibian biology. Vol. 4: Palaeontology, the evolutionary history of amphibians. Chipping Norton, Australia: Surrey Beatty & Sons, pp. 1274–1282.

Roček, Z., and Rage, J-C. 2000b. Tertiary Anura of Europe, Africa, Asia, North America, and South America. *In* H. Heatwole and R. L. Carroll, eds., Amphibian biology. Vol. 4: Palaeontology, the evolutionary history of amphibians. Chipping Norton, Australia: Surrey Beatty & Sons, pp. 1332–1387.

Rogers, K. L. 1976. Herpetofauna of the Beck Ranch Local Fauna (Upper Pliocene: Blancan) of Texas. Paleontological Series (East Lansing) 1:167–200.

Rogers, K. L. 1982. Herpetofaunas of the Courland Canal and Hall Ash local faunas (Pleistocene:early Kansan) of Jewell Co., Kansas. Journal of Herpetology 16:174–177.

Rogers, K. L. 1984. Herpetofaunas of the Big Springs and Hornet's Nest quarries (northeastern Nebraska, Pleistocene: late Blancan). Transactions of the Nebraska Academy of Sciences 12:81–94.

Rogers, K. L. 1987. Pleistocene high altitude amphibians and reptiles from Colorado (Alamosa Local Fauna; Pleistocene, Irvingtonian). Journal of Vertebrate Paleontology 7:82–95.

Rogers, K. L., Repenning, C. A., Forester, R. M., Larson, E. E., Hall, S. A., Smith, G. R., Anderson, E. M., and Brown, T. J. 1985. Middle Pleistocene (late Irvingtonian) climatic changes in south-central Colorado. National Geographic Research 1:535–563.

Romer, A. S. 1959. The vertebrate story. 4th ed. Chicago: University of Chicago Press.

Sanchiz, B. 1998. Handbuch der Paläoherpetologie. Part 4: Salientia. München: Verlag Dr. Friedrich Pfeil, 275 pp.

Sanchiz, B., Schleich, H. H., and Esteban, M. 1993. Water frogs (Ranidae) from the Oligocene of Germany. Journal of Herpetology 27:486–489.

Saunders, J. J. 1977. Late Pleistocene vertebrates of the western Ozark highlands. Reports of Investigation, Illinois State Museum, No. 33, 118 pp.

Shubin, N. H., and Jenkins, F. A., Jr. 1995. An Early Jurassic jumping frog. Nature (London) 377:49–52.

Špinar, Z. V. 1972. Tertiary frogs from Central Europe. The Hague: Academia Prague and D. W. Junk.

Špinar, Z. V. 1976. Endolymphatic sacs and dorsal endocranial pattern: their significance for systematics and phylogeny of frogs. Věstnik Ústředního Ustavu Geologického 51:285–290.

Stebbins, R. C. 1985. A field guide to western reptiles and amphibians. Boston: Houghton Mifflin.

Stock, C. 1956. Rancho La Brea: A record of Pleistocene life in California. 6th ed. Science Series (Los Angeles), No. 20, 81 pp.

Taylor, E. H. 1939. "1938." A new anuran amphibian from the Pliocene of Kansas. Kansas University Science Bulletin 25:407–419.

Taylor, E. H. 1942. Extinct toads and frogs from the Upper Pliocene deposits of Meade County, Kansas. Kansas University Science Bulletin 28:199–235.

Tihen, J. A. 1951. Anuran remains from the Miocene of Florida with the description of a new species of *Bufo*. Copeia 1951:230–235.

Tihen, J. A. 1952. *Rana grylio* from the Pleistocene of Florida. Herpetologica 8: 107.

Tihen, J. A. 1954. A Kansas Pleistocene herpetofauna. Copeia 1954:217–221.

Tihen, J. A. 1960a. On *Neoscaphiopus* and other Pliocene pelobatid frogs. Copeia 1960:89–94.

Tihen, J. A. 1960b. Notes on late Cenozoic hylid and leptodactylid frogs from Kansas, Oklahoma, and Texas. Southwestern Naturalist 5:66–70.

Tihen, J. A. 1962a. Osteological observations on New World *Bufo*. American Midland Naturalist 67:157–183.

Tihen, J. A. 1962b. A review of New World fossil bufonids. American Midland Naturalist 68:1–50.

Trueb, L. 1973. Bones, frogs, and evolution. *In* J. L. Vial, ed., Evolutionary biology of the anurans, major problems. Columbia: University of Missouri Press, pp. 65–132.

Trueb, L. 1993. Patterns of cranial diversity among the Lissamphibia. *In* J. Hanken and B. K. Hall, eds., The skull. Chicago: University of Chicago Press, pp. 255–343.

Trueb, L., and Cannatella, D. C. 1982. The cranial osteology and hyolaryngeal apparatus of *Rhinophrynus dorsalis* (Anura: Rhinophrynidae) with comparisons to Recent pipid frogs. Journal of Morphology 171:11–40.

Trueb, L., and Cloutier, R. 1991. A phylogenetic investigation of the inter- and intrarelationships of the Lissamphibia (Amphibia: Temnospondyli). *In* H. P. Schulze and L. Trueb, eds., Origins of the higher groups of tetrapods: Controversy and consensus. Ithaca: Comstock, pp. 223–313.

Van Dam, G. H. 1978. Amphibians and reptiles (in the late Pleistocene Baker Bluff site, Tennessee). Bulletin of Carnegie Museum of Natural History 11: 19–25.

Van Devender, T. R., and Lowe, C. H., Jr. 1977. Amphibians and reptiles of Yepomera, Chihuahua, Mexico. Journal of Herpetology 11:41–50.

Van Devender, T. R., and Mead, J. I. 1978. Early Holocene and late Pleistocene amphibians and reptiles in Sonora Desert packrat middens. Copeia 1978: 464–475.

Van Devender, T. R., and Worthington, R. D. 1977. The herpetofauna of the Howell's Ridge Cave and the paleoecology of the northwestern Chihuahuan desert. *In* D. H. Wauer and D. H. Riskind, eds., Transactions of the Symposium on the biological resources of the Chihuahuan Desert region, United States and Mexico. U.S. Department of the Interior National Park Service Transactions and Proceedings Series, No. 3, pp. 85–106.

Van Devender, T. R., Rea, A. M., and Smith, M. L. 1985. The Sangamon interglacial vertebrate fauna from Rancho la Brisca, Sonora, Mexico. Transactions of the San Diego Society of Natural History 21:23–55.

Vergnaud-Grazzini, C. 1966. Les amphibiens de Miocene de Beni-Mellal. Notes du Service géologique du Maroc 27:43–74.

Voorhies, M. R. 1990. Vertebrate paleontology of the proposed Norden Reservoir area, Brown, Cherry, and Keya Paha counties, Nebraska. Division of Archeological Research, Department of Anthropology, University of Nebraska, Lincoln, Technical Report 82-09, 731 pp.

Voorhies, M. R., Holman, J. A., and Xue Xiang-Xu. 1987. The Hottell Ranch rhino quarries (basal Ogallala: medial Barstovian), Banner County, Nebraska. Part I: Geologic setting, faunal lists, lower vertebrates. Contributions to Geology, University of Wyoming 25:55–69.

Walker, C. F. 1938. The structure and systematic relationships of the genus *Rhinophrynus*. Occasional Papers of the Museum of Zoology, University of Michigan, No. 372, 11 pp.

Wang, Y., and Gao, K. 1999. Earliest Asian discoglossid frog fossil from western Liaoning. Chinese Science Bulletin 10: 636–641.

Weigel, R. D. 1962. Fossil vertebrates of Vero, Florida. Special Publication, Florida Geological Survey, No. 10, 59 pp.

Wellstead, C. F. 1981. Sedimentology of Norden Bridge and Egelhoff fossil quarries (Miocene) of north-central Nebraska. Transactions of the Nebraska Academy of Sciences 9:67–85.

Wilson, R. L. 1968. Systematics and faunal analysis of a Lower Pliocene vertebrate assemblage from Trego County, Kansas. Contributions from the Museum of Paleontology, University of Michigan 22:75–126.

Wilson, V. V. 1975. The systematics and paleoecology of two Pleistocene herpetofaunas of the southeastern United States. Ph.D. dissertation, Michigan State University, East Lansing.

Winkler, D. A., Murray, P. A., and Jacobs, L. L. 1990. Early Cretaceous (Comanchean) vertebrates of central Texas. Journal of Vertebrate Paleontology 10: 89–96.

Wood, H. H., II, Chaney, R. W., Clark, J., Colbert, E. H., Jepson, G. L., Reedside, J. B., Jr., and Stock, C. 1941. Nomenclature and correlation of the North

American continental Tertiary. Geological Society of America Bulletin 52: 1–48.

Woodburne, M. O., ed. 1987. Cenozoic mammals in North America: Geochronology and biostratigraphy. Berkeley: University of California Press.

Wright, A. H., and Wright, A. A. 1949. Handbook of frogs and toads of the United States and Canada. Ithaca, New York: Comstock.

Zweifel, R. G. 1954. A new *Rana* from the Pliocene of California. Copeia 1954: 85–87.

Zweifel, R. G. 1956. Two pelobatid frogs from the Tertiary of North America and their relationships to fossil and recent forms. American Museum Novitates, No. 1762, 45 pp.

FIGURE CREDITS

Figure 23. From Henrici (1991), courtesy of the Carnegie Museum of Natural History.

Figure 24. From Henrici (1991), courtesy of the Carnegie Museum of Natural History.

Figure 27. From Henrici (1998), courtesy of the *Journal of Vertebrate Paleontology*.

Figure 32. From Estes (1970), courtesy of the Museum of Comparative Zoology, Harvard University.

Figure 33. From Holman (1975), courtesy of the Museum of Paleontology, University of Michigan.

Figure 35. From Estes (1970), courtesy of the Museum of Comparative Zoology, Harvard University.

Figure 39. From Kluge (1966), courtesy of the Los Angeles County Museum of Natural History.

Figure 40. From Henrici (1994), courtesy of the Carnegie Museum of Natural History.

Figure 43. From Tihen (1962b), courtesy of the *American Midland Naturalist*.

Figure 49. From Tihen (1962b), courtesy of the *American Midland Naturalist*.

Figure 50. From Parmley (1992), courtesy of the *Journal of Herpetology*.

Figure 51. From Holman (1973b), courtesy of the *Michigan Academician*.

Figure 52. From Tihen (1962b), courtesy of the *American Midland Naturalist*.

Figure 53. From Parmley (1992), courtesy of the *Journal of Herpetology*.

Figure 55. From Holman (1967a), courtesy of the *Florida Academy of Sciences, Quarterly Journal*.

Figure 58. From Tihen (1962b), courtesy of the *American Midland Naturalist*.

Figure 59. From Parmley (1992), courtesy of the *Journal of Herpetology*.

Figure 60. From Tihen (1962b), courtesy of the *American Midland Naturalist*.

Figure 61. From Tihen (1962b), courtesy of the *American Midland Naturalist*.

Figure 63. From Auffenberg (1957), courtesy of the *Florida Academy of Sciences, Quarterly Journal*.

Figure 64. From Estes and Tihen (1964), courtesy of the *American Midland Naturalist*.

Figure 67. From Holman (1967a), courtesy of the *Florida Academy of Sciences, Quarterly Journal*.

Figure 76. From Holman (1966b), courtesy of the *Florida Academy of Sciences, Quarterly Journal*.

Figure 77. From Holman (1967a), courtesy of the *Florida Academy of Sciences, Quarterly Journal*.

Figure 79. From Holman (1968), courtesy of the *Florida Academy of Sciences, Quarterly Journal*.

Figure 83. From Holman (1987), courtesy of the *Journal of Vertebrate Paleontology*.

Figure 100. From Holman (2001b), courtesy of the Michigan State University Press.

Figure 101. From Holman (1992c), courtesy of the *Michigan Academician*.

GENERAL INDEX

Note: The general index covers chapters 1 and 3. The taxonomic and site indices to follow cover chapter 2 only.

Hemisus, 11
Henrici, A. C., 16
#Hesperotestudo, 209
#Hesperotestudo crassiscutata, 213
Hibbard, C. W., 14
History, of North American fossil anuran studies, 13–16
Human overkill hypothesis, relative to Pleistocene frogs, 209
Hyla, 8–9, 11, 192–193, 195–197, 200, 202
Hyla arenicolor, 202
*Hyla baderi, 202, 206–207
Hyla chrysoscelis, 202, 207
Hyla chrysoscelis or versicolor, 200, 202, 210
Hyla cinerea, 200, 202
Hyla femoralis, 202–203
Hyla gratiosa, 194, 203
*Hyla miocenica, 195
*Hyla miofloridana, 192, 194
Hyla squirella, 203
*Hyla swanstoni, 192–193
Hyla versicolor, 202, 206–207

Ice sheet, general effects of, 204–206
Identification, of fossil anurans, 30–33
Irvingtonian land-mammal age, of Pleistocene, 36

Jenkins, F. A., 16
Jurassic anurans, 187–189

Lag time in community development, in Pleistocene, 206
Lake and sea levels, of Pleistocene, 206
Laurentide Ice Sheet, 204
Larval stages, as aspects of anuran evolution, 27–28
Leiopelma, 12, 30, 188, 190–191
Lepidobatrachus, 6
Leptobrachium, 8
Leptodactylus, 201
Leptodactylus cf. labialis, 201
Leptodactylus melanonotus, 201
Lissamphibia, definition and discussion of, 1–3
Lynch, J. D., 15

Mead, J. I., 16
#Megalania, 208
Megistolotis, 11–12
Mertensophryne, 11–12
Mertensophryne micranotis, 9
Microhyla, 12
Miocene anurans, 194–197
#Miopelodytes, 197
#Miopelodytes gilmorei, 195
"Montana discoglossids I, II, and III," 190

Nectophrynoides, 12
#Nezpercius dodsoni, 190

Nomen vanum, definition of, 188–189
North American land-mammal ages, 33, 191
Notobatrachus degiustoi, 188
Nyctimystes, 8

Ocala Island, of Florida Pleistocene, 206
Oligocene anurans, 193

Paedomorphosis, in anurans, 26–27
#Palaeobatrachus occidentalis, 190–191
Paleocene anurans, 191
#Paradiscoglossus, 191
#Paradiscoglossus americanus, 190
Pedicellate teeth, of lissamphibians, 2
Pelobates fuscus, 10
Phyllodytes, 8
Phyllomedusa, 6
Physalaemus, 11–12
Physiological adaptations to cold, in Pleistocene anurans, 206
Pipa, 11
Pipa carvalhoi, 12
Pipa pipa, 12
Plaid hypothesis, of Pleistocene, 205–206
Pleistocene anurans, 199–204
Pleistocene frogs, an overview, 204–215
Pleistocene, of North America, 35–37
Postcranial elements useful in the identification of anurans: astragalus, 26; calacaneum, 26; cervical vertebra, 20–21; clavicle, 23; coracoid, 23; femur, 24; humerus, 23–24; ilium, 24; presacral vertebra, 21–22; puboischium, 24; radio-ulna, 24; sacrum, 22; scapula, 22; suprascapula, 22; tibiofibula, 24, 26; urostyle, 22
Primary postglacial invasion of anura, 212
#Proacris, 196
#Proacris mintoni, 194, 196
"Progressive evolution," in anurans, 27
#Prosalirus, 187–188
#Prosalirus bitis, 29, 187, 189
Pseudacris, 11, 195, 197–198, 200, 202–203
Pseudacris cf. clarkii, 200, 210
Pseudacris cf. triseriata, 210
Pseudacris clarkii, 203
Pseudacris crucifer, 200, 203
Pseudacris nigrita, 203
#Pseudacris nordensis, 195–197
Pseudacris ornata, 203
Pseudacris regilla, 203
Pseudacris streckeri, 203, 210

Pseudacris triseriata, 200, 203, 206, 210, 214
Pseudophryne, 11
Pternohyla, 202–203
Pternohyla fodiens, 203
Pyxicephalus, 8

Rana, 9, 11–12, 17, 31–32, 196, 200, 203, 214
Rana areolata, 197–198, 200, 203
Rana aurora, 203
Rana catesbeiana, 9, 11, 196–198, 200, 203, 210, 214–215
Rana cf. areolata, 196
Rana cf. aurora, 198
Rana cf. blairi, 200
Rana cf. catesbeiana, 198
Rana cf. clamitans, 195
Rana clamitans, 9, 195, 197, 200, 203–204, 210, 214–215
Rana grylio, 204
Rana palustris, 198–199, 204
Rana pipiens, 8, 204, 214
Rana pipiens complex, 194–198, 200, 203, 207, 214–215
*Rana pliocenica, 196–197
Rana sphenocephala, 204
Rana sylvatica, 199–200, 204, 206, 210, 212, 214–215
Rancholabrean land-mammal age, of Pleistocene, 36–37
Range adjustment, of Pleistocene anurans, 209–211
Recolonization of postglacial Michigan by anurans, a model, 211–213
#Rhadinosteus parvus, 188–189
Rhinoderma, 12
Rhinophrynus, 6, 13, 192–193
*Rhinophrynus canadensis, 192–193
Rhinophrynus dorsalis, 7, 192, 200
Rogers, K. L., 16

Scaphiopus, 192–193, 197, 200
*Scaphiopus cf. (Scaphiopus) alexanderi, 194
*Scaphiopus cf. (Spea) bombifrons, 198
*Scaphiopus cf. (Spea) hammondii, 198
*Scaphiopus (Scaphiopus) alexanderi, 196
Scaphiopus (Scaphiopus) couchii, 201
*Scaphiopus (Scaphiopus) guthriei, 192–193
*Scaphiopus (Scaphiopus) hardeni, 194–195
Scaphiopus (Scaphiopus) holbrookii, 201
*Scaphiopus (Scaphiopus) skinneri, 192–193
*Scaphiopus (Scaphiopus) wardorum, 194, 196–197
Scaphiopus (Spea) bombifrons, 194, 197–199, 201

TAXONOMIC INDEX

SITE INDEX

Fossil anuran sites are listed alphabetically by provinces and states, chronologically by time units from oldest to youngest, and then alphabetically by site names. Sites are abbreviated in this index because terms such as "fauna," "local fauna," "prospect," "locality," and "quarry" are often used inconsistently by authors for the same site. Abbreviations: **JU** = Jurassic; **CR** = Cretaceous; **PA** = Paleocene; **EO** = Eocene; **OL** = Oligocene; **MI** = Miocene; **PLI** = Pliocene; and **PLE** = Pleistocene.

ALABAMA
PLE: Bell Cave, 120–121, 158, 173, 176

ARIZONA
JU: Gold Springs, 40
MI: White Cone, 136, 147, 150, 172
PLI: Benson, 119, 147
PLE: Deadman Cave, 83, 138, 148
  Wolcott Peak, 138

ARKANSAS
PLE: Conard Fissure, 126
  Peccary Cave, 90, 121, 173, 176, 179

BAJA CALIFORNIA (MEXICO)
CR: El Gallo, 57

CALIFORNIA
MI: Rodeo, 181
PLE: Costeau Pit, 122
  Newport Beach Mesa, 122, 167, 174
  Potter Creek Cave, 122
  Rancho La Brea, 122, 174
  Redtail Peaks, 138
  Tunnel Ridge, 138

CHIHUAHUA (MEXICO)
MI: Patterson Field, 123

COLORADO
MI: Quarry A (Logan County), 150
PLE: Hansen Bluff, 98, 125, 147, 169, 173, 176
  Haystack Cave, 122

FLORIDA
MI: Haile VIA, 144, 172
  Thomas Farm, 137, 150–151, 159, 162, 172, 185
PLE: Arredondo, 90, 143, 155–156, 158, 166, 176, 182, 185
  Bradenton, 126, 143
  Cutler Hammock, 90
  Devil's Den (Chamber 3), 90, 126, 143, 176, 182
  Haile (Rancholabrean complex), 126, 143, 157–158, 160, 178
  Hornsby Spring, 143
  Ichetucknee River, 126, 143
  Inglis IA, 156, 173, 176
  Kanapaha I, 126, 144
  Kendrick IA, 126
  Leisey Shell Pit, 143
  Orange Lake, 90, 144
  Reddick I, 91, 126, 139, 144, 156, 165, 173, 178, 182, 185
  Sabertooth Cave, 91, 182, 185
  Vero Beach (strata 2 and 3), 91, 176, 182
  Williston IIIA, 91, 144, 157, 182, 185
  Winter Beach, 144

GEORGIA
PLE: Isle of Hope, 166, 176
  Kingston Saltpeter Cave, 90, 121, 126, 164, 176–177, 185
  Ladds, 121, 144, 156, 164, 173, 182, 185

IDAHO
PLI: Hagerman, 99, 150, 172, 174

INDIANA
MI: Pipe Creek Sinkhole, 172, 175
PLE: Christensen Bog, 173
  Kolarik Mastodon, 173
  Prairie Creek D, 173, 176, 178, 183

KANSAS
MI: Edson, 128
  WaKeeney, 49, 87–88, 124, 129, 133, 135, 146, 150, 172–173
  Quarry E near Long Island, 118, 124, 142
PLI: Borchers, 147
  Fox Canyon, 140, 142–143
  Rexroad, 124, 142, 150
  Sanders, 150
  Wedell Fox Pasture, 142
  White Rock, 125, 147, 150, 172, 175
PLE: Butler Spring, 125, 152
  Courland Canal-Hall Ash, 127, 172, 175
  Cragin, 98, 125, 169
  Cudahy, 127, 169
  Duck Creek, 148, 173, 183
  Jinglebob, 98, 125, 148, 169, 173, 176
  Jones, 125, 148, 169
  Kanapolis, 153, 156, 173, 176
  Mount Scott, 173
  Nash, 98
  Sandahl, 125, 148, 169, 173
  Williams, 148, 153, 173, 176

MARYLAND
PLE: Cumberland Cave, 120, 126, 164, 169, 172–173, 177, 183

MICHIGAN
PLE: Meskill Road, 121
  Shelton, 178

MISSOURI
PLE: Boney Spring, 176
  Brynjulfson Cave I, 176
  Zoo Cave, 126, 173

MONTANA
CR: Bug Creek, 56–57, 63
  Judith River Formation type area, 44
PA: Tongue River Formation, 57
  Tullock Formation, 57
MI: Cabbage Patch, 107

NEBRASKA

MI: Achilles, 97
Annies Geese Cross, 98
Carrot Top, 172
Devil's Nest Airstrip, 130
Driftwood Creek, 124
Egelhoff, 88, 97, 128, 146, 150, 165, 172
Glad Tidings, 150, 172
Hottell Ranch, 88, 97, 128, 146, 150, 172, 177
Kuhre, 172
Lemoyne, 129, 136, 146, 150, 172
Mailbox, 172, 175
Mouth of McCann's Canyon, 101
Niobrara River (UC V3218), 128
Norden Bridge, 82, 89, 95, 97, 128, 132, 145, 150, 165, 172
Notre Dame University (locality ND V337), 145
Railway A, 146
Santee, 95, 124, 130, 141, 172, 175
Welke, 97
PLI: Big Springs, 124, 141–142, 172
Hornets Nest, 124, 172, 175, 183
PLE: Albert Ahrens, 98, 120, 125, 155–156, 163, 175, 177, 183

NEVADA

EO: Core 8, Standard Oil Co. of California, 49
MI: Elko Shales, 106
Fish Lake, 82
PLE: Gypsum Cave, 138
Hidden Cave, 100
Smith Creek Cave, 99–100, 122, 148

NEW MEXICO

CR: Fruitland Formation, 57
PA: Tullock Formation, 57
PLE: Dark Canyon Cave, 83, 100, 113, 147–148, 169, 173
Dry Cave, 98–99, 138, 148, 169, 173
Howell's Ridge Cave, 83, 154, 173
Shelter Cave, 83

NORTH DAKOTA

OL: Leo Fitterer Ranch, 92

NOVA SCOTIA (CANADA)

PLE: East Milford Mammoth, 173

OHIO

PLE: Sheriden Pit Cave, 121, 169, 173, 178, 180

OKLAHOMA

PLE: Doby Springs, 169

ONTARIO (CANADA)

PLE: Kelso Cave, 121

PENNSYLVANIA

PLE: Bootlegger Sink, 90, 121, 126, 173
Frankstown Cave, 121, 156, 164, 173, 183
Hanover No. 1 fissure, 120, 173
New Paris 4, 121, 126, 164, 173, 177, 179, 183

QUEBEC (CANADA)

PLE: Eardley 2, 180

SASKATCHEWAN (CANADA)

EO: Calf Creek, 72, 93, 161
MI: Kleinfelder Farm, 82, 146

SONORA (MEXICO)

PLE: El Golfo, 119
Rancho la Brisca, 83, 113, 117, 119, 125, 131, 134, 154, 170, 173, 186

SOUTH DAKOTA

OL: Wounded Knee, 101
MI: Bijou Hills, 150
Glenn Olson, 150
Mission, 136
PLE: Java, 147, 172, 175

TAMAULIPAS (MEXICO)

PLE: Cueva de Abra, 75, 114, 117, 173

TENNESSEE

PLE: Baker Bluff Cave, 120, 126, 183
Cheek Bend Cave, 121, 126, 143, 164
Guy Wilson Cave, 121, 126, 173
Robinson Cave, 121

TEXAS

MI: Trinity River, 159, 172
PLI: Beck Ranch, 124, 141, 146–147, 152, 172–173, 179
PLE: Avenue, 176
Ben Franklin, 176
Clear Creek, 147, 152, 173, 186
Easley Ranch, 115, 148, 152, 156, 163, 173
Fowlkes Cave, 146, 148, 173
Friesenhahn Cave, 113, 125, 148
Gilliland, 115, 147, 173
Howard Ranch, 125, 141, 153, 156, 168, 173
Ingleside, 176
Lubbock Lake, 125, 148, 153, 173, 176, 179
Miller's Cave, 168, 173, 176
Schulze Cave, 83, 113, 115, 148, 156
Sims Bayou, 173
Slaton, 125, 141, 148, 153, 163, 173
Vera, 115, 127, 152

UTAH

JU: Rainbow Park, 76
PLE: Bechan Cave, 100

VIRGINIA

PLE: Clark's Cave, 90, 121, 126, 164, 173, 176–177, 179, 183
Natural Chimneys, 90, 121, 126, 173, 176–177, 179, 183
Saltville, 126, 173
Strait Canyon Fissure, 121, 126, 176, 178

WEST VIRGINIA

PLE: Hamilton Cave, 120, 126, 156, 164, 173, 176, 183
New Trout Cave, 121, 164, 173, 177, 183
Trout Cave, 120, 173, 183
Worm Hole Cave, 183

WISCONSIN

PLE: Moscow Fissure, 121

WYOMING

JU: Como Bluff Quarry Nine, 55
CR: Bushy Tailed Blowout, 46, 59, 62
Lull 2, 59, 62
EO: Lysite Mountain, 65
Tabernacle Butte, 5, 70
"Unamed locality"(Fremont County), 84